Horatio N. Robinson

Conic Sections and Analytical Geometry

Theoretically and practically illustrated

Horatio N. Robinson

Conic Sections and Analytical Geometry
Theoretically and practically illustrated

ISBN/EAN: 9783337105167

Printed in Europe, USA, Canada, Australia, Japan

Cover: Foto ©berggeist007 / pixelio.de

More available books at **www.hansebooks.com**

ROBINSON'S MATHEMATICAL SERIES.

CONIC SECTIONS

AND

ANALYTICAL GEOMETRY;

THEORETICALLY AND PRACTICALLY ILLUSTRATED.

BY

HORATIO N. ROBINSON, LL. D.,

LATE PROFESSOR OF MATHEMATICS IN THE U. S. NAVY, AND AUTHOR OF A FULL COURSE
OF MATHEMATICS.

NEW YORK:
IVISON, PHINNEY & COMPANY,
48 & 50 WALKER STREET.
CHICAGO:
S. C. GRIGGS & COMPANY,
39 & 41 LAKE STREET.
1863.

ROBINSON'S

Series of Mathematics,

The most COMPLETE, *most* PRACTICAL, *and most* SCIENTIFIC SERIES *of* MATHEMATICAL TEXT-BOOKS *ever issued in this country.*

(IN TWENTY-TWO VOLUMES.)

I.	Robinson's Progressive Table Book,	-$ 12
II.	Robinson's Progressive Primary Arithmetic,	15
III.	Robinson's Progressive Intellectual Arithmetic,	25
IV.	Robinson's Rudiments of Written Arithmetic,	25
V.	Robinson's Progressive Practical Arithmetic,	56
VI.	Robinson's Key to Practical Arithmetic,	50
VII.	Robinson's Progressive Higher Arithmetic,	75
VIII.	Robinson's Key to Higher Arithmetic,	75
IX.	Robinson's New Elementary Algebra,	75
X.	Robinson's Key to Elementary Algebra,	75
XI.	Robinson's University Algebra,	1 25
XII.	Robinson's Key to University Algebra,	1 00
XIII.	Robinson's New University Algebra,	1 50
XIV.	Robinson's Key to New University Algebra,	1 25
XV.	Robinson's New Geometry and Trigonometry,	1 50
XVI.	Robinson's Surveying and Navigation,	1 50
XVII.	Robinson's Analyt. Geometry and Conic Sections,	1 50
XVIII.	Robinson's Differen. and Int. Calculus, (in preparation,)	1 50
XIX.	Robinson's Elementary Astronomy,	75
XX.	Robinson's University Astronomy,	1 75
XXI.	Robinson's Mathematical Operations,	2 25
XXII.	Robinson's Key to Geometry and Trigonometry, Conic Sections and Analytical Geometry,	1 50

Entered, according to Act of Congress, in the year 1860, by
HORATIO N. ROBINSON, LL.D.,
In the Clerk's Office of the District Court of the United States for the Northern
District of New York.

PREFACE.

In the preparation of the following work the object has been to bring within the compass of one volume of convenient size an elementary treatise on both Conic Sections and Analytical Geometry.

In the first part, the properties of the curves known as the Conic Sections are demonstrated, principally by geometrical methods ; that is, in the investigations, the curves and parts connected with them are constantly kept before the mind by their graphic representations, and we reason directly upon them.

In the purely Analytical Geometry the process is quite different. Here the geometrical magnitudes, themselves, or those having certain relations to them, are represented by algebraic symbols, and we seek to express properties and imposed conditions by means of these symbols. The mind is thus relieved, in a great measure, of the necessity of holding in view the often-times complex figures required in the intermediate steps of the first method. It is, mainly, at the beginning and end of our investigations that we have to deal with concrete quantity. That is, after we have expressed known and imposed conditions, analytically, our reasoning is independent of the kind of quantity involved, until the conclusion is reached in the form of an algebraic expression, which must then receive its geometrical interpretation.

Much of the value of Analytical Geometry, as a disciplinary study, will be derived from a careful consideration, in each case, of this process of passing from the concrete to the abstract and the

(iii)

converse, and both teacher and student are earnestly recommended to give it a large share of their attention.

In both divisions of the work the object has been to present the subjects in the simplest manner possible, and hence, in the first, analytical methods have been employed in several propositions when results could be thereby much more easily obtained; and for the same reason, in the second division, a few of the demonstrations are almost entirely geometrical.

The analytical part terminates, with the exception of some examples, with the Chapter on Planes. Three others might have been added; one on the transformation of Co-ordinates in Space, another on Curves in Space, and a third on Surfaces of Revolution and curved surfaces in general: but the work, as it is, covers more ground than is generally gone over in Schools and Colleges, and is sufficiently extensive for the wants of elementary education. Numerous examples are given under the several divisions in the second part to illustrate and impress the principles.

The Author has great pleasure in acknowledging his obligations to Prof. I. F. Quinby, A. M., of the University of Rochester, N. Y., formerly Assistant Prof. of Mathematics in the United States Military Academy, at West Point, for valuable services rendered in the preparation of this treatise, as well as for the contribution to it of much that is valuable both in matter and arrangement. His thorough scholarship, as well as his long and successful experience as an instructor in the class-room, preëminently qualified him to perform such labor.

December, 1861.

CONTENTS.

CONIC SECTIONS.

DEFINITIONS.

Conical Surfaces,.................................PAGE 9
Conic Sections, ... 10

THE ELLIPSE.

Definitions and Explanations,............................ 11
Propositions relating to the Ellipse,....................... 13

THE PARABOLA.

Definitions and Explanations,............................ 41
Propositions relating to the Parabola,.................... 43

THE HYPERBOLA.

Definitions and Explanations,............................ 65
Propositions relating to the Hyperbola,................... 67

ASYMPTOTES.

Definition, ... 91
Propositions establishing relations between the Hyperbola and
its Asymptotes, ... 91

1* (v)

ANALYTICAL GEOMETRY.

General Definitions and Remarks, ... 96

GENERAL PROPERTIES OF GEOMETRICAL MAGNITUDES.

CHAPTER I.

OF POSITIONS AND STRAIGHT LINES IN A PLANE AND THE TRANSFORMATION OF CO-ORDINATES.

Definitions and Explanations, ... 97
Propositions relating to Straight Lines in a Plane, ... 100
Transformation of Co-ordinates, ... 119
Polar Co-ordinates, ... 122

CHAPTER II.

THE CIRCLE.

LINES OF THE SECOND ORDER.

Propositions relating to the Circle 124
Polar equation of the Circle, ... 132
Application in the solution of Equations of the second degree, 134
Examples, ... 139

CHAPTER III.

THE ELLIPSE.

The description of the Ellipse and Propositions establishing
    its properties, ... 140
Example, ... 167

CHAPTER IV.

THE PARABOLA.

The description of the Parabola and propositions establishing
    its properties, ... 169
Polar equation of the Parabola, ... 183
Application in the solution of equations of the second degree, 185
Examples 187

CHAPTER V.

THE HYPERBOLA.

The Description of the Curve, and Propositions Establishing its Properties,.................................... 188

ASYMPTOTES OF THE HYPERBOLA.

Definition and Explanation,.............................. 201
The Equation of the Hyperbola referred to its Asymptotes, and Properties deduced therefrom,...................... 202

CHAPTER VI.

ON THE GEOMETRICAL REPRESENTATION OF EQUATIONS OF THE SECOND DEGREE BETWEEN TWO VARIABLES.

Object of the Discussion,............................... 210
Solution and Discussion of the General Equation,.......... 211
Criteria for the Interpretation of any Equation of the Second Degree between two Variables,...................... 221

APPLICATIONS.

First, $B^2-4AC<0$, the Ellipse,........................ 222
Second, $B^2-4AC>0$, the Hyperbola,..................... 226
Third, $B^2-4AC=0$, the Parabola,...................... 231
Examples, ... 233

CHAPTER VII.

On the Intersection of Lines, and the Geometrical Solution of Equations,... 237
Remarks on the Interpretation of Equations,.............. 244

CHAPTER VIII.

STRAIGHT LINES IN SPACE.

Co-ordinate Planes and Axes,............................ 249
The Equations and Relations of Straight Lines in Space,.... 250

CHAPTER IX.

ON THE EQUATION OF A PLANE.

The Equations and Relations of Planes,................... 258
Examples Relating to Straight Lines in Space and to Planes,. 269
Miscellaneous Examples,............................. 273

CONIC SECTIONS.

DEFINITIONS.

1. **A Conical Surface**, or a **Cone** is, in its general accept-ation, the surface that is generated by the motion of a straight line of indefinite extent, which in its different positions constantly passes through a fixed point and touches a given curve.

The moving line is called the *generatrix*, the curve that it touches the *directrix*, the fixed point the *vertex*, and the generatrix in any of its positions an *element*, of the cone.

The generatrix in all its positions extending without limit beyond the vertex on either side, will by its motion generate two similar surfaces separated by the vertex, called the *nappes* of the cone.

2. The **Axis** of a cone is the indefinite line passing through the vertex and the center of the directrix.

3. The intersection of the cone by any plane not pass-ing through its vertex, that cuts all its elements, may be taken as the directrix; and when we regard the cone as limited by such intersection, it is called the *base* of the cone. If the axis is perpendicular to the plane of the base, the cone is said to be *right;* and if in addition the base is a circle, we have a *right cone with a circular base.* This is the same as the cone defined in Geometry, (Book VII, Def. 16), and in the following pages it is to be understood that all references are made to it, unless otherwise stated.

4. Conic Sections are the figures made by a plane cutting a cone.

5. There are *five* different figures that can be made by a plane cutting a cone, namely: a *triangle*, a *circle*, an *ellipse*, a *parabola*, and an *hyperbola*.

REMARK. The three last mentioned are commonly regarded as embracing the whole of conic sections; but with equal propriety the triangle and the circle might be admitted into the same family. On the other hand we may examine the properties of the ellipse, the parabola, and the hyperbola, in like manner as we do a triangle or a circle, without any reference whatever to a cone.

It is important to study these curves, on account of their extensive application to astronomy and other sciences.

6. If a plane cut a cone through its vertex, and terminate in any part of its base, the section will evidently be a *triangle*.

7. If a plane cut a cone parallel to its base, the section will be a *circle*.

8. If a plane cut a cone obliquely through all of the elements, the section will represent a curve called an *ellipse*.

9. If a plane cut a cone parallel to one of its elements, or what is the same thing, if the cutting plane and an element of the cone make equal angles with the base, then the section will represent a *parabola*.

10. If a plane cut a cone, making a greater angle with the base than the element of the cone makes, then the section is an *hyperbola*.

11. And if the plane be continued to cut the other nappe of the cone, this latter intersection will be the opposite hyperbola to the former.

12. The **Vertices** of any section are the points where the cutting plane meets the opposite elements of the cone, or the sides of the vertical triangular section, as *A* and *B*.

Hence, the ellipse and the opposite hyperbolas have each two vertices; but the parabola has only one, unless we consider the other as at an infinite distance.

13. The **Axis**, or **Transverse Diameter** of a conic section, is the line or distance AB between the vertices.

Hence, the axis of a parabola is infinite in length, AB being only a part of it.

The properties of the three curves known as the Conic Sections will first be investigated without any reference to the cone whatever; and afterward it will be shown that these curves are the several intersections of a cone by a plane.

THE ELLIPSE.

DEFINITIONS.

1. The **Ellipse** is a plane curve described by the motion of a point subjected to the condition that the sum of its distances from two fixed points shall be constantly the same.

2. The two fixed points are called the *foci*. Thus F, F', are *foci*.

3. The **Center** is the point C, the middle point between the foci.

4. A **Diameter** is a straight line through the center, and terminated both ways by the curve.

5. The extremities of a diameter are called its *vertices*.

Thus, DD' is a diameter, and D and D' are its *vertices*.

6. The **Major**, or **Transverse Axis**, is the diameter which passes through the foci. Thus, AA' is the major axis.

7. The **Minor**, or **Conjugate Axis** is the diameter at right angles to the major axis. Thus, CE is the *semi* minor axis.

8. The distance between the center and either focus is called the *eccentricity* when the semi major axis is unity.

That is, the eccentricity is the ratio between CA and CF; or it is $\dfrac{CF}{CA}$; hence, it is always less than unity.

The less the eccentricity, the nearer the ellipse approaches the circle.

9. A **Tangent** is a straight line which meets the curve in one point only; and, being produced, does not cut it.

10. A **Normal** to a curve at any point is a perpendicular to the tangent at that point.

11. An **Ordinate to a Diameter** is a straight line drawn from any point of the curve to the diameter, *parallel to a tangent* passing through one of the vertices of *that* diameter.

REMARK.—A diameter and its ordinate are not at right angles, unless the diameter be either the *major* or *minor* axis.

12. The parts into which a diameter is divided by an ordinate, are called *abscissas*.

13. Two diameters are said to be *conjugate*, when either is parallel to the tangent lines at the vertices of the other.

14. The **Parameter** of a diameter is a third proportional to that diameter and its conjugate.

15. The paramater of the major axis is called the *principal parameter*, or *latus rectum;* and, as will be proved, is equal to the double ordinate through the focus. Thus $F'G$ is one half of the principal parameter.

16. A **Sub-tangent** is that part of the axis produced, which is included between a tangent and the ordinate, drawn from the point of contact.

17. A **Sub-normal** is that part of the axis which is included between the normal and the ordinate, drawn from the point of contact.

PROPOSITION I. PROBLEM.

To describe an Ellipse.

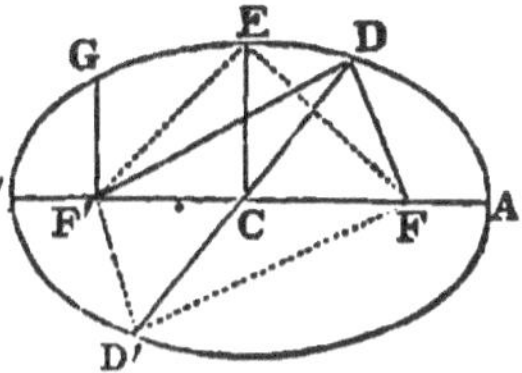

Assume any two points, as F and F' and take a thread longer than the distance between these points, fastening one of its extremities at the point F and the other at the point F'. Now if the point of a pencil be placed in the loop and moved entirely around the points F and F', the thread being constantly kept tense, it will describe a curve as represented in the adjoining figure, and, by definition 1, this curve is an ellipse.

PROPOSITION II.—THEOREM.

The major axis of an ellipse is equal to the sum of the two lines drawn from any point in the curve to the foci.

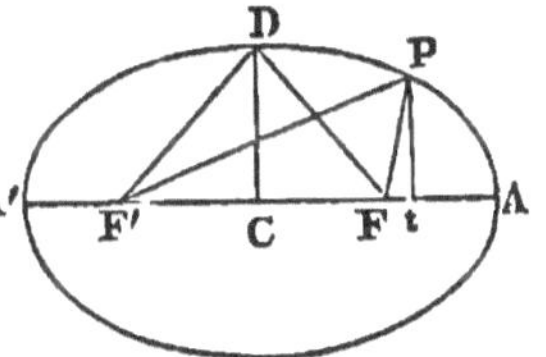

Suppose the point of a pencil at D to move along in the loop, holding the threads $F'D$ and FD at equal tension; when D arrives at A, there will be *two lines* of threads between F and A. Hence, the entire length of the threads will be measured by $F'F+2FA$. Also, when D arrives at A', the length of the threads is measured by $FF'+2F'A'$.

Therefore, $\qquad FF'+2FA=FF'+2F'A'$

Hence, $\qquad\qquad FA=F'A'$

From the expression $FF'+2FA$, take away FA, and add $F'A'$, and the sum will not be changed, and we have

$$FF'+2FA=A'F'+FF'+FA=A'A$$

Therefore, $\qquad F'D+FD=A'A$

Hence the theorem ; *the major axis of an ellipse, etc.*

2

PROPOSITION III.—THEOREM

An ellipse is bisected by either of its axes.

Let F, F' be the foci, AA' the major and BB' the minor axis of an
ellipse; then will either of these
axis divide the ellipse into equal
parts.

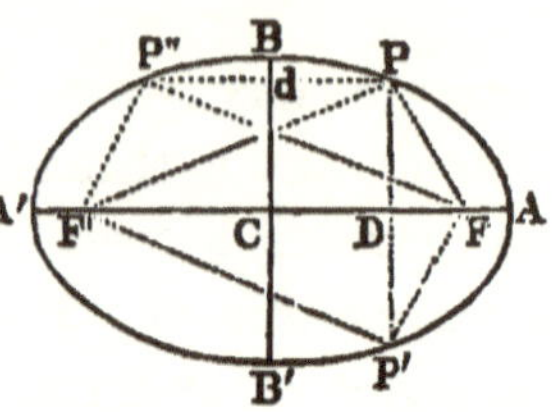

Take any point, as P in the el-
lipse, and from this point draw ordinates, one to the major and another to the minor axis, and produce these ordinates, the first to P', the second to P'', making the parts
produced equal to the ordinates themselves. It is evident
that the proposition will be established when we have
proved that P' and P'' are points of the curve.

First. F is a point in the perpendicular to PP' at its
middle point; therefore $FP'=FP$ (Scho. 1, Th. 18, B. 1
Geom.) for the same reason $F'P'=F'P$.

Whence, by addition,

$$FP'+F'P'=FP+F'P.$$

That is, the sum of the distances from P' to the foci is
equal to the sum of the distances from P to the foci; but
by hypothesis P is a point of the ellipse; therefore P' is
also a point of the ellipse, (Def. 1).

Second. The trapezoids $P''dCF'$, $PdCF$ are equal, because $F'C=FC$, $dP''=dP$ by construction, and the angles
at d and C in each are equal, being right angles; these
figures will therefore coincide when applied, and we have
$P''F'$ equal to PF and the angle $P''F'F$ equal to the angle
PFF'. Hence the triangles $P''F'F$, PFF' are equal having the two sides $P''F'$, $F'F$ and the included angle $P''F'F$
in the one equal, each to each to the two sides PF, FF'
and the included angle PFF' in the other.

Therefore, $P''F'+P''F=PF'+F'P$

That is, the sum of the distances from P'' to the foci is

equal to the sum of the distances from P to the foci, and since P is a point of the ellipse P'' must also be found on the ellipse.

Hence the theorem; *an ellipse is bisected, etc.*

PROPOSITION IV.—THEOREM.

The distance from either focus of an ellipse to the extremity of the minor axis is equal to the semi-major axis.

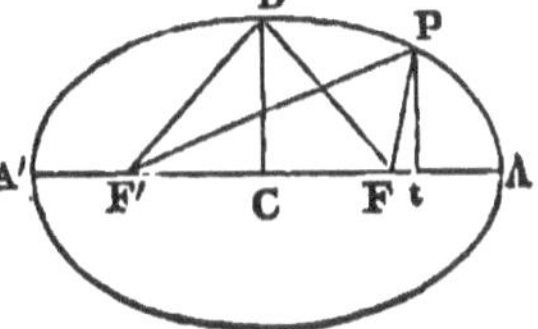

Let AA' be the major axis, F and F' the foci, and CD the semi-minor axis of an ellipse; then will $FD=F'D$ be equal to CA.

Because $F'C=CF$ and CD is at right angles to $F'F$, we have $F'D=FD$.

But, $$F'D+FD=A'A$$
Or, $$2FD=A'A$$
Therefore, $$FD=\tfrac{1}{2}A'A, \text{ or } CA.$$

Hence the theorem; *the distance from either focus, etc.*

SCHOLIUM.—The half of the minor axis is a mean proportional between the distance from either focus to the principal vertices.

In the right-angled triangle FCD we have
$$\overline{CD}^2=\overline{FD}^2-\overline{FC}^2$$
But, $$FD=AC$$
Therefore, $$\overline{CD}^2=\overline{AC}^2-\overline{FC}^2$$
$$=(AC+FC)(AC-FC)$$
$$=AF'\times AF$$
Or, $$AF:CD=CD:FA'$$

PROPOSITION V.—THEOREM

Every diameter of an ellipse is bisected at the center.

Let D be any point in the curve, and C the center. Draw DC, and produce it. From F' draw $F'D'$ parallel

to *FD;* and from *F* draw *FD'* parallel to *F'D.* The figure *DFD'F''* is
a parallelogram by construction; and
therefore its opposite sides are equal.

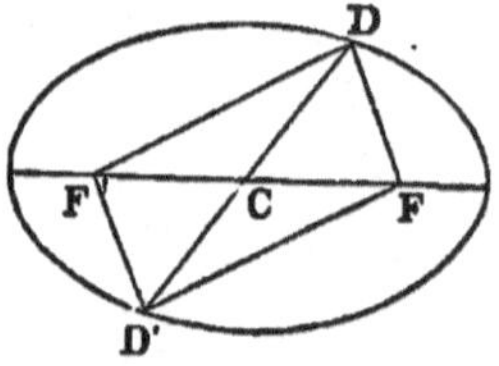

Hence, the sum of the two sides
F'D' and *D'F* is equal to *F'D* and *DF;* therefore, by def-
inition 1, the point *D'* is in the ellipse. But the two di-
agonals of a parallelogram bisect each other; therefore,
DC=CD', and the diameter *DD'* is bisected at the center,
C, and *DD'* represents any diameter whatever.

Hence the theorem; *every diameter, etc.*

Cor. The quadrilateral formed by drawing lines from
the extremities of a diameter to the foci of an ellipse, is
a parallelogram.

PROPOSITION VI.—THEOREM.

*A tangent to the ellipse makes equal angles with the two
straight lines drawn from the point of contact to the foci.*

Let *F* and *F'* be the foci and
D any point in the curve. Draw
F'D and *FD*, and produce *F'D*
to *H*, making *DH=DF*, and draw
FH. Bisect *FH* in *T.* Draw *TD*
and produce it to *t.*

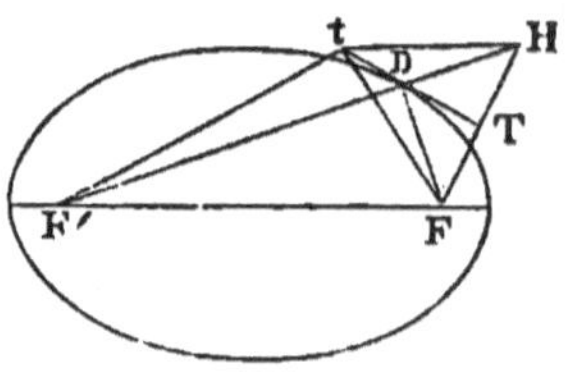

Now, (by Cor. 2, Th. 18, B. I, Geom.), the angle *FDT=*
the angle *HDT*, and *HTD=*its vertical angle *F'Dt.*

Therefore, $\qquad FDT=F'Dt.$

It now remains to be shown that *Tt* meets the curve
only at the point *D*, and is, therefore, a tangent.

If possible, let it meet the curve in some other point,
as *t*, and draw *Ft*, *tH*, and *F't.*

(*By* Scholium 1, Th. 18, B. I, Geom.) *Ft=tH.*

To each of these add *F't;*

Then, $\qquad F''t+tH=F't+Ft$

But $F't$ and tH are, together, greater than $F'H$, because a straight line is the shortest distance between two points; that is, $F''t$ and Ft, the two lines from the foci, are, together, greater than FH, or greater than $F'D+FD$; therefore, the point t is without the ellipse, and t is any point in the line Tt, except D. Therefore, Tt is a tangent, touching the ellipse at D; and it makes equal angles with the lines drawn from the point of contact to the foci.

Hence the theorem; *a tangent, etc.*

Cor. The tangents at the vertices of either axis are perpendicular to that axis; and, as the ordinates are parallel to the tangents, it follows that all ordinates to either axis must cut that axis at right angles, and be parallel to the other axis.

Scholium 1.—From this proposition we derive the following simple rule for drawing a tangent line to an ellipse at any point: *Through the given point draw a line bisecting the angle included between the line connecting this point with one of the foci and the line produced connecting it with the other focus.*

Scholium 2. Any point in the curve may be considered as a point in a tangent to the curve at that point.

It is found by experiment that rays of *light, heat* and *sound* are incident upon, and reflected from surfaces under equal angles; that is, for a ray of either of these principles the angles of incidence and reflection are equal. Therefore, if a reflecting surface be formed by turning an ellipse about its major axis, the light, heat, or sound which proceeds from one of the foci of this surface will be concentrated in the other focus.

Whispering galleries are made on this principle, and all theaters and large assembly rooms should more or less approximate this figure. The concentration of the rays of heat from one of these points to the other, is the reason why they are called the *foci* or burning points.

2*   B

PROPOSITION VII.—THEOREM.

Tangents to the ellipse, at the vertices of a diameter, are parallel to each other.

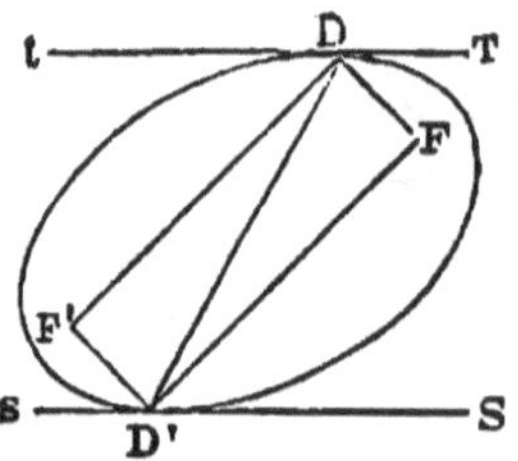

Let DD' be the diameter, and F' and F the foci. Draw $F'D$, $F'D'$, FD, and FD'.

Draw the tangents, Tt and Ss, one through the point D, the other through the point D'. These tangents will be parallel.

By Cor. Prop. 5, $F'D'FD$ is a parallelogram, and the angle $F'D'F$ is equal to its opposite angle, $F'DF$.

But the sum of all the angles that can be made on one side of a line is equal to two right angles. Therefore, by leaving out the equal angles which form the opposite angles of the parallelogram, we have

$$sD'F' + SD'F = tDF' + TDF$$

But (by Prop. 6) $sD'F' = SD'F$; and also $tDF' = TDF$; therefore, the sum of the two angles in either member of this equation is double either of the angles, and the above equation may be changed to

$$2SD'F = 2tDF' \qquad \text{or} \qquad SD'F = tDF'$$

But DF' and $D'F$ are parallel; therefore $SD'F$ and tDF' are, in effect, alternate angles, showing that Tt and Ss are parallel.

Cor. If tangents be drawn through the vertices of any two conjugate diameters, they will form a parallelogram circumscribing the ellipse.

PROPOSITION VIII.—THEOREM.

If, from the vertex of any diameter of an ellipse, straight lines are drawn through the foci, meeting the conjugate diameter, the part of either line intercepted by the conjugate, is equal to one half the major axis.

Let DD' be the diameter, and Tt the tangent. Through the center draw EE' parallel to Tt. Draw $F'D$ and DF, and produce DF to K; and from F draw FG parallel to EE' or Tt.

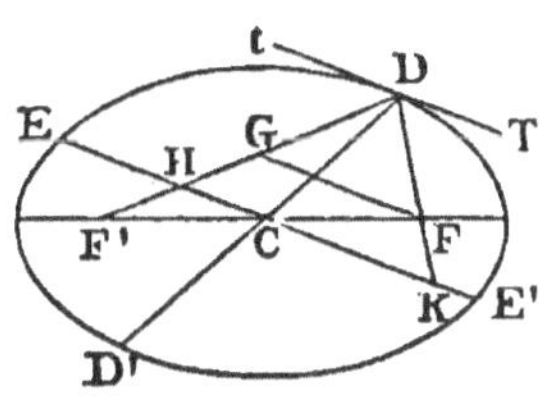

Now, by reason of the parallels, we have the following equations among the angles:

$$\left. \begin{array}{l} tDG = DGF \\ TDF = DFG \end{array} \right\} \text{Also,} \left\{ \begin{array}{l} tDG = DHK \\ TDF = DKH \end{array} \right.$$

But (Prop. 6) $tDG = TDF;$
Therefore, $DGF = DFG;$
And, $DHK = DKH$

Hence, the triangles DGF and DHK are isosceles. Whence, $DG = DF$, and $DH = DK$.

Because HC is parallel to FG, and $F'C = CF$,

therefore, $F'H = HG$
Add, $DF = DG$

and we have $F'H + DF = DH$

But the sum of the lines in both members of this equation is $F'D + DF$, which is equal to the major axis of the ellipse; therefore, either member is one half the major axis; that is, DH, and its equal, DK, are each equal to one half the major axis.

Hence the theorem; *if from the vertex of any diameter, etc.*

PROPOSITION IX.—THEOREM.

Perpendiculars from the foci of an ellipse upon a tangent, meet the tangent in the circumference of a circle whose diameter is the major axis.

Let F', F be the foci, C the center of the ellipse, and D a point through which passes the tangent Tt. Draw $F'D$

and FD, produce $F'D$ to H, making $DH=FD$, and produce FD to G, making $DG=F'D$. Then $F'H$ and FG are each equal to the major axis, $A'A$.

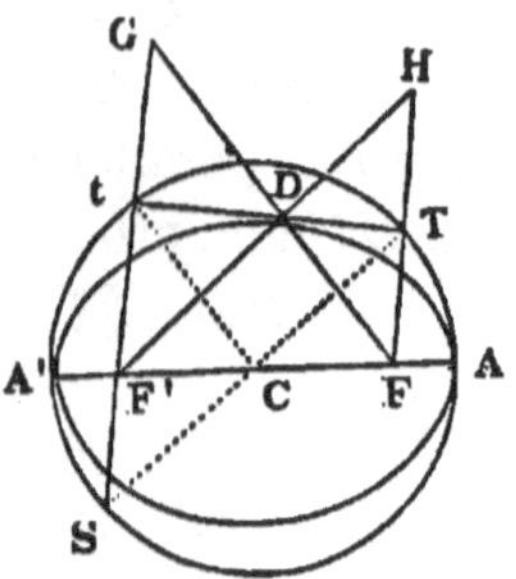

Draw FH meeting the tangent in T and $F'G$ meeting it in t. Draw the dotted lines, CT and Ct.

By Prop. 6, the angle $FDT=$ the angle $F'Dt$; and since opposite or vertical angles are equal, it follows that the four angles formed by the lines intersecting at D, are all equal.

The triangles $DF'G$ and DHF are isosceles by construction; and as their vertical angles at D are bisected by the line Tt, therefore $F't=tG$, $FT=TH$, and FT and $F't$ are perpendicular to the tangent Tt.

Comparing the triangles $F'GF$ and $F'Ct$, we find that $F'C$ is equal to the half of $F'F$, and $F't$, the half of $F'G$; therefore, Ct is the half of FG; but $A'A=FG$; hence, $Ct=\frac{1}{2}A'A=CA$.

Comparing the triangles $FF'H$ and FCT, we find the sides FH and FF' cut proportionally in T and C; therefore, they are equi-angular and similar, and CT is parallel to $F'H$, and equal to one half of it. That is, CT is equal to CA; and CA, CT, and Ct are all equal; and hence a circumference described from the center C, with the radius CA, will pass through the points T and t.

Hence the theorem: *perpendiculars from the foci, etc.*

PROPOSITION X.—THEOREM.

The product of the perpendiculars from the foci of an ellipse upon a tangent, is equal to the square of one half the minor axis.

Produce TC and GF', and they will meet in the circumference at S; for FT and $F't$ are both perpendicular to

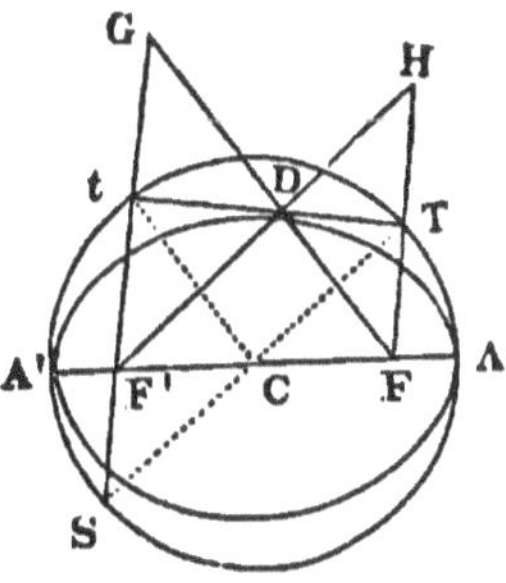

the same line Tt, they are therefore parallel; and the two triangles, CFT and $CF'S$, having a side, FC, of the one, equal to the side, CF', of the other, and their angles equal, each to each, are themselves equal. Therefore, $CS = CT$, S is in the circumference, and $SF' = FT$.

Now, since $A'A$ and St are two lines that intersect each other in a circle, therefore (Th. 17, B. III, Geom.),

$$SF' \times F't = A'F' \times F'A \; ;$$

Or, $$FT \times F't = A'F' \times F'A.$$

But, by the Scholium to Prop. 4, it is shown that

$A'F' \times F'A =$ the square of one half the minor axis.

Therefore, $FT \times F't =$ the square of one half the minor axis.

Hence the theorem; *The product of the perpendiculars, etc.*

Cor. The two triangles, FTD and $F'tD$, are similar, and from them we have $TF : F't = FD : DF'$; that is, *perpendiculars let fall from the foci upon a tangent, are to each other as the distances of the point of contact from the foci.*

PROPOSITION XI.—THEOREM.

If a tangent, drawn to an ellipse at any point, be produced until it meets either axis, and from the point of tangency an ordinate be drawn to the same axis, one half of the axis will be a mean proportional between the distances from the center to the intersections of these lines with the axis.

Let Tt be a tangent at any point in the ellipse, as P.

Draw $F'P$ and FP, F and F' being the foci, and produce $F'P$ to Q, making $PQ = PF$; join T, Q, and draw PG perpendicular to the axis AA'.

The triangles PFT and PTQ are equal, because PT is common, $PQ=PF$ by construction, and the $\llcorner TPF=$ the angle $\llcorner TPQ$ (Th. 6).

Therefore, TP bisects the angle FTQ, and $QT=FT$.

As the angle at T is bisected by TP, the sides about this angle in the triangle $F'TQ$ are to each other, as the segments of the third side, (Th. 24, B. II, Geom.)

That is, $\qquad F'T : TQ : : F'P : PQ$

Or, $\qquad F'T : FT : : F'P : PF$

From this last proportion we have (Th. 9, B. II, Geom.),

$$F'T+FT : F'T-FT : : F'P+PF : F'P-PF$$

Or, since $\quad F'T+FT=2CT$ and $F'P+PF=2CA,$ by substitution we have

$$2CT : F'F : : 2CA : F'P-PF \qquad (1)$$

Again, because PG is drawn perpendicular to the base of the triangle $F'PF$, the base is to the sum of the two sides, as the difference of the sides is to the difference of the segments of the base, (Prop. 6, Pl. Trig.)

Whence, $\quad F'F : F'P+PF : : F'P-PF : 2CG \quad (2)$

If we multiply proportions (1) and (2), term by term, omitting in the resulting proportion the factor $F'F$, common to the terms of the first couplet, and the factor $F'P-PF$, common to the terms of the second couplet, we shall have

$$2CT : 2CA : : 2CA : 2CG$$

Or, $\qquad CT : CA : : CA : CG$

In like manner it may be proved that

$$Ct : CB : : CB : Cg$$

Hence the theorem ; *If a tangent, drawn to an ellipse, etc.*

PROPOSITION XII.—THEOREM.

The sub-tangent on either axis of an ellipse is equal to the corresponding sub-tangent of the circle described on that axis as a diameter.

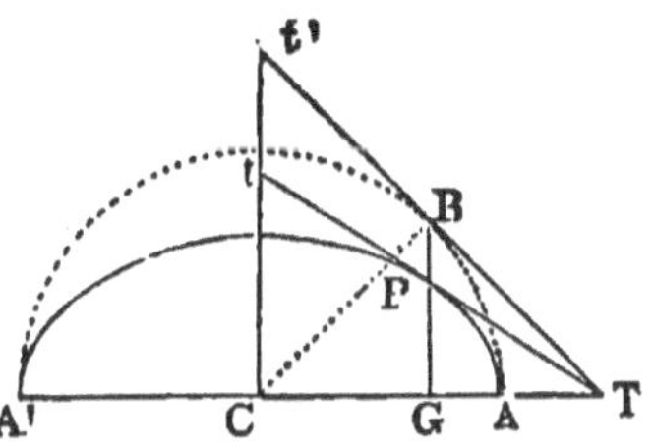

Let P be the point of tangency of the tangent line Tt to the ellipse, of which AA' is the major axis and C the center. Draw the ordinate PG to this axis, and produce it to meet the circumference of the circle described on AA' as a diameter, at B, and draw BC and BT, T being the intersection of the tangent with the major axis; then will the line BT be a tangent to the circumference, at the point B.

By the preceding theorem we have

$$CT : CA :: CA : CG$$

And since $CA = CB$, this proportion becomes

$$CT : CB :: CB : CG$$

Hence, the triangles CBT and CBG have the common angle C, and the sides about this angle proportional; they are therefore similar (Cor. 2 Th. 17, B. II, Geom.). But CBG is a right-angled triangle; therefore, CBT is also right-angled, the right angle being at B. Now, since the line BT is perpendicular to the radius CB at its extremity, it is tangent to the circumference, and GT is therefore a common sub-tangent to the ellipse and circle.

If a circumference be described on the minor axis as a diameter, it may be proved in like manner that the corresponding sub-tangents of the ellipse and circle are equal.

Hence, the theorem; *The sub-tangent on either axis, etc.*

Scholium 1.—This proposition furnishes another easy rule for drawing a tangent line to an ellipse, at any point.

Rule. *On the major axis as a diameter, describe a semi-circumference, and from the given point on the ellipse draw an ordinate to the major axis; draw a tangent to the semi-circumference at the point in which the ordinate produced meets it. The line that connects the point in which this tangent intersects the major axis with the given point on the ellipse, will be the required tangent.*

SCHOLIUM 2.—Because CBT is a right-angled triangle,
$$CG \cdot GT = \overline{BG}^2 \text{; but } A'G \cdot AG = \overline{BG}^2$$
Therefore, $CG \cdot GT = A'G \cdot AG$

PROPOSITION XIII.—THEOREM.

The square of either semi-axis of an ellipse is to the square of the other semi-axis, as the rectangle of any two abscissas of the former axis is to the square of the corresponding ordinate.

From any point, as P, of the ellipse of which C is the center, AA' the major, and BB' the minor axis, draw the ordinate PG to the major axis; then it is to be proved that

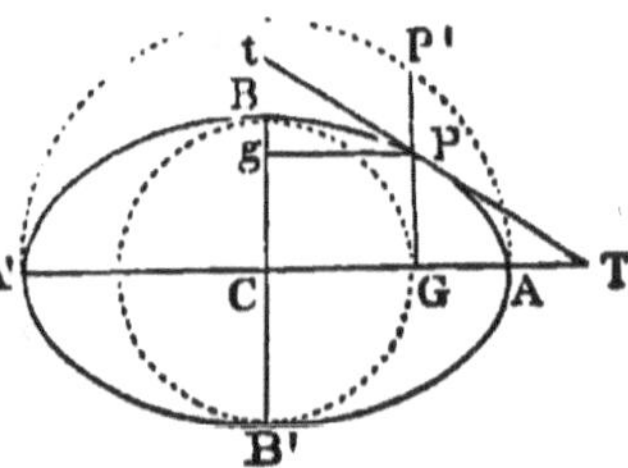

$$\overline{CA}^2 : \overline{CB}^2 : : AG \cdot GA' : \overline{PG}^2$$

Through P draw a tangent line intersecting the axes at T and t; then, by Prop. 11, we have
$$CT : : CA : : CA : CG$$
Whence, $CT \cdot CG = \overline{CA}^2$

and by multiplying both members of this equation by CG, it becomes
$$CT \cdot \overline{CG}^2 = \overline{CA}^2 \cdot CG$$
which may be resolved into the proportion
$$\overline{CA}^2 : \overline{CG}^2 : : CT : CG \quad .$$

From this we find, (Cor. Th. 8, B. II, Geom.),
$$\overline{CA}^2 : \overline{CA}^2 - \overline{CG}^2 : : CT : GT \quad (1)$$
Again, drawing the ordinate Pg to the minor axis, we have
$$Ct : CB : : CB : Cg \text{ or } PG$$
Whence, $Ct \cdot PG = \overline{CB}^2$

Multiplying both members of this equation by PG, it becomes

$$Cl \cdot \overline{PG}^2 = \overline{CB}^2 \cdot PG$$

from which we have the proportion

$$\overline{CB}^2 : \overline{PG}^2 :: Cl : PG$$

By similar triangles we have

$$Cl : PG :: CT : GT$$

And, since the first couplet in this proportion is the same as the second couplet in the preceding, the terms of the other couplets are proportional.

That is, $\qquad \overline{CB}^2 : \overline{PG}^2 :: CT : GT \qquad$ (2)

By comparing proportions (1) and (2), we obtain

$$\overline{CB}^2 : \overline{PG}^2 :: \overline{CA}^2 : \overline{CA}^2 - \overline{CG}^2 \qquad (3)$$

But $\overline{CA}^2 - \overline{CG}^2 = (CA + CG)(CA - CG) = A'G \cdot AG$;

Whence, by inverting the means in proportion (3) and substituting the values of $\overline{CA}^2 - \overline{CG}^2$, we have finally

$$\overline{CB}^2 : \overline{CA}^2 :: \overline{PG}^2 : A'G \cdot AG$$

or, $\qquad \overline{CA}^2 : \overline{CB}^2 :: A'G \cdot AG : \overline{PG}^2$

By a process in all respects similar to the above, we will find that

$$\overline{CB}^2 : \overline{CA}^2 :: Bg \cdot B'g : \overline{(Pg)}^2$$

Hence the theorem; *the square of either semi-axis, etc.*

Scholium 1.—From the theorem just demonstrated is readily deduced what is called, in Analytical Geometry, the *equation of the ellipse referred to its center and axes.* If we take any point, as P, on the curve, and can find a general relation between AG and PG, or between CG and PG, the equation expressing such relation will be the equation of the curve. Let us represent CA, one half of the major axis, by A, and CB, one half of the minor axis, by B; that is, the symbols A and B denote the numerical values of these semi-axes, respectively. Also, denote the CG by x, and PG by y, then $A'G = A + x$, and $AG = A - x$; and by the theorem we have

$$A^2 : B^2 :: (A + x)(A - x) : y^2$$

Whence, $\qquad A^2 y^2 = A^2 B^2 - B^2 x^2$

Or, $\qquad A^2 y^2 + B^2 x^2 = A^2 B^2$

3

This is the required equation in which the variable quantities, x and y, are called the *co-ordinates* of the curve, the first, x, being the *abscissa*, and the second, y, the *ordinate;* the center C from which these variable distances are estimated, is called the *origin* of co-ordinates, and the major and minor axes are the *axes* of co-ordinates.

Had we denoted $A'G$ by x, without changing y, then we should have $\qquad A G = 2A - x,$

And $\qquad A^2 : B^2 : : (2A - x)\, x : y^2$

Whence, $\quad y^2 = \dfrac{B^2}{A^2}(2Ax - x^2)$, which is the equation of the ellipse when the origin of co-ordinates is on the curve at A'.

SCHOLIUM 2.—*If a circle be described on either axis of an ellipse as a diameter, then any ordinate of the circle to this axis is to the corresponding ordinate of the ellipse, as one half of this axis is to one half of the other axis.*

Retaining the notation in Scholium 1, and producing the ordinate PG to meet the circumference described on $A'A$ as a diameter, at P, we have, by the theorem,

$$A^2 : B^2 : : (A+x)\,(A-x) : y^2$$

But $\qquad (A+x)\,(A-x) = \overline{GP'}^2$

Whence, $\qquad A^2 : B^2 : : \overline{GP'}^2 : y^2$

Or, $\qquad A \ : B : : GP' \ : y$

That is, $\qquad GP' : y : : A : B$

By describing a circle on BB' as a diameter, we may in like manner prove that $\qquad pg : Pg : : B : A$

PROPOSITION XIV.—THEOREM.

The squares of the ordinate to either axis of an ellipse are to each other, as the rectangles of the corresponding abscissas.

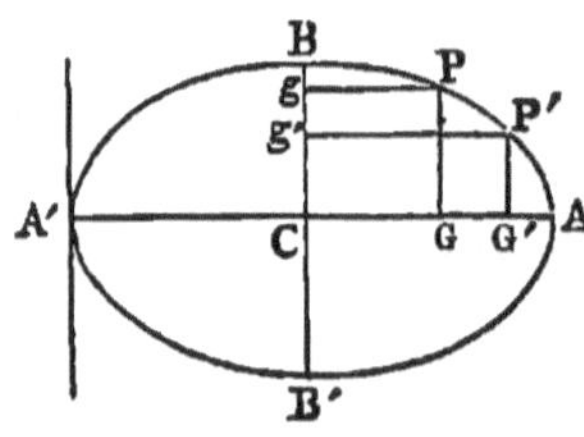

Let AA' be the major, and BB' the minor axis of the ellipse, and PG, $P'G'$ any two ordinates to the first axis. Denoting CG by by x, CG' by x', PG by y and $P'G'$ by y', we have, by Scho. 1,

Prop. 13, $\qquad A^2y^2 + B^2x^2 = A^2B^2$

and $\qquad\qquad A^2y'^2 + B^2x'^2 = A^2B^2$

Whence, $\qquad y^2 = \dfrac{B^2}{A^2}(A^2 - x^2) = \dfrac{B^2}{A^2}(A+x)(A-x)$ (1)

and $\qquad y'^2 = \dfrac{B^2}{A^2}(A^2 - x'^2) = \dfrac{B^2}{A^2}(A+x')(A-x')$ (2)

Dividing equation (1) by equation (2), member by member, and omitting the common factors in the numerator and denominator of the second member of the resulting equation, it becomes

$$\frac{y^2}{y'^2} = \frac{(A+x)\ (A-x)}{(A+x')(A-x')}$$

By simply inspecting the figure, we perceive that $A+x$ and $A-x$ represent the abscissas of the axis AA', corresponding to the ordinate y; and $A+x'$, and $A-x'$ those corresponding to the ordinate y'.

By placing the two equations first written above, under the form

$$x^2 = \frac{A^2}{B^2}(B^2 - y^2)$$

$$x'^2 = \frac{A^2}{B}(B^2 - y'^2)$$

and proceeding as before, we should find

$$\frac{x^2}{x'^2} = \frac{(B+y)\ (B-y)}{(B+y')\ (B-y')}$$

in which $B+y$, $B-y$ are the abscessas of the axis BB', corresponding to the ordinate $x = CG = Pg$; and $B+y'$, $B-y'$ are those corresponding to the ordinate $x' = CG' = P'g'$.

Hence the theorem; *the squares of the ordinates, etc.*

PROPOSITION XV.—THEOREM.

The parameter of the transverse axis of an ellipse, or, the latus rectum, is the double ordinate to this axis through the focus.

Let F and F' be the foci of an ellipse of which AA' and BB' respectively are the major and minor axes.

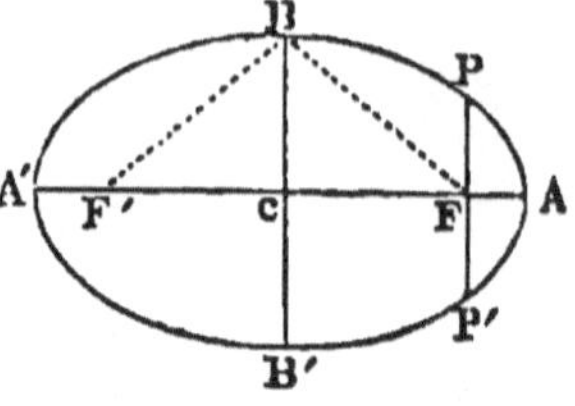

Through the focus F draw the double ordinate PP'. Then will PP' be the parameter of the major axis.

We will denote the semi-major axis by A, the semi-minor axis by B, the ordinate through the focus by P, and and the distance from the center to the focus by c.

The equation of the curve referred to the center and axis, is

$$A^2y^2+B^2x^2=A^2B^2.$$

If in this equation we substitute c for x, y will become P, and we have

$$A^2P^2+B^2c^2=A^2B^2.$$

Transposing the term B^2c^2, and factoring the second member of the resulting equation, it becomes

$$A^2P^2=B^2\,(A^2-c^2) \qquad (1)$$

In the right-angled triangle BCF, since $BF=A$ (Prop. 4) and $Bc=B$, we have $A^2-c^2=B^2$.

Replacing A^2-c^2 in eq. (1) by its value, that equation becomes

$$A^2 \cdot P^2=B^2 \cdot B^2$$

Or, by taking the square roots of both members,

$$A\cdot P=B\cdot B$$

Whence, $\qquad A:B::B:P$

Or, $\qquad 2A:2B::2B:2P$

$2P$ is therefore a third proportional to the major and minor axes, and (Def. 14) it is the parameter of the former axis.

Hence the theorem; *the parameter, etc.*

PROPOSITION XVI.—THEOREM.

The area of an ellipse is a mean proportional between two circles described, the one on the major, and the other on the minor axis as diameters.

On the major axis AA' of the ellipse represented in the figure, describe a circle, and suppose this axis to be divided into any number of equal parts.

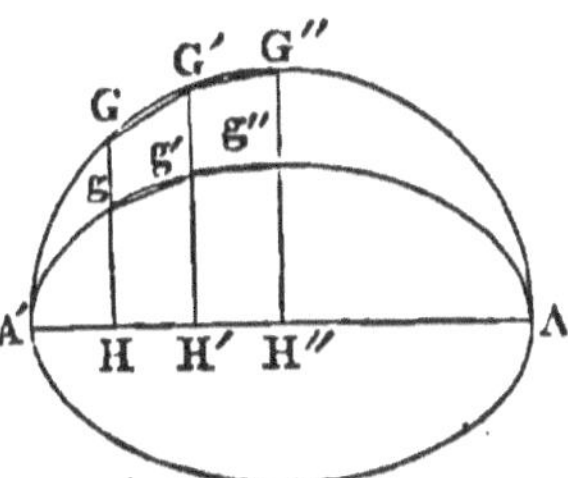

Through the points of division draw ordinates to the circle, and join the extremities of these consecutive ordinates, and also those of the corresponding ordinates of the ellipse, by straight lines. We shall thus form in the semi-circle a number of trapezoids, and a like number in the semi-ellipse.

Let GH, $G'H'$ be two adjacent ordinates of the circle, and gH $g'H'$ those of the ellipse answering to them; and let us denote GH by Y, $G'H'$ by Y', gH by y, $g'H'$ by y', and the part HH' of the axis by x.

The trapezoidal areas, $GHH'G'$, $gHH'g'$, are respectively measured by

$$\frac{Y+Y'}{2}\cdot x \text{ and } \frac{y+y'}{2}\cdot x \text{ (Th. 34, B. I, Geom.)}$$

But (Prop. 13, Scho. 2)

$$A : B :: Y : y$$
$$:: Y' : y'$$

Hence (Th. 7, B. II, Geom.)

$$A : B :: Y+Y' : y+y' :: \frac{Y+Y'}{2} : \frac{y+y'}{2}$$

or,

$$A : B :: \frac{Y+Y'}{2}\cdot x : \frac{y+y'}{2}\cdot x$$

If the ordinates following Y', y' in order, be represented by Y'', y'', etc., we shall also have

3*

$$A : B : : \frac{Y' + Y''}{2} \cdot x : \frac{y' + y''}{2} \cdot x$$

That is, any trapezoid in the circle will be to the corresponding trapezoid in the ellipse, constantly in the ratio of A to B; and therefore the sum of the trapezoids in the circle will be to the sum of the trapezoids in the ellipse as A is to B; and this will hold true, however great the number of trapezoids in each.

Calling the first sum S, and the second s, we shall then have

$$A : B : : S : s$$

But, when the number of equal parts into which the axis AA' is divided, is increased without limit, S becomes the area of the semi-circle and s that of the semi-ellipse.

Therefore, $A : B : :$ area semi-circle : area semi-ellipse.

Or, $\qquad A : B : :$ area circle : area ellipse.

By substituting in this last proportion for area circle, its value πA^2, it becomes

$$A : B : : \pi A^2 : \text{area ellipse.}$$

Whence $\qquad$ area ellipse$= \pi A B$,

which is a mean proportional between πA^2 and πB^2.

Hence the theorem; *the area of an ellipse, etc.*

SCHOLIUM.—This theorem leads to the following rule in mensuration for finding the area of an ellipse.

RULE.=*Multiply the product of the semi-major and semi-minor axes by* 3.1416.

PROPOSITION XVII.—THEOREM.

If a cone be cut by a plane making an angle with the base less than that made by an element of the cone, the section is an ellipse.

Let V be the vertex of a cone, and suppose it to be cut by a plane at right-angles to the plane of the opposite elements, *VN VB*, these elements being cut by the first plane at *A* and *B*. Then, if the secant plane be not parallel to the base of the cone, the section will be an ellipse, of which *AB* is the major axis.

Through any two points, *F* and *H*, on *AB*, draw the lines *KL, MN*, parallel to the base of the cone, and

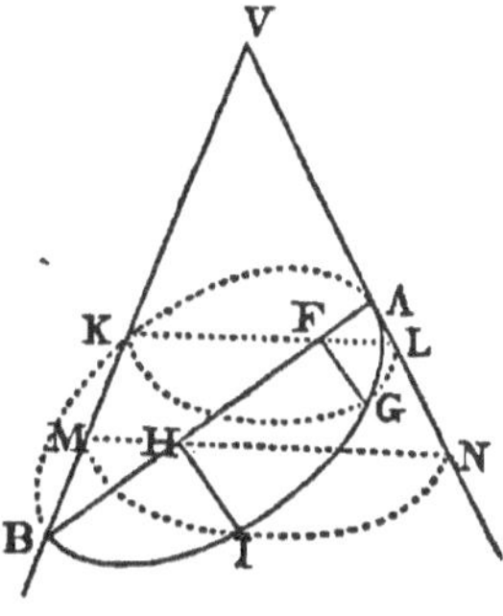

through these lines conceive planes to be passed also parallel to this base. The sections of the cone made by these planes will be circles, of which *KGL* and *MIN* are the semi-circumferences, passing the first through *G*, and the second through *I*, the extremities of the perpendiculars to *BA*, lying in the section made by the oblique plane.

The triangles *AFL, AHN*, are similar; so also are the triangles *BMH, BKF;* and from them we derive the following proportions:

$$AF : FL :: AH : HN$$
$$BF : KF :: BH : HM$$

By multiplication, $AF{\cdot}BF : FL{\cdot}KF :: AH{\cdot}BH : HN{\cdot}HM$

Because *KL* is a diameter of a circle, and *FG* an ordinate to this diameter, we have

$$KF{\cdot}FL = \overline{FG}^2,$$

and for a like reason, $HM{\cdot}HN = \overline{HI}^2$

Therefore, $AF{\cdot}BF : \overline{FG}^2 :: AH{\cdot}HB : \overline{HI}^2$

or, $AF{\cdot}BF : AH{\cdot}HB :: \overline{FG}^2 : \overline{HI}^2$

This proportion expresses the property of the ellipse proved in (Prop. 14); and the section *A GIB* is, therefore, an ellipse.

Hence the theorem; *if a cone be cut, etc.*

SCHOLIUM.—The proportion $AF{\cdot}BF : AH{\cdot}HB :: \overline{FG}^2 : \overline{HI}^2$ would still hold true, were the line *AB* parallel to the base of the cone, and the section a circle; the ratios would then become equal to unity. The circle may therefore be regarded as a particular case of the ellipse.]

PROPOSITION XVIII.—THEOREM.

If, from one of the vertices of each of two conjugate diameters of an ellipse, ordinates be drawn to either axis, the sum of the squares of these ordinates will be equal to the square of the other semi-axis.

Let $APP'A'QQ'$ be an ellipse, of which AA' is the major and BB' the minor axis; also let PQ, $P'Q'$ be any two conjugate diameters. Through the vertices of these

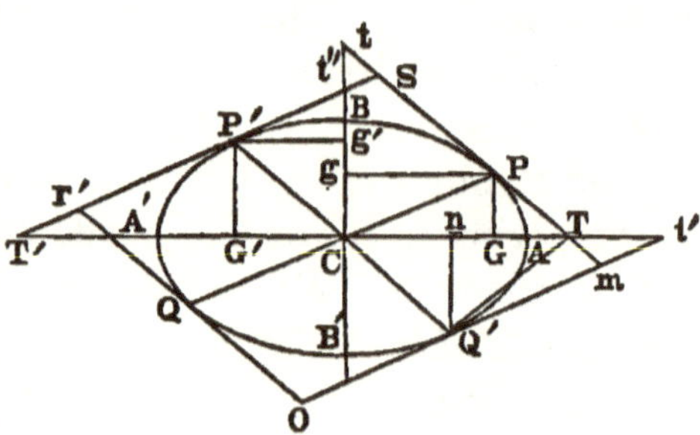

diameters draw the tangents to the ellipse and the ordinates to the axes, as represented in the figure. Then we are to prove that

$$\overline{CA}^2=(Pg)^2+(P'g')^2=\overline{CG}^2+\overline{CG'}^2$$

and $$\overline{CB}^2=(PG)^2+(P'G')^2=(Cg)^2+(Cg')^2$$

Now (by Prop. 11) we have

$$OT:CA::CA:CG,$$

also, $$Ct:CA::CA:Cn$$

Whence, $$\overline{CA}^2=CT\cdot CG, \qquad (1)$$

and $$\overline{CA}^2=Ct'\cdot Cn.$$

Therefore, $$OT\cdot CG=Ct\cdot Cn,$$

which, resolved into a proportion, gives

$$Ct:CT::CG:Cn \qquad (2)$$

By the construction, it is evident that the triangles CPT, $CQ't$, are similar, as are also the triangles PGT and $CQ'n$.

From these triangles we derive the proportions

$$Ct' : CT :: CQ' : PT$$
$$CQ : PT :: Cn : GT$$

Whence, $\qquad Ct : CT :: Cn : GT$

Comparing the last proportion with proportion (2) above, we have

$$CG : Cn :: Cn : GT$$

Whence, $\qquad (Cn)^2 = CG \cdot GT$

But $\quad GT = CT - CG;$ then $(Cn)^2 = CG\,(CT - CG),$ from which we get

$$(Cn)^2 + \overline{CG}^2 = CG \cdot CT = \overline{CA}^2 \quad \text{(See eq. 1.)}$$

Substituting, in this equation, for $(Cn)^2$, its equal $\overline{CG'}^2$, it becomes

$$\overline{CA}^2 = \overline{CG}^2 + \overline{CG'}^2$$

In a similar manner it may be proved that

$$\overline{CB}^2 = \overline{PG}^2 + \overline{P'G'}^2$$

Hence the theorem; *if from one of the vertices of each, etc.*

PROPOSITION XIX.—THEOREM.

The sum of the squares of any two conjugate diameters of an ellipse is a constant quantity, and equal to the sum of the squares of the axes.

The annexed figure, being the same as that employed in the preceding proposition, by that proposition we have

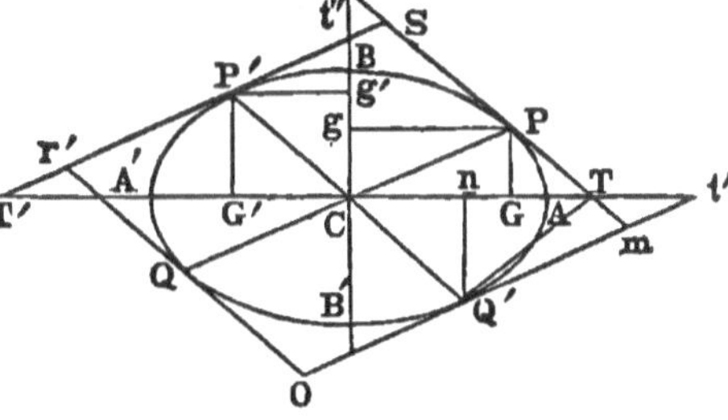

$$\overline{CA}^2 = \overline{CG}^2 + \overline{CG'}^2$$

and

$$\overline{CB}^2 = \overline{PG}^2 + \overline{P'G'}^2$$

By addition, $\overline{CA}^2 + \overline{CB}^2 = \overline{CG}^2 + \overline{PG}^2 + \overline{CG'}^2 + \overline{P'G'}^2$

But CG and PG are the two sides of the right-angled triangle CPG, and CG' and $P'G'$ are the two sides of the right-angled triangle $CP'G'$;

Therefore, $\quad \overline{CA}^2 + \overline{CB}^2 = \overline{CP}^2 + \overline{CP'}^2$

Whence, $\quad \overline{4CA}^2 + \overline{4CB}^2 = \overline{4CP}^2 + \overline{4CP'}^2$

The first member of this equation expresses the sum of the squares of the axes, and the second member the sum of the squares of the two conjugate diameters.

Hence the theorem; *the sum of the squares of any two, etc.*

PROPOSITION XX.—THEOREM.

The parallelogram formed by drawing tangents through the vertices of any two conjugate diameters of an ellipse, is equal to the rectangle of the axes.

Employing the figure of the last two propositions, we have, from proposition 18,

$$\overline{CA}^2 = \overline{CG}^2 + \overline{CG}'$$

from which, by transposition and factoring the second member, we get

$$\overline{CG}^2 = (CA + CG')\,(CA - CG') = AG'\cdot A'G'$$

But $\quad \overline{CA}^2 : \overline{CB}^2 :: AG'\cdot A'G' : \overline{P'G'}^2;$ $\quad$ (Prop. 13.)

Whence, $\overline{CA}^2 : \overline{CB}^2 :: \overline{CG}^2 : \overline{P'G'}^2$

Or, $\quad CA : CB :: CG : P'G' = Q'n$ (1)

But, $\quad CT : CA :: CA : CG$ (2) $\quad\quad$ (Prop. 11.)

Multiplying proportions (1) and (2), term by term, omitting, in the first couplet of the resulting proportion, the common factor CA, and in the second couplet the common factor CG, we find

$$CT : CB :: CA : Q'n$$

Whence, $\qquad CT \cdot Q'n = CA \cdot CB$

Or, $\qquad 4CT \cdot Q'n = 4CA \cdot CB$

The first member of this equation measures eight times the area of the triangle $CQ'T$, and this triangle is equivalent to one half of the parallelogram $CQ'mP$, because it has the same base, CQ', as the parallelogram, and its vertex is in the side opposite the base. This parallelogram is obviously one fourth of that formed by the tangent lines through the vertices of the conjugate diameters; $4CT.Q'n$ therefore, measures the area of this parallelogram. Also, $4CA \cdot CB$ is the measure of the rectangle that would be formed by drawing tangent lines through the vertices of the major and minor axes of the ellipse.

Hence, the theorem; *the parallelogram formed, etc.*

PROPOSITION XXI.—THEOREM.

If a normal line be drawn to an ellipse at any point, and also an ordinate to the major axis from the same point, then will the square of the semi-major axis be to the square of the semi-minor axis, as the distance from the center to the foot of the ordinate is to the sub-normal on the major axis.

Let P be the assumed point in the ellipse, and through this point draw the tangent PT, the normal PD, and the ordinate PG, to the major axis; then C being the center of the ellipse, and A denoting the semi-major, and B the semi-minor axis, it is to be proved that

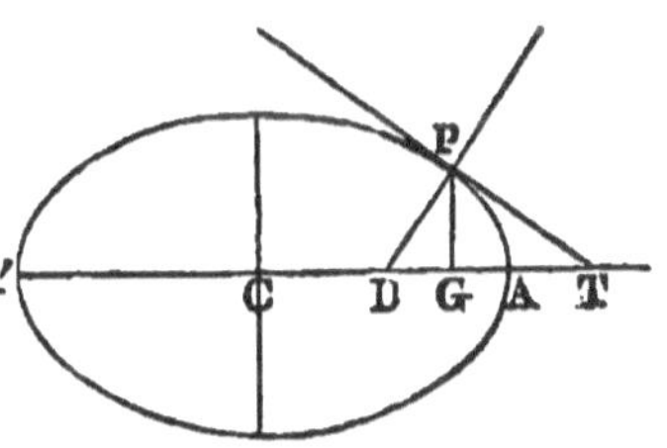

$$A^2 : B^2 :: CG : DG$$

By (Prop. 13) we have

$$A^2 : B^2 :: A'G \cdot AG : \overline{PG}^2 \qquad (1)$$

and because DPT is a right-angled triangle, and PG is a

perpendicular let fall from the vertex of the right-angle upon the hypotenuse, we also have

(Th. 25, B. II, Geom.) $\overline{PG}^2 = DG \cdot GT$

But $A'G \cdot AG = CG \cdot GT$ (Scho. 2, Prop. 12)

Substituting in proportion (1), for the terms of the second couplet, their values, it becomes

$$A^2 : B^2 :: CG \cdot GT : DG \cdot GT$$

or $\qquad A^2 : B^2 :: CG : DG.$

Hence the theorem; *if a normal line be drawn, etc.*

Cor. If $CG = x$, then this theorem will give for the subnormal, DG, the value $\dfrac{B^2}{A^2} x$, which is its analytical expression.

PROPOSITION XXII.—THEOREM.

If two tangents be drawn to an ellipse, the one through the vertex of the major axis and the other through the vertex of any other diameter, each meeting the diameter of the other produced, the two tangential triangles thus formed will be equivalent.

Let PP' be any diameter of the ellipse whose major axis is AA'. Draw the tangents AN and PT, the first meeting the diameter produced at N, and the second the axis produced at 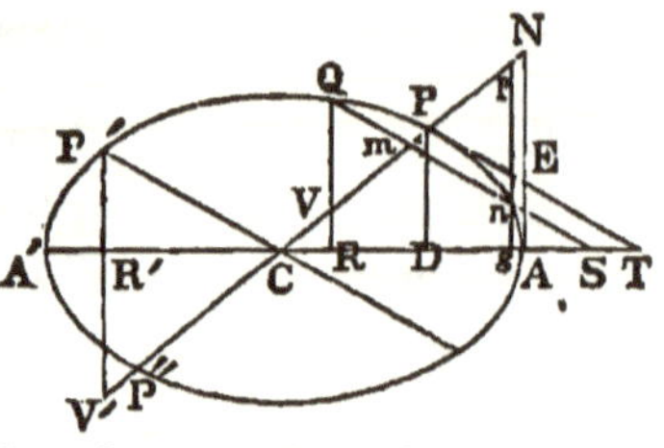 T; the triangles CAN and CPT thus formed are equivalent.

Draw the ordinate PD; then by similar triangles we have

$$CD : CA :: CP : CN$$

But $\qquad CD : CA :: CA : CT$ (Prop. 11)

Whence $\quad CP : CN :: CA : CT$

Therefore, $\quad CP \cdot CT = CN \cdot CA$

Multiplying both members of this equation by sin. C, it becomes

$$CP \cdot CT \sin. C = CN \cdot CA \sin. C$$

or, $\quad \frac{1}{2}CT \cdot CP \sin. C = \frac{1}{2}CA \cdot CN \sin. C \qquad (1)$

But $\quad CP \cdot \sin. C = PD$, and $CN \cdot \sin. C = AN;$ therefore the first member of equation (1) measures the area of the triangle CPT, and the the second member measures that of the triangle CAN.

Hence the theorem; *if two tangents be drawn to an, etc.*

Cor. 1. Taking the common area $CAEP$, from each triangle, and there is left $\triangle PEN = \triangle AET$.

Cor. 2. Taking the common $\triangle CDP$, from each triangle, and there is left $\triangle PDT =$ trapezoidal area $PDAN$.

PROPOSITION XXIII.—THEOREM.

The supposition of Proposition 22 being retained, then, if a secant line be drawn parallel to the second tangent, and ordinates to the major axis be drawn from the points of intersection of the secant with the curve, thus forming two other triangles, these triangles will be equivalent each to each to the corresponding trapezoids cut off, by the ordinates, from the triangle determined by the tangent through the vertex of the major axis.

Draw the secant QnS parallel to the tangent PT, and also the ordinates QR, ng, producing the latter to p. Then is $\triangle SQR =$ trapezoid $ANVR$, and $\triangle Sng =$ trapezoid $ANpg$.

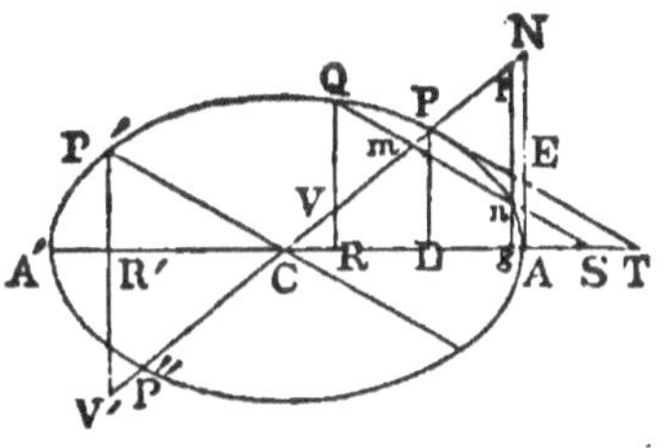

The three triangles, CVR, CPD, CNA are similar, by construction; therefore,

$$\triangle CNA : \triangle CPD : : \overline{CA}^2 : : \overline{CD}^2$$

Whence,

$$\text{trapezoid } ANPD : \triangle CNA : : \overline{CA}^2 - \overline{CD}^2 : \overline{CA}^{2\,(1)}$$

(Th. 8, B. II, Geom.)

4

In like manner,

$$\text{trapezoid } ANVR : \triangle CNA :: \overline{CA}^2 - \overline{CR}^2 : \overline{CA}^2 \quad (2)$$

Dividing proportion (1) by (2), term by term, we get

$$\frac{\text{trapezoid } ANPD}{\text{trapezoid } ANVR} : 1 :: \frac{\overline{CA}^2 - \overline{CD}^2}{\overline{CA}^2 - \overline{CR}^2} : 1$$

Whence,

$$\text{trapez. } ANPD : \text{trapez. } ANVR :: \overline{CA}^2 - \overline{CD}^2 : \overline{CA}^2 - \overline{CR}^2$$

But $\overline{PD}^2 : \overline{QR}^2 :: A'D \cdot DA : A'R \cdot RA$, (Prop. 14); and since

$$A'D = CA + CD, \ A'R = CA + CR, \ DA = CA - CD \text{ and}$$

$RA = CA - CR$, we have

$$\overline{PD}^2 : \overline{QR}^2 :: (CA + CD)(CA - CD) : (CA + CR)$$

$$(CA - CR) :: \overline{CA}^2 - \overline{CD}^2 : \overline{CA}^2 - \overline{CR}^2$$

Therefore,

$$\text{trapezoid } ANPD : \text{trapezoid } ANVR :: \overline{PD}^2 : \overline{QR}^2,$$

But the trapezoid $ANPD = \triangle TPD$, (Cor. 2, Prop. 22); whence,

$$\triangle TPD : \text{trapezoid } ANVR :: \overline{PD}^2 :: \overline{QR}^2 \quad (3)$$

and since the triangles TPD and SQR are similar, we have

$$\triangle TPD : \triangle SQR :: \overline{PD}^2 : \overline{QR}^2 \quad (4)$$

By comparing proportions (3) and (4) we find

$$\triangle TPD : \text{trapezoid } ANVR :: \triangle TPD : \triangle SQR$$

Whence, trapezoid $ANVR = \triangle SQR$;

and by a similar process we should find that

$$\text{trapezoid } ANpg = \triangle Sng.$$

Hence the theorem ; *if a secant line be drawn parallel, etc.*

Cor. 1. Taking the trapezoid $ANpg$ from the trapezoid $ANVR$, and the $\triangle Sng$ from the $\triangle SQR$, we have

$$\text{trapezoid } gpVR = \text{trapezoid } gnQR.$$

Cor. 2. The spaces $ANVR$, $TPVR$, and SQR are equivalent, one to another.

Cor. 3. Conceive QR and QS to move parallel to their present positions, until R coincides with C; then QR

becomes the semi-minor axis, the space $ANVR$ the triangle ANC, and the $\triangle QRS$ equivalent to the $\triangle CPT$.

PROPOSITION XXIV.—THEOREM.

Any diameter of the ellipse bisects all of the chords of the ellipse drawn parallel to the tangent through the vertex of the diameter.

By Cor. 1 to the preceding proposition we have trapez. $gpVR$=trapez. $gnQR$. If from each of these equals we subtract the common area $gnmVR$, there will remain the 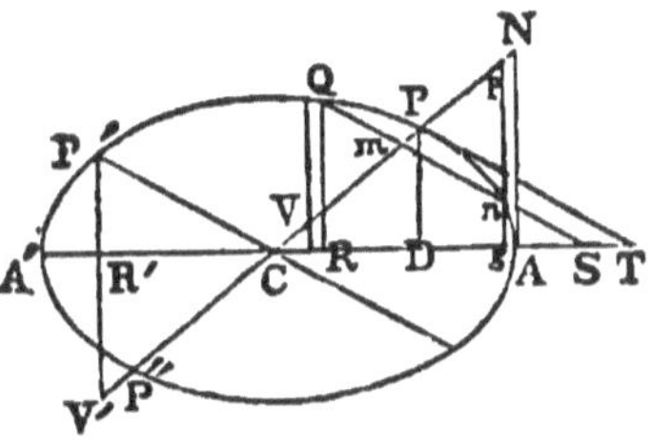 $\triangle mnp$, equivalent to the $\triangle QmV$; and as these triangles are also equi-angular, they are absolutely equal.

Therefore, $Qm=mn$.

Hence the theorem ; *any diameter of the ellipse bisects, etc.*

REMARK.—The property of the ellipse demonstrated in this proposition is merely a generalization of that previously proved in Prop. 3.

PROPOSITION XXV.—THEOREM.

The square of any semi-diameter of an ellipse is to the square of its semi-conjugate, as the rectangle of any two abscissas of the former diameter is to the square of the corresponding ordinate.

Let AA' be the major axis of the ellipse, CP any semi-diameter and CP' its semi-conjugate. Draw the tangents TP and AN, the ordinate Qm, producing it to meet 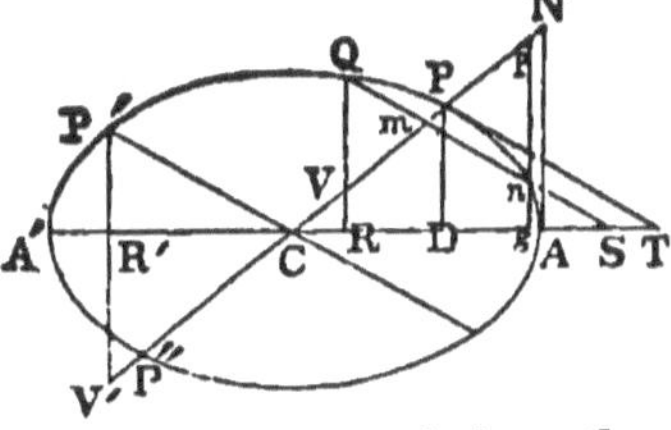 the axis at $S;$ and $P'V'$, parallel to AN, and in other respects make the construction as indicated in the figure. It is then to be proved that

$$\overline{CP}^2 : \overline{CP'}^2 :: Pm \cdot mP'' : \overline{Qm}^2$$

Now in the present construction, the triangles $CP'R'$ and $CV'R'$ take the place of the triangles SQR and CVR respectively, in Prop. 23; and hence by that proposition, the triangles $CP'V'$, CAN, and CPT are equivalent one to another.

The triangles CPT and CmS are similar; therefore,

$$\triangle CPT : \triangle CmS :: \overline{CP}^2 : \overline{Cm}^2$$

Whence,

$$\triangle CPT : \triangle CPT - \triangle CmS :: \overline{CP}^2 : \overline{CP}^2 - \overline{Cm}^2$$

Or, $\triangle CPT :$ trapez. $mPTS :: \overline{CP}^2 : \overline{CP}^2 - \overline{Cm}^2$ (1)

From the similar triangles, $CP'V'$ and mQV, we have

$$\triangle CP'V' : \triangle mQV :: \overline{CP'}^2 : \overline{mQ}^2$$

But area $SmVR + \triangle CVR + \triangle mQV =$ area $SmVR + \triangle CVR +$ trapez. $mPTS$, (Prop. 23.); therefore, $\triangle mQV =$ trapez. $mPTS$; also $\triangle CP'V' = \triangle CPT$.

Substituting these values in the preceding proportion, it becomes

$$\triangle CPT : \text{trapez. } mPTS :: \overline{CP'}^2 : \overline{mQ}^2 \quad (2)$$

By comparing proportions (1) and (2), we get

$$CP^2 : \overline{CP}^2 - \overline{Cm}^2 :: \overline{CP'}^2 : \overline{mQ}^2$$

Or, $\qquad CP^2 : \overline{CP'}^2 :: \overline{CP}^2 - \overline{Cm}^2 : \overline{mQ}^2$

Whence, $\overline{CP}^2 : \overline{CP'}^2 :: (CP + Cm)(CP - Cm) : \overline{mQ}^2$

Or, $\qquad CP^2 : \overline{CP'}^2 :: P''m \cdot mP : \overline{mQ}^2$

Hence the theorem; *the square of any semi-diameter, etc.*

Remark. The property of the ellipse relating to conjugate diameters, established by this proposition, is but the generalization of that before demonstrated in reference to the axes, in Prop. 13.

THE PARABOLA.

DEFINITIONS.

1. The **Parabola** is a plane curve, generated by the motion of a point subjected to the condition that its distances from a fixed point and a fixed straight line shall be constantly equal.

2. The fixed point is called the *focus* of the parabola, and the fixed line the *directrix*.

Thus, in the figure, F is the focus and BB'' the directrix of the parabola $PVP'P''$, etc.

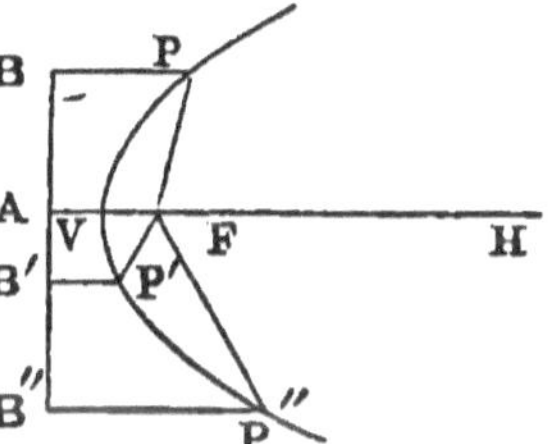

3. A **Diameter** of the parabola is a line drawn through any point of the curve, in a direction from the directrix, and at right-angles to it.

4. The **Vertex** of a diameter is the point of the curve through which the diameter is drawn.

5. The **Principal Diameter**, or the **Axis**, of the parabola is the diameter passing through the focus. The vertex of the axis is called the *principal vertex*, or simply *the vertex* of the parabola.

The vertex of the parabola bisects the perpendicular distance from the focus to the directrix, and all the diameters of the parabola are parallel lines.

6. An **Ordinate** to a diameter is a straight line drawn from any point of the curve to the diameter, parallel to the

4*

tangent line through its vertex. Thus, *PD*, drawn parallel to the tangent $V'T$, is an ordinate to the diameter $V'D$. It will be shown that $DP=.DG$; and hence *PG* is called a *double ordinate*.

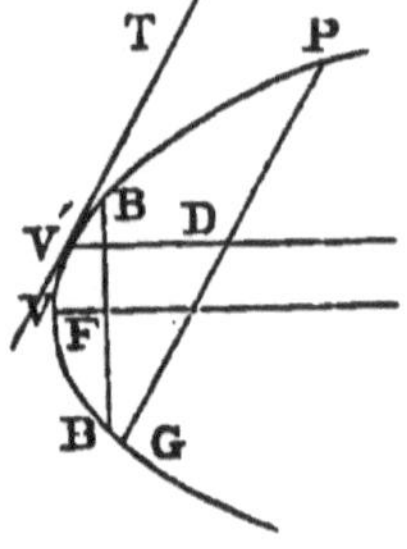

7. An **Abscissa** is the part of the diameter between the vertex and an ordinate. Thus, $V'D$ is the abscissa corresponding to the ordinate *PD*.

8. The **Parameter** of any diameter of the parabola is one of the extremes of a proportion, of which any ordinate to the diameter is the mean, and the corresponding abscissa the other extreme.

9. The parameter of the axis of the parabola is called the *principal parameter*, or simply the *parameter of the parabola*. It will be shown to be equal to the double ordinate to the axis through the focus. Thus, BB', the chord drawn through the focus at right-angles to the axis, is the parameter of the parabola.

The principal parameter is sometimes called the *latus-rectum*.

10. A **Sub-tangent**, on any diameter, is the distance from the point of intersection of a tangent line with the diameter produced to the foot of that ordinate to this diameter that is drawn from the point of contact.

11. A **Sub-normal**, on any diameter, is the part of the diameter intercepted between the normal to the curve, at any point, and the ordinate from the same point to the diameter. Thus, in the figure, $V'N$ being any diameter, *PT* a tangent, and *PN* a normal at the point *P*, and *PQ* an ordinate to the diameter; then *TQ* is a sub-tangent and *QN* a sub-normal on this diameter.

When the terms, sub-tangent and sub-normal, are used without reference to the diameter on which they are taken, the axis will always be understood.

PROPOSITION I.—PROBLEM.

To describe a parabola mechanically.

Let CD be the given line, and F the given point. Take a square, as DBG, and to one side of it, GB, attach a thread, and let the thread be of the same length as the side GB of the square. Fasten one end of the thread at the point G, the other end at F. 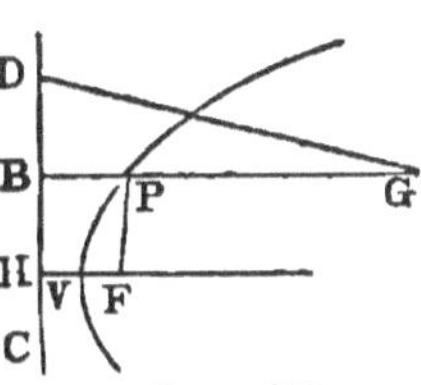

Put the other side of the square against the given line, CD, and with the point of a pencil, in the thread, bring the thread up to the side of the square. Slide the side BD of the square along the line CD, and at the same time keep the thread close against the other side, permitting the thread to slide round the point of the pencil. As the side BD of the square is moved along the line CD, the pencil will describe the curve represented as passing through the points V and P.

For $\quad GP+PF=$ the length of the thread,

and $\quad GP+PB=$ the length of the thread.

By subtraction, $PF-PB=0$, or $PF=PB$.

This result is true at any and every position of the point P; that is, it is true for every point on the curve corresponding to definition 1.

Hence, $\quad\quad\quad FV=VH.$

If the square be turned over and moved in the opposite direction, the other part of the parabola, on the other side of the line FH, may be described.

Cor. It is obvious that chords of the curve which are perpendicular to the axis, are bisected by it.

PROPOSITION II.—THEOREM.

Any point within the parabola, or on the concave side of the curve, is nearer to the focus than to the directrix; and any point without the parabola, or on the convex side of the curve, is nearer to the directrix than to the focus.

Let F be the focus and HB' the directrix of a parabola.

First.—Take A, any point within the curve. From A draw AF to the focus, and AB perpendicular to the directrix; then will AF be less than AB.

Since A is within the curve, and B is without it, the line AB must cut the curve at some point, as P. Draw PF. By the definition of the parabola, $PB=PF$; adding PA to each member of this equation, we have

$$PB+PA=BA=PA+PF$$

But PA and PF being two sides of the triangle APF, are together greater than the third side AF; therefore their equal, BA, is greater than AF.

Second.—Now let us take any point, as A', without the curve, and from this point draw $A'F$ to the focus, and $A'B'$ perpendicular to the directrix.

Because A' is without the curve and F is within it, $A'F$ must cut the curve at some point, as P. From this point let fall the perpendicular, BP, upon the directrix, and draw $A'B$.

As before, $PB=PF$; adding $A'P$ to each member of this equation, and we have $A'P+PB=A'P+PF=A'F$. But $A'P$ and PB being two sides of the triangle $A'PB$, are together greater than the third side, $A'B$; therefore their equal, $A'F$, is greater than $A'B$. Now $A'B$, the hypotenuse of the right-angled triangle $A'BB'$ is greater than either side; hence, $A'B$ is greater than $A'B'$; much more then is $A'F$ greater than $A'B'$.

Hence the theorem; *any point within the parabola, etc.*

Cor. Conversely: *If the distance of any point from the directrix is less than the distance from the same point to the focus, such point is without the parabola; and, if the distance from any point to the directrix is greater than the distance from the same point to the focus, such point is within the parabola.*

First.—Let A' be a point so taken that $A'B' < A'F$. Now A' is not a point on the curve, since the distances $A'B'$ and $A'F$ are unequal; and A' is not within the curve, for in that case $A'B'$ would be greater than $A'F$ according to the proposition, which is contrary to the hypothesis. Therefore A' being neither on nor within the parabola, must be without it.

Second.—Let A be a point so taken that $AB > AF$. Then, as before, A is not on the curve, since AF and AB are unequal; and A is not without the curve, for in that case AB would be less than AF, which is contrary to the hypothesis. Therefore, since A is neither on nor without the parabola, it must be within it.

PROPOSITION III.—THEOREM.

If a line be drawn from the focus of a parabola to any point of the directrix, the perpendicular that bisects this line will be a tangent to the curve.

Let F be the focus, and HD the directrix of a parabola.

Assume any point whatever, as B, in the directrix, and join this point to the focus by the line BF; then will tA, the perpendicular to BF through its middle point t, be a tangent to the parabola. Through B draw BL perpendicular to the directrix, and join P, its intersection with tP, to the focus. Then, since P is a point in the perpendicular to BF at its middle point, it is equally distant from the extremities of BF; that is, $PB = PF$. P is therefore a point in the parabola, (Def. 1). Hence, the line tP meets the curve at the point P.

We will now prove that all other points in the line tP are without the parabola. Take A, any point except P in the line tP, and draw AF, AB; also draw AD perpendicular to the directrix. AF is equal to AB, because A is a point in the perpendicular to BF at its middle point; but AB, the hypotenuse of the right-angled triangle ABD, is greater than the side AD; therefore AD is less than AF, and the point A is without the parabola. (Cor., Prop. 2). The line tA and the parabola have then no point in common except the point P. This line is therefore tangent to the parabola.

SCHOLIUM 1.—The triangles BPt and FPt are equal; therefore the angles FPt and BPt are equal. Hence, to draw a tangent to the parabola at a given point, we have the following

RULE.—*From the given point draw a line to the focus, and another perpendicular to the directrix, and through the given point draw a line bisecting the angle formed by these two lines. The bisecting line will be the required tangent.*

SCHOLIUM 2.—Just at the point P the tangent and the curve coincide with each other; and the same is true at every point of the curve. Now, because the angles BPt and FPt are equal, and the angles BPt and LPA are vertical, it follows that the angles LPA and FPt are equal. Hence it follows, from the law of reflection, that if rays of light parallel to the axis VF be incident upon the curve, they will all be reflected to the focus F. If therefore a reflecting surface were formed, by turning a parabola about its axis, all the rays of light that meet it parallel with the axis, will be reflected to the focus; and for this reason many attempts have been made to form perfect parabolic mirrors for reflecting telescopes.

If a light be placed at the focus of such a mirror, it will reflect all its rays in one direction; hence, in certain situations, parabolic mirrors have been made for lighthouses, for the purpose of throwing all the light seaward.

Cor. 1. The angle BPF continually increases, as the

pencil P moves toward V, and at V it becomes equal to two right angles; and the tangent at V is perpendicular to the axis, which is called the *vertical tangent.*

Cor. 2. The vertical tangent bisects all the lines drawn from the focus of a parabola to the directrix.

Let Vt be the vertical tangent; then because the two right-angled triangles FVt and FHB are similar, and $VF = VH$, we have $Ft = tB$.

PROPOSITION IV.—THEOREM.

The distance from the focus of a parabola to the point of contact of any tangent line to the curve, is equal to the distance from the focus to the intersection of the tangent with the axis.

Through the point P of the parabola of which F is the focus and BH the directrix, draw the tangent line PT, meeting the axis produced at the point T; then will FP be equal to FT 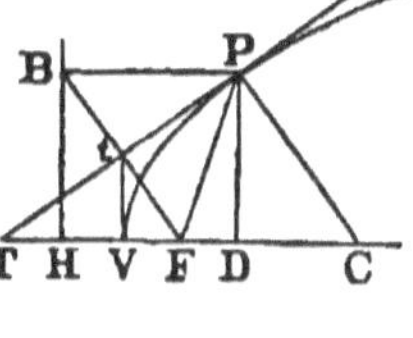

Draw PB perpendicular to the directrix, and join F, B.

The angles BPT and TPF are equal, (Scho. 1, Prop. 3); and since PB is parallel to TC, the alternate angles BPT, and PTC are also equal. Hence the angle TPF is equal to the angle PTF, and the triangle PFT is isosceles; therefore $FP = FT$.

Hence the theorem; *the distance from the focus to, etc.*

Scholium.—To draw a tangent line to a parabola at a given point, we have the following

Rule.—*Produce the axis, and lay off on it from the focus a distance equal to the distance from the focus to the point of contact. The line drawn through the point thus determined and the given point will be the required tangent.*

PROPOSITION V.—THEOREM.

The perpendicular distance from the focus of a parabola to any tangent to the curve, is a mean proportional between the distance from the focus to the vertex and the distance from the focus to the point of contact.

In the figure of the preceding proposition draw in addition the vertical tangent Vt; then we are to prove that $\overline{Ft}^2 = VF \cdot FP$. Because TtF and VFt are similar right-angled triangles, we have

$$TF : Ft : : Ft : VF. \quad \text{But } TF = PF, (\text{Prop. 4});$$

therefore, $PF : Ft : : Ft : VF$

Whence, $\overline{Ft}^2 = PF \cdot VF$

Hence, the theorem; *the perpendicular distance from, etc.*

PROPOSITION VI.—THEOREM.

The sub-tangent on the axis of the parabola is bisected at the vertex.

In the figure which is constructed as in the two preceding propositions, draw in addition the ordinate PD, from the point of contact to the axis; then we are to prove that TD is bisected at the vertex V.

The two right-angled triangles TFt and tFP have the side Ft common, and the angle FTt equal to the angle FPt; hence the remaining angles are equal, and the triangles themselves are equal; therefore $tT = tP$. From the similar triangles TDP, TVt, we have the proportion

$$Tt : tP : : TV : VD$$

But $tT = tP$; whence $TV = VD$

Hence the theorem; *the sub-tangent on the axis, etc.*

Cor. Since $TV=\frac{1}{2}TD$, it follows that $Vt=\frac{1}{2}PD$. That is, *The part of the vertical tangent included between the vertex and any tangent line to the parabola, is equal to one half of the ordinate to the axis from the point of contact.*

PROPOSITION VII.—THEOREM.

The sub-normal is equal to twice the distance from the focus to the vertex of the parabola.

In the figure (which is the same as that of the last three propositions), PC is a normal to the parabola at the point C, and DC is the sub-normal; it is to be proved that $DC=2FV$.

Because BH and PD are parallel lines included between the parallel lines BP and HD, they are equal. BF and PC are also parallel, since each is perpendicular to the tangent PT; hence $BF=PC$, and also the two triangles HBF and DPC are equal.

Therefore $\qquad HF=DC;$
but $\qquad HF=2FV;$
whence $\qquad DC=2FV.$

Hence the theorem; *the sub-normal is equal to twice, etc.*

SCHOLIUM.—This proposition suggests another easy process for constructing a tangent to a parabola at a given point.

RULE.—*Draw an ordinate to the axis from a given point, and from the foot of this ordinate lay off on the axis, in the opposite direction of the vertex, twice the distance from the focus to the vertex. Through the point thus determined and the given point draw a line, and it will be the required tangent.*

PROPOSITION VIII.—THEOREM.

Any ordinate to the axis of a parabola is a mean proportional between the corresponding sub-tangent and sub-normal.

5 D

Assume any point, as P, in the parabola of which F is the focus and HB the directrix. Through this point draw the tangent PT, the normal PC, and the ordinate PD to the axis. Then in reference to the point P, TD is the sub-tangent, and DC the sub-normal on the axis; and we are to prove that

$$TD : PD :: PD : DC$$

The triangle TPC is right-angled at P, and PD is a perpendicular let fall from the vertex of this angle upon the hypotenuse. Therefore, PD is a mean proportional between the segments of the hypotenuse, (Th. 25, B. II, Geom.)

Hence the theorem; *any ordinate to the axis, etc.*

Scholium 1.—For a given parabola, the fourth term of the proportion, $TD : PD :: PD : DC$, is a constant quantity, and equal to twice the distance from the focus to the vertex, (Prop. 7). By placing the product of the means of this proportion equal to the product of the extremes, we have

$\overline{PD}^2 = TD \cdot DC = \frac{1}{2} TD \cdot 2DC$, which may be again resolved into the proportion

$$\tfrac{1}{2} TD : PD :: PD : 2DC$$

Or, $$VD : PD :: PD : 2DC$$

But VD is the abscissa, and PD is the ordinate of the point P; hence (Def. 8) $2DC$ is the parameter of the parabola, and is equal to four times the distance from the focus to the vertex, or to twice the distance from the focus to the directrix.

Scholium 2.—If we designate the ordinate PD by y, the abscissa VD by x, and the parameter by $2p$, the above proportion becomes

$$x : y :: y : 2p$$

Whence, $$y^2 = 2px.$$

This equation expresses the general relation between the abscissa and ordinate of any point of the curve, and is called, in Analytical Geometry, the equation of the parabola referred to its principal vertex as an origin.

Cor. *The sub-normal in the parabola is equal to one-half of the parameter.*

PROPOSITION IX.—THEOREM.

The parameter, or latus rectum, of the parabola is equal to twice that ordinate to the axis which passes through the focus.

Let F be the focus, and BB' the directrix of a parabola; and through the focus draw a perpendicular to the axis intersecting the curve at P and P'. From P and P' let fall the perpendiculars PB, $P'B'$, on the directrix. Then will $2PF$ be equal to $2FH$, or to the parameter of the parabola.

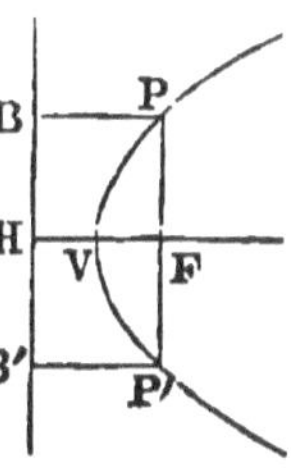

By the definition of the parabola, $PF=PB$; and because PP' and BB' are parallel, and the parallels PB and FH are included between them, we have $PB=FH$.

Hence $PF=FH$, or $2PF=2FH=$ the parameter. Scho. 1, Prob. 8.

Cor. Since the axis bisects those chords of the parabola which are perpendicular to it, $FP=FP'$. That is, FP'; therefore $PP'=2FH$. That is,

The parameter of the parabola is equal to the double ordinate through the focus.

PROPOSITION X.—THEOREM.

The squares of any two ordinates to the axis of a parabola are to each other as their corresponding abscissas.

Let y and y' denote the ordinates, and x and x' the abscissas of any two points of the parabola; then, by Scho. 2, Prop. 8, we have the two following equations:

$$y^2=2px \text{ and } y'^2=2px'$$

Dividing the first of these equations by the second, member by member, we have

$$\frac{y^2}{y'^2}=\frac{2px}{2px'}=\frac{x}{x'}$$

Whence $\qquad y^2 : y'^2 :: x : x'$

Hence the theorem; *the squares of any two ordinates, etc.*

PROPOSITION XI.—THEOREM.

*If a perpendicular be drawn from the focus of a parabola
to any tangent line to the curve, the intersection of the perpendicular with the tangent will be on the vertical tangent.*

Let F be the focus, and BH the directrix of the parabola, and PT a tangent to the curve at the point P.   From
F draw FB perpendicular to the tangent,
intersecting it at t, and the directrix at B.   We will now
prove that the point t is also the intersection of the vertical tangent with the tangent PT.

Because the triangle TFP is isosceles, the perpendicular Ft bisects the base PT; therefore $tP=tT$.   Again,
since Vt and DP are both perpendicular to the axis, they
are parallel, and the vertical tangent divides the sides of
the triangle TDP proportionally.

Hence, $TV: VD :: Tt : tP;$ but $TV= VD$ (Prop. 6)
therefore, $Tt=tP.$

That is, the tangent PT is bisected by both the perpendicular let fall upon it from the focus, and the vertical
tangent.   Therefore the tangent PT, the vertical tangent
and the perpendicular FB, meet in the common point t.

Hence the theorem ; *if a perpendicular be drawn, etc.*

PROPOSITION XII. THEOREM.

*The parameter of the parabo'a is to the sum of any two or-
dinates to the axis, as the difference of those ordinates is to the
difference of the corresponding abscissas.*

Take any two points, as P and Q, in the parabola represented in the following figure, and through these points
draw the double ordinates Pp and Qq. VD and VE are
the corresponding abscissas.

Draw PS and pt parallel to the axis.   Then, since

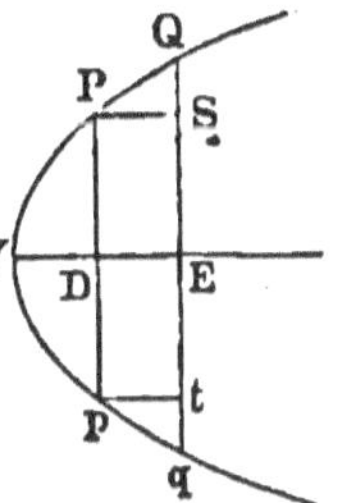

$PD = Dp$ and $QE = Eq$, we have $QE + PD = Qt$, equal to the sum of the two ordinates; and $QE - PD = QS$, equal to their difference; also $VE - VD = DE$, equal to the difference of the corresponding abscissas. We are now to prove that

$$2p : Qt :: QS : DE$$

in which $2p$ denotes the parameter of the parabola.

Because PD and QE are ordinates to the axis, we have (Scho. 2, Prop. 8)

$$\overline{PD}^2 = 2p \cdot VD \qquad (1)$$

and
$$\overline{QE}^2 = 2p \cdot VE \qquad (2)$$

Whence
$$\overline{QE}^2 - \overline{PD}^2 = 2p\,(VE - VD) = 2p \cdot DE \qquad (3)$$

But
$$\overline{QE}^2 - \overline{PD}^2 = (QE + PD)\,(QE - PD) = Qt \cdot QS,$$

therefore
$$Qt \cdot QS = 2p \cdot DE \qquad (4)$$

Whence
$$2p : Qt :: QS : DE$$

Hence the theorem; *the parameter of the parabola, etc.*

Cor. By dividing eq. (4) by eq. (2), member by member, we obtain

$$\frac{Qt \cdot QS}{\overline{QE}^2} = \frac{DE}{VE}$$

Whence
$$VE : DE :: \overline{QE}^2 : Qt \cdot QS$$

PROPOSITION XIII.—THEOREM.

If a tangent line be drawn to a parabola at any point, and from any point of the tangent a line be drawn parallel to the axis terminating in the double ordinate from the point of contact, this line will be cut by the curve into parts having to each other the same ratio as the segments into which it divides the double ordinate.

5*

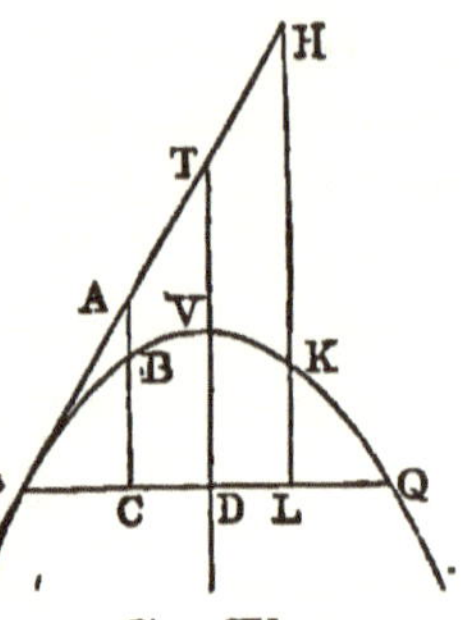

Take any point as P in the parabola represented in the figure, and of which VD is the axis, and through this point draw the tangent PT to the curve, and the double ordinate PQ to the axis. Assume a point in the tangent at pleasure, as A, and through it draw AC parallel to the axis, cutting the curve at B and the double ordinate at C. Then we are to prove that

$$AB : BC : : PC : CQ$$

By similar triangles we have

$$PC : CA : : PD : DT; \text{ but } DT=2DV \text{ (Prop. 6)}$$

therefore $\qquad PC : CA : : PD : 2DV \qquad\qquad$ (1)

But $\qquad\qquad DV : PD : : PD : 2p$ (Scho. 2, Prop. 8)

or $\qquad\qquad 2DV : PD : : 2PD : 2p.$

Inverting terms, $PD : 2DV : : 2p : 2PD=PQ$ (2)

By comparing proportions (1) and (2), we get

$$PC : CA : : 2p : PQ$$

But $\qquad\qquad 2p : CQ : : PC : BC \qquad$ (Prop. 12)

Multiplying the last two proportions, term by term, we have

$$2p{\cdot}PC : CA{\cdot}CQ : : 2p{\cdot}PC : BC{\cdot}PQ$$

The first and third terms of this proportion are equal; therefore the second and fourth are also equal. Hence we have the proportion

$$CA : BC : : PQ : CQ$$

Whence by division, $\quad CA{-}BC : BC : : PQ{-}CQ : CQ$

or $\qquad\qquad AB : BC : : PC : CQ$

If we take any other point, H, on the tangent, and through it draw the line HL parallel to the axis, intersecting the curve at K and the ordinate at L, we will have, in like manner,

$$HK : KL : : PL : LQ$$

Hence the theorem ; *if a tangent be drawn, etc.*

PROPOSITION XIV.—THEOREM.

If any two points be taken on a tangent line to a parabola, and through these points lines parallel to the axis be drawn to meet the curve, such lines will be to each other as the squares of the distances of the points from the point of contact.

The figure and construction being the same as in the foregoing proposition, we are to prove that

$$AB : HK :: \overline{PA}^2 : \overline{PH}^2$$

We have

$$AB : BC :: PC : CQ \quad \text{(1) (Prop. 13.)}$$

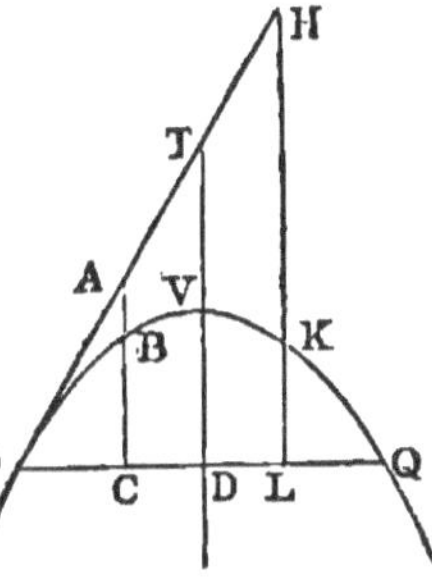

Multiplying the terms of the second couplet of this proportion by PC, it becomes

$$AB : BC :: \overline{PC}^2 : PC \cdot CQ \quad (2)$$

But, (Cor. Prop. 12) $\quad VD : BC :: \overline{PD}^2 : PC \cdot CQ \quad (3)$

Dividing proportion (2) by proportion (3), term by term, we have

$$\frac{AB}{VD} : 1 :: \frac{\overline{PC}^2}{\overline{PD}^2} : 1$$

Whence, $\qquad AB : VD :: \overline{PC}^2 : \overline{PD}^2 \quad (4)$

From the similar triangles, APC and TPD, we get the proportion

$$\overline{PA}^2 : \overline{PT}^2 :: \overline{PC}^2 : \overline{PD}^2 \quad (5)$$

By comparing proportions (4) and (5) we find

$$AB : VD :: \overline{PA}^2 : \overline{PT}^2 \quad (6)$$

In like manner we can prove that

$$HK : VD :: \overline{PH}^2 : \overline{PT}^2 \quad (7)$$

Dividing proportion (6) by proportion (7), term by term, we have

$$\frac{AB}{HK} : 1 :: \frac{\overline{PA}^2}{\overline{PH}^2} : 1$$

Whence, $\qquad AB : HK :: \overline{PA}^2 : \overline{PH}^2$

Hence the theorem; *if any two points be taken, etc.*

APPLICATION.—Conceive PH to be the direction in which a body thrown from the surface of the earth, would move, if it were undisturbed by the resistance of the air and by the force of gravity. It would then move along the line PH, passing over equal spaces in equal times. When a body falls under the action of gravity, one of the laws of its motion is, that *the spaces are proportional to the squares of the times of descent;* hence, if we suppose gravity to act upon the body in the direction AC, the lines AB, TV, HK, etc., must be to each other as the squares PA^2, PT^2, PH^2, etc.; that is, the real path of a projectile in vacuo, possesses the property of the parabola that has been demonstrated in this proposition. In other words,

The path of a projectile, undisturbed by the resistance of the air, is a parabola, more or less curved, depending upon the direction and intensity of the projectile force.

PROPOSITION XV.—THEOREM.

The abscissas of any diameter of the parabola are to each other as the squares of their corresponding ordinates.

Let P be any point on a parabola, PL a tangent line, and PF a diameter through this point. From the points B, V, K, etc., assumed at pleasure on the curve, draw ordinates and parallels to the diameter, forming the quadrilaterals $PCBA$, $PDVT$, etc.

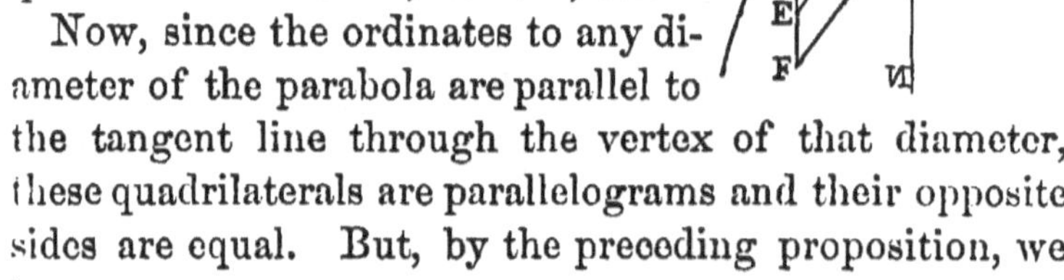

Now, since the ordinates to any diameter of the parabola are parallel to the tangent line through the vertex of that diameter, these quadrilaterals are parallelograms and their opposite sides are equal. But, by the preceding proposition, we have

$$AB : TV : HK, \text{ etc.,} : : \overline{PA}^2 : \overline{PT}^2 : \overline{PH}^2, \text{ etc.}$$

or $\quad PC : PD : PE, \text{ etc.,} : : \overline{BC}^2 : \overline{VD}^2 : \overline{KE}^2, \text{ etc.}$

By definition 6, PC is the ordinate and BC the abscissa of the point B, and so on.

Hence the theorem; *the abscissas of any diameter, etc.*

PROPOSITION XVI.—THEOREM.

If a secant line be drawn-parallel to any tangent line to the parabola, and ordinates to the axis be drawn from the point of contact and the two intersections of the secant with the curve, these three ordinates will be in arithmetical progression.

Let CT be the tangent line to the parabola, and EH the parallel secant. Draw the ordinates EG, CD, and HI, to the axis VI, and through E draw EK parallel to VI.

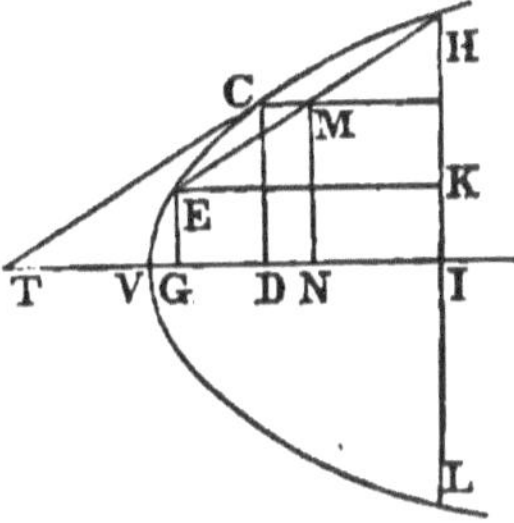

We are now to prove that

$$EG + HI = 2CD$$

The similar triangles, HKE and CDT, give the proportion

$$HK : KE :: CD : DT = 2VD$$

and, by proposition 12, we have

$$2p : KL :: HK : KE.$$

Therefore $\quad 2p : KL :: CD : 2VD,$ $\qquad\qquad$ (1)

and from the equation, $y^2 = 2px$, we get, by making $y = CD$ and $x = VD,$

$$2p : 2CD :: CD : 2VD \qquad\qquad (2)$$

By dividing proportion (1) by (2), term by term, we shall have

$$1 : \frac{KL}{2CD} :: 1 : 1$$

Whence $\qquad\qquad KL = 2CD$

But $\qquad\qquad KL = HI + KI = HI + EG;$

therefore $\qquad\qquad HI + EG = 2CD$.

Hence the theorem; *if a secant line be drawn, etc.*

Scholium 1.—If we draw CM parallel, and MN perpendicular to VI, then $2CD=2MN=EG+HI$; and since MN is parallel to each of the lines EG and HI, the point M bisects the line EH. That is, the diameter through C bisects its ordinate EH; and as HE is any ordinate to this diameter, it follows that

A diameter of the parabola divides into equal parts all chords of the curve parallel to the tangent through the vertex of the diameter.

Scholium 2.—Hence, as the abscissas of any diameter of the parabola and their ordinates have the same relations as those of the axis, namely; that the ordinates are bisected by the diameter, and their squares are proportional to the abscissas; so all the other properties of this curve, before demonstrated in reference to the abscissas and ordinates of the axis, will likewise hold good in reference to the abscissas and ordinates of any diameter.

PROPOSITION XVII.—THEOREM.

The square of an ordinate to any diameter of the parabola is equal to four times the product of the corresponding abscissa and the distance from the vertex of that diameter to the focus.

Let VX be the axis of a paraola, and through any point, as P, of the curve, draw the tangent PT, and the diameter PW; also draw the secant Qq, parallel PT, and produce the ordinate QN, and the diameter PW, to meet at D.

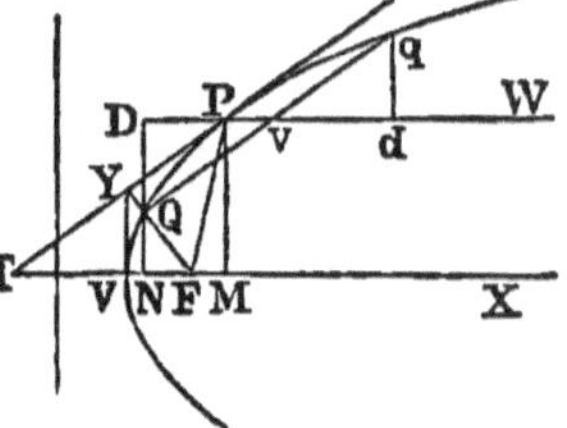

From the focus let fall the perpendicular FY upon the tangent, and draw FP and VY. We are now to prove that

$$\overline{Qv}^2=4PF\cdot Pv$$

Because FY is perpendicular to PT, Qv parallel to PT and DQ parallel to each of the lines PM and VY, the triangles DQv, PMT, TYV and TYF are all similar.

Whence　$\overline{Qv}^2 : \overline{QD}^2 :: \overline{TF}^2 : \overline{YF}^2$　(1)

But　$\overline{TF}^2=\overline{PF}^2$ and $\overline{YF}^2=PF\cdot VF.$　(Prop. 5)

Substituting these values in proportion (1) and dividing the third and fourth terms of the result by PF, it becomes

$$\overline{Qv}^2 : \overline{QD}^2 :: PF : VF \qquad (2)$$

Again, from the triangles QDv and PMT we get

$$QD : Dv :: PM : MT = 2\,VM$$
$$:: \overline{PM}^2 : 2PM \cdot VM$$

But (Scho. 2, Prop. 8) $\overline{PM}^2 = 4\,VF \cdot VM$

Whence $QD : Dv :: 4\,VF \cdot VM : 2PM \cdot VM;$
$$:: 4\,VF : 2PM$$

therefore $2PM \cdot QD = 4\,VF \cdot Dv \qquad (3)$

By subtracting the equation $\overline{QN}^2 = 4\,VF \cdot VN$ from the equation $\overline{PM}^2 = 4\,VF \cdot VM$, member from member, we have

$$\overline{PM}^2 - \overline{QN}^2 = 4\,VF \cdot (VM - VN)$$
$$= 4\,VF \cdot NM$$
$$= 4\,VF \cdot DP$$

Whence
$$(PM + QN)(PM - QN) = (PM + QN)\,DQ = 4\,VF \cdot DP \quad (4)$$

Subtracting eq. (4) from eq. (3), member from member, we obtain

$$(PM - QN)\,DQ = 4\,VF\,(Dv - DP) = 4\,VF \cdot Pv$$

and because $PM - QN = DQ$, this last equation becomes

$$\overline{DQ}^2 = 4\,VF \cdot Pv$$

Substituting this value of $\overline{DQ}^2$ in proportion (2), we have

$$\overline{Qv}^2 : 4\,VF \cdot Pv :: PF : VF$$

or $\overline{Qv}^2 : 4Pv :: PF : 1$

Whence $\overline{Qv}^2 = 4PF \cdot Pv$

Hence the theorem ; *the square of an ordinate, etc.*

Cor. If, in the course of this demonstration, we had used the triangle vdq in the place of vDQ, to which it is similar, we would have found that $\overline{qv}^2 = 4PF \cdot Pv$; whence $Qv = qv$. And since the same may be proved for any ordinate, it follows that

All the ordinates of the parabola to any of its diameters are bisected by that diameter.

Scholium.—The parameter of any diameter of the parabola has been defined (Def. 8) to be one of the extremes of a proportion, of which any ordinate to the diameter is the mean and the corresponding abscissa the other extreme.

Now, we have just shown that $\overline{Qv}^{2}=\overline{qv}^{2}=4PF\cdot Pv$.

Whence, $Pv : Qv :: Qv : 4PF$. $4PF$, which remains constant for the same diameter, is therefore the parameter of the diameter PW. And as the same may be shown for any other diameter, we conclude that

The parameter of any diameter of the parabola is equal to four times the distance from the vertex of that diameter to the focus.

PROPOSITION XVIII.—THEOREM.

The parameter of any diameter of the parabola is equal to the double ordinate to this diameter that passes through the focus.

Through any point, as P, of the parabola of which F is the focus and V the vertex, draw the diameter PW, the tangent PT, and, through the focus the double ordinate BD, to the diameter. It is now to be proved that $4PF$, or the parameter to this diameter, is equal to BD.

Because PW is parallel to TX, and BD to TP, $TPvF$ is a parallelogram, and $Pv=TF$. But $PF=FT$ (Prop. 4), hence $Pv=PF$.

By the preceding proposition, $\overline{Bv}^{2}=4PF\cdot Pv=4PF\cdot PF$

Whence, $Bv=2PF$; therefore, $2Bv=BD=4PF$.

Hence the theorem; *the parameter of any diameter, etc.*

PROPOSITION XIX.—THEOREM.

The area of the portion of the parabola included between the curve, the ordinate from any of its points to the axis, and

the corresponding abscissa, is equivalent to two thirds of the rectangle contained by the abscissa and ordinate.

Let VD be the axis of a parabola, and VIP any portion of the curve. Draw the extreme ordinate PD to the axis, and complete the rectangle $VAPD$; then will the area included between the curve 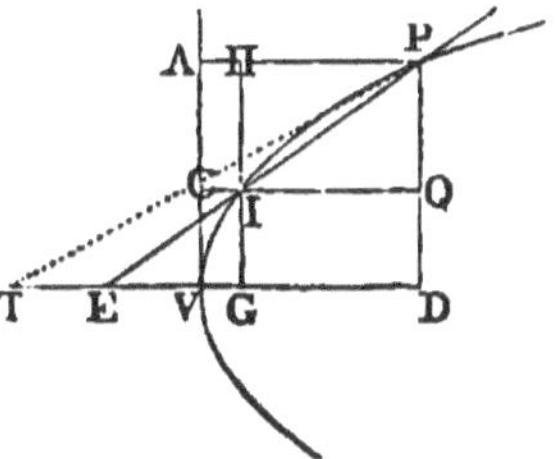 VIP, the ordinate PD, and the abscissa VD, be equivalent to two thirds of the rectangle $VAPD$.

Take any point I, between P and the vertex, and draw PI, producing it to meet the axis produced at E.

Now, from the similar triangles, PQI and PDE, we get the proportion

$$PQ : QI : : PD : DE.$$

Whence $\quad PQ \cdot DE = QI \cdot PD = GD \cdot PD.$ $\qquad$ (1)

If we suppose the point I to approach P, the secant line PE will, at the same time, approach the tangent PT; and finally, when I comes indefinitely near to P, the secant will sensibly coincide with the tangent PT, and DE may then be replaced by $DT = 2DV = 2PA$. Under this supposition, eq. (1) becomes

$$2PQ \cdot PA = PD \cdot GD.$$

That is, when the rectangles $GDPH$ and $APQC$ become indefinitely small, we shall have

$$\text{Rect. } GDPH = 2 \text{ Rect. } APQC.$$

We will call Rect. $GDPH$ the interior rectangle, and Rect. $APQC$ the exterior rectangle. If another point be taken very near to I, and between it and the vertex, and with reference to it the interior and exterior rectangles be constructed as before, we should again have the interior equivalent to twice the exterior rectangle. Let us conceive this process to be continued until all possible interior and exterior rectangles are constructed; then would we have

Sum interior rectangles=2 sum exterior rectangles.

But, under the supposition that these rectangles are indefinitely small, the sum of the interior rectangles becomes the interior curvilinear area, and the sum of the exterior rectangles the exterior curvilinear area, and the two sums make up the rectangle $APDV$. Therefore, if this rectangle were divided into three equal parts, the interior area would contain two of these parts.

Hence the theorem; *the area of the portion of the, etc.*

PROPOSITION XX.—THEOREM.

If a parabola be revolved on its axis, the solid generated will be equivalent to one half of its circumscribing cylinder.

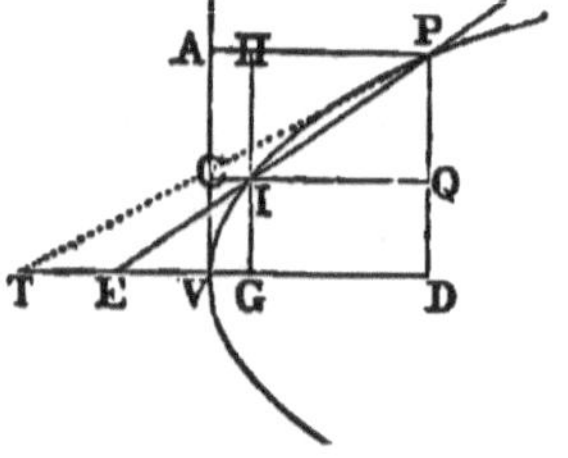

Conceive the parabola in the figure, which is constructed as in the last proposition, to revolve on its axis VD. We are then to find the measure of the volume generated.

The rectangle ID will generate a cylinder having DQ for the radius of its base, and DG for its axis; and the rectangle AI will generate a cylindical band, whose length is CI, and thickness PQ.

The solidity of the cylinder $=\pi \overline{DQ}^2 \cdot DG$

The solidity of the band $=\pi(\overline{PD}^2 - \overline{DQ}^2) \cdot VG =$
$\pi[PD^2 - (PD - PQ)^2] \cdot VG = \pi[2PD \cdot PQ - \overline{PQ}^2] \cdot VG$

Now, under the supposition that the point I is indefinitely near to P, DQ may be replaced by PD, VG by VD, and $\overline{PQ}^2$ may be neglected as insensible in comparison with $2PD \cdot PQ$. These conditions being introduced in the above expressions for the solidities of the cylinder and band, they become

The solidity of the cylinder $= \pi \overline{PD}^2 \cdot DG$
The solidity of the band $= 2\pi PD \cdot PQ \cdot VD$

Whence,

sol. of cylinder : sol. of band : : $\overline{PD}^2 \cdot DG : 2PD \cdot PQ \cdot VD$ (1)

But, when I and P are sensibly the same point,

$$PQ : GD : : PD : 2VD$$

therefore,

$$2VD \cdot PQ = PD \cdot GD, \text{ or } 2VD \cdot PQ \cdot PD = \overline{PD}^2 \cdot DG$$

The terms in the last couplet of proportion (1) are therefore equal, and we have

$$\text{sol. of cylinder : sol. of band : : } 1 : 1$$

or sol. of cylinder=sol. of band.

In the same manner we may prove that any other interior cylinder is equivalent to the corresponding exterior band. Hence the sum of all the possible interior solids is equivalent to the sum of the exterior solids. But the two sums make up the cylinder generated by the rectangle $VDPA$; therefore either sum is equivalent to one half of the cylinder.

Hence the theorem ; *if a parabola be revolved, etc.*

REMARK.—The body generated by the revolution of a parabola about its axis is called a *Paraboloid of Revolution.*

PROPOSITION XXI.—THEOREM.

If a cone be cut by a plane parallel to one of its elements, the section will be a parabola.

Let MVN be a section of a cone by a plane passing through its axis, and in this section draw AH parallel to the element $VM.$ Through AH conceive a plane to be passed perpendicular to the plane $MVN;$ then will the section $DAGI$ of the cone made by this last plane, be a parabola. In the plane $MVN,$ draw MN and KL perpendicular to the axis of the cone, and through them, pass planes perpendicular to this axis. The sections of the cone, by these planes, will be circles, of which MN and KL, respectively, are the diameters. Through the points F and H, in which AH meets KL and MN, draw in the section $DAGI$ the lines FG and HI, perpendicular to AH. Because the planes DAI and MVN are at right angles to each other, FG is perpendicular to KL, and HI is perpendicular to MN.

Now, from the similar triangles AFL, AHN, we have

$$AF : AH :: FL : HN \qquad (1)$$

By reason of the parallels, $KF = MH$; multiplying the first term of the second couplet of proportion (1) by KF, and the second term by MH, it becomes

$$AF : AH :: FL \cdot KF : HN \cdot MH \qquad (2)$$

But FG is an ordinate of the circle of which KL is the diameter, and HI an ordinate of the circle of which MN is the diameter: therefore

$FL \cdot KF = \overline{FG}^2$, and $HN \cdot MH = \overline{HI}^2$ (Cor., Th. 17, B. III, Geom.)

Substituting, for the terms of the second couplet, in proportion (2), these values, it becomes

$$AF : AH :: \overline{FG}^2 : \overline{HI}^2$$

This proportion expresses the property that was demonstrated in proposition 15 to belong to the parabola.

Hence the theorem; *if a cone be cut by a plane, etc.*

Cor. From the proportion, $AF : AH :: \overline{FG}^2 : \overline{HI}^2$ we get $\dfrac{\overline{FG}^2}{AF} = \dfrac{\overline{HI}^2}{AH}$; that is, $\dfrac{\overline{FG}^2}{AF}$ or $\dfrac{\overline{HI}^2}{AH}$, which is a third proportional to any abscissa and the corresponding ordinate of the section, is constant, and (by Def. 8) is the parameter of the section.

THE HYPERBOLA.

DEFINITIONS.

1. The **Hyperbola** is a plane curve, generated by the motion of a point subjected to the condition that the difference of its distances from two fixed points shall be constantly equal to a given line.

REMARK 1.—The distance between the foci is also supposed to be known, and the given line must be less than the distance between the fixed points; that is, less than the distance between the foci.

REMARK 2.—The ellipse is a curve confined by two fixed points called the foci; and the sum of two lines drawn from any point in the curve is constantly equal to a given line. In the hyperbola, the *difference* of two lines drawn from any point in the curve, to the fixed points, is equal to the given line. The ellipse is but a single curve, and the foci are within it; but it will be shown in the course of our investigation, that

The *hyperbola consists of two equal and opposite branches, and the least distance between them is the given line.*

2. The **Center** of the hyperbola is the middle point of the straight line joining the foci.

3. The **Eccentricity** of the hyperbola is the distance from the center to either focus.

4. A **Diameter** of the hyperbola is a straight line passing through the center, and terminating in the opposite branches of the curve. The extremities of a diameter called its *vertices*.

6*   E

5. The **Major**, or **Transverse Axis**, of the hyperbola is the diameter that, produced, passes through the foci.

6. The **Minor**, or **Conjugate Axis**, of the hyperbola bisects the major axis at right-angles; and its half is a mean proportional between the distances from either focus to the vertices of the major axis.

7. An **Ordinate** to a diameter of the hyperbola is a straight line, drawn from any point of the curve to meet the diameter produced, and is parallel to the tangent at the vertex of the diameter.

8. An **Abscissa** is the part of the diameter produced that is included between its vertex and the ordinate.

9. Conjugate Hyperbolas are two hyperbolas so related that the major and minor axes of the one are, respectively, the minor and major axes of the other.

10. Two diameters of the hyperbola are *conjugate*, when either is parallel to the tangent lines drawn through the vertices of the other.

The conjugate to a diameter of one hyperbola will terminate in the branches of the conjugate hyperbola.

11. The **Parameter** of any diameter of the hyperbola is a third proportional to that diameter and its conjugate.

12. The parameter of the major axis of the hyperbola is called the *principal parameter*, the *latus-rectum*, or simply the *parameter;* and it will be proved to be equal to the chord of the hyperbola through the focus and at right-angles to the major axis.

EXPLANATORY REMARKS.—Thus, let $F'F$ be two fixed points. Draw a line between them, and bisect it in C. Take CA, CA', each equal to one half the given line, and CA may be any distance less than CF; $A'A$ is the given line, and is called the *major axis* of the hyperbola. Now, let us suppose the curve already found and represented by ADP. Take any point, as P, and join P, F and P, F'; then, by Def. 1, the difference between PF'

and PF must be equal to the given line $A'A$; and conversely, if $PF''-PF=A'A$, then P is a point in the curve.

By taking any point, P, in the curve, and joining P, F and P, F' a triangle PFF' is always formed, having $F'F$ for its base, and $A'A$ for the difference of the sides; and these are all the conditions necessary to define the curve.

As a triangle can be formed *directly opposite* $PF'F$, which shall be in all respects exactly equal to it, the two triangles having $F'F$ for a common side; the difference of the other two sides of this opposite triangle will be equal to $A'A$, and correspond with the condition of the curve.

Hence, a curve can be formed about the focus F', exactly similar and equal to the curve about the focus F.

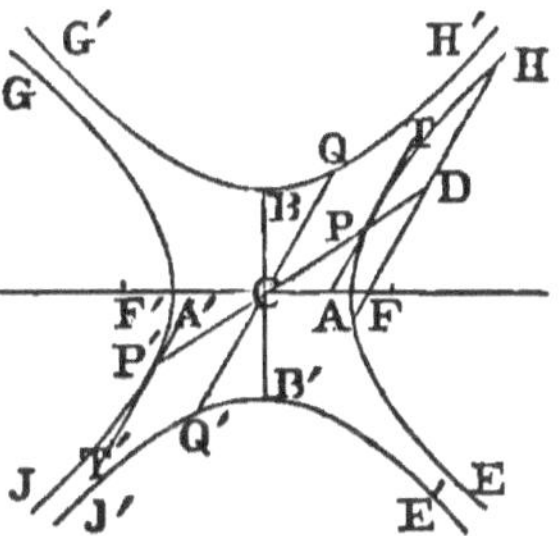

We perceive, then, that the hyperbola is composed of two equal curves called *branches*, the one on the right of the center and curving around the right-hand focus, and the other on the left of the center and curving around the left-hand focus. In like manner, by making CB equal to a mean proportional between FA and FA', and constructing above and below the center the branches of the hyperbola 'of which $BB'=2CB$ is the major, and AA' the minor axis, we have the hyperbola which is conjugate to the first. PP'' is a diameter of the hyperbola, PT a tangent line through the vertex of the diameter, and QQ', parallel to PT and terminating in the branches of the conjugate hyperbola, is conjugate to the diameter PP'. HD is the ordinate from the point H to the diameter CP, and PD is the corresponding abscissa.

PROPOSITION I.—PROBLEM.

To describe an hyperbola mechanically.

Take a ruler, $F'H$, and fasten one end at the point F', on which the ruler may turn as a hinge. At the other end, attach a thread, the length of which is less than that of the

ruler by the given line $A'A$. Fasten the other end of the thread at F.
With the pencil, P, press the thread
against the ruler, and keep it at
equal tension between the points H
and F. Let the ruler turn on the
point F', keeping the pencil close

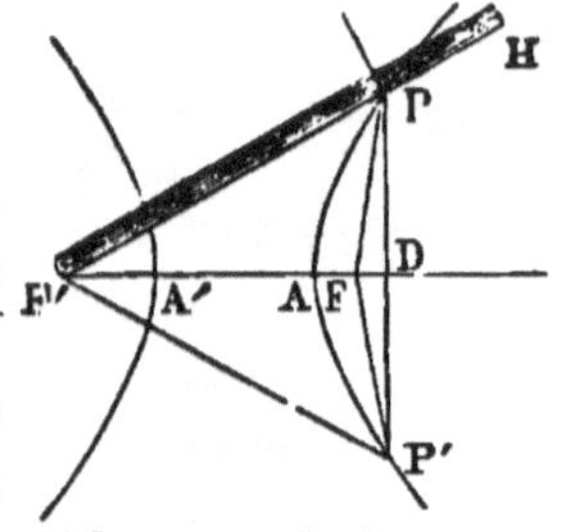

to the ruler and letting the thread slide round the pencil; the pencil will thus describe a curve on the paper.

If the ruler be changed, and made to revolve about the
other focus as a fixed point, the opposite branch of the
curve can be described.

In all positions of P, except when at A or A', PF' and
PF will be two sides of a triangle, and the difference of
these two sides is constantly equal to the difference between the ruler and the thread; but that difference was
made equal to the given line $A'A$; hence, by Definition
1, the curve thus described must be an hyperbola.

Cor. From any point, as P, of the hyperbola, draw the
ordinate PD to the major axis, and produce this ordinate
to P', making DP' equal to PD; and draw FP, FP',
$F'P$ and $F'P'$. Then, because $F'D$ is a perpendicular to
PP at its middle point, we have $FP=FP'$, and $F'P=
F'P'$; whence

$F'P-FP=F'P'-FP'$, and P' is a point of the hyperbola. Therefore, PP' is a chord of the hyperbola at right
angles to the major axis, and is bisected by this axis; and
as the same may be proved for any other chord drawn at
right angles to the major axis, we conclude that

*All chords of the hyperbola which are drawn at right angles
to the major axis are bisected by that axis.* It may be proved,
in like manner, that

*All chords of the hyperbola which are drawn at right angles
to the conjugate axis are bisected by that axis.*

PROPOSITION II.—THEOREM.

If a point be taken within either branch of the hyperbola, or on the concave side of the curve, the difference of its distances from the foci will be greater than the major axis; and if a point be taken without both branches, or on the convex side of both curves, the difference of its distances from the foci will be less than the major axis.

Let AA' be the major axis, and F and F' the foci of an hyperbola. Within the branch APX take any point, Q, and draw FQ and $F'Q$; then we are to prove

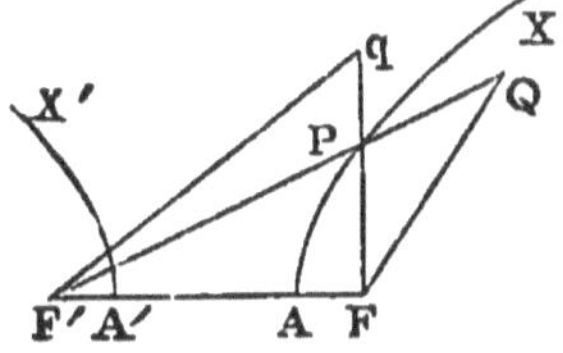

First.—That $F'Q—FQ$ is greater than AA'.

Since Q is within the branch APX, the line $F'Q$ must cut the curve at some point, as P. Draw PF and FQ.

By the definition of the hyperbola, $F'P—PF=AA'$. Adding $PQ+PF$ to both members of this equation, it becomes

$$F'P—PF+PQ+PF=AA'+PQ+PF$$
or, $$F'Q=AA'+PQ+PF.$$

But PQ and PF being two sides of the triangle FPQ, are together greater than the third side FQ. Therefore $F'Q>AA'+FQ;$ and, by taking FQ from both members of this inequality, we have

$$F'Q—FQ>AA'.$$

Second.—Take any point, q, without both branches of the hyperbola, and join this point to either focus, as F. Then since q is without the branch APF, the line qF must cut the curve at some point, P. Draw qF, qF', and PF'.

Because P is a point on the curve, we have $PF'—PF=AA'$. Adding $Pq+PF$ to the members of this equation it becomes

$$PF'—PF+Pq+PF=AA'+PF+Pq$$
or, $$PF'+Pq=AA'+PF+Pq=AA'+qF.$$

But PF' and Pq, being two sides of the triangle $F'Pq$, are together greater than the third side qF'. Whence $qF' < AA' + qF$; and by taking qF from both members of this inequality, we have $qF' - qF < AA'$.

Hence the theorem; *if a point be taken, etc.*

Cor. Conversely: *If the difference of the distances from any point to the foci of an hyperbola be greater than the major axis, the point will be within one of the branches of the curve; and if this difference be less than the major axis, the point will be without both branches.*

For, let the point Q be so taken that $F'Q - FQ > AA'$; then the point Q cannot be on the curve; for in that case we should have $F'Q - FQ = AA'$. And it cannot be without both branches of the curve, for then we should have $F'Q - FQ < AA'$, from what is proved above. But it is contrary to the hypothesis that $F'Q - FQ$ is either equal to or less than AA'; hence the point Q must be within one of the branches of the hyperbola.

In like manner we prove that, if the point q be so chosen that $qF' - qF < AA'$, this point must be without both branches of the hyperbola.

PROPOSITION III.—THEOREM.

A tangent to the hyperbola bisects the angle contained by lines drawn from the point of contact to the foci.

Let F', F be the foci, and P any point on the curve; draw PF', PF and bisect the angle $F'PF$ by the line TT'; this line will be a tangent at P.

If TT' be a tangent at P, every other point on this line will be without the curve.

Take $PG=PF$ and draw GF; TT' bisects GF, and any point in the line TT' is at equal distances from F and G (Scho. 1, Th. 18, B. I, Geom). By the definition of the curve, $F'G=A'A$ the given line. Now take any other point than P in TT', as E, and draw EF', EF and EG.

Because EF is equal to EG we have

$$EF'-EF=EF'-EG.$$

But $EF'-EG$, is less than $F'G$, because the difference of any two sides of a triangle is less than the third side. That is, $EF'-EF$ is less than $A'A$; consequently the point E is without the curve (Prop. 2), and as E is any point on the line TT', except P, therefore, the line TT', which bisects the angle at P, is a tangent to the curve at that point.

Hence the theorem; *a tangent to the hyperbola, etc.*

Scholium.—It should be observed that by joining the *variable* point, P, in the curve, to the two *invariable* points, F' and F, we form a triangle; and that the tangent to the curve at the point P, bisects the angle of that triangle at P.

But when any angle of a triangle is bisected, the bisecting line cuts the base into segments proportional to the other sides. (Th. 24, B. II, Geom).

Therefore, $\qquad F'P : PF=F'T : TF$

Represent $F'P$ by r' and PF by r;
then $\qquad r' : r=F'T : TF$

But as r' must be greater than r by a given quantity, a, therefore, $\qquad r+a : r=F'T : TF$

Or, $\qquad 1+\dfrac{a}{r} : 1=F'T : TF$

Let it be observed that a is a constant quantity, and r a variable one which can increase without limit; and when r is *immensely* great in respect to a, the fraction $\dfrac{a}{r}$ is *extremely minute*, and the first term of the above proportion would not in any practical sense differ from the second; therefore, in that case, the third term would not essentially differ from the fourth; that is, $F'T$ does not essentially differ from FT when r, or the distance of P from F' is immensely great. *Hence, the tangent at any point P, of the hyperbola, can never cross the line FF' at its middle point*, but it may approach within the *least imaginable* distance to that point.

If, however, we conceive the point P to be removed to an infinite distance on the curve, the tangent at that point would cut AA' at its middle point C, and the tangent itself is then called an *asymptote*.

PROPOSITION IV.—THEOREM.

Every diameter of the hyperbola is bisected at the center.

Let F and F' be the foci, and AA' the major axis of an hyperbola. Take any point, as P, in one of the branches of the curve; draw PF and PF', and complete the parallelogram $PFP'F'$.

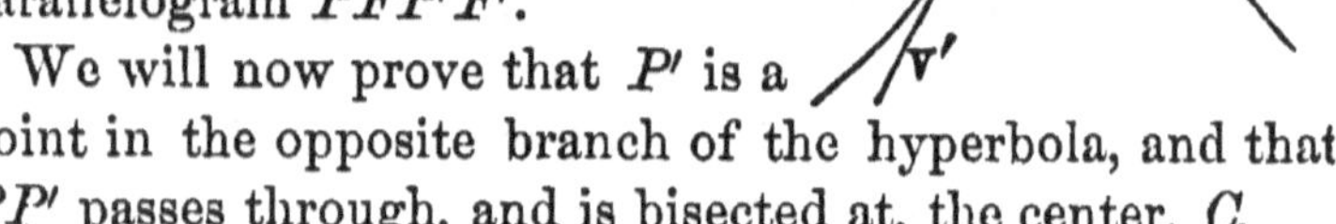

We will now prove that P' is a point in the opposite branch of the hyperbola, and that PP' passes through, and is bisected at, the center, C.

Because $PFP'F'$ is a parallelogram, the opposite sides are equal; therefore $F'P-PF=FP'-P'F'$; but since F is, by hypothesis, a point of the hyperbola, $F'P-PF=AA'$; hence $FP'-P'F'=AA'$, and P' is also a point of the hyperbola.

Again, the diagonals, $F'F$, $P'P$ of the parallelogram, mutually bisect each other; hence C is the middle point of the line joining the foci, and (Def. 2) is the center of the hyperbola. PP' is therefore a diameter, and is bisected at the center, C.

Hence, the theorem; *every diameter of the hyperbola, etc.*

PROPOSITION V.—THEOREM.

Tangents to the hyperbola at the vertices of a diameter are parallel to each other.

At the extremities of the diameter, PP', of the hyperbola represented in the figure, draw the tangents TT' and VV'. We are now to prove that these tangents are parallel. By proposition (Prop. 3)

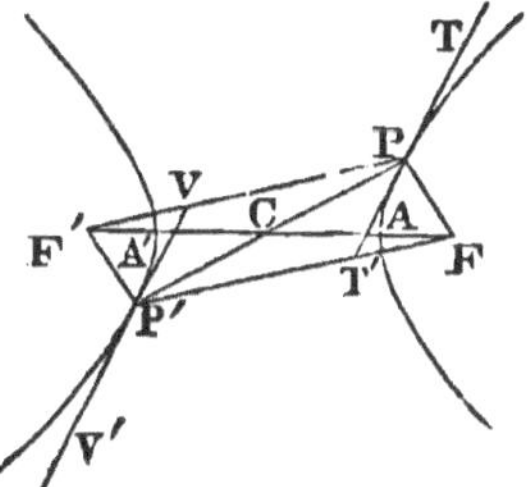

TT' bisects the angle FPF', and VV' also bisects the angle $F'P'F$. But these angles being the opposite angles of the parallelogram $FPF'P'$, are equal; therefore the $\angle T'PF =$ the $\angle PT'F =$ the $\angle VP'F$. But the $\angle$'s $PT'F$, $VP'F$, formed by the line FP' meeting the tangents, are opposite exterior and interior angles. The tangents are therefore parallel (Cor. 1, Th. 7, B. I, Geom).

Hence the theorem; *tangents to the hyperbola, etc.*

PROPOSITION VI.—THEOREM.

The perpendiculars let fall from the foci of an hyperbola on any tangent line to the curve, intersect the tangent on the circumference of the circle described on the major axis as a diameter.

In the hyperbola of which AA' is the major axis, F and F' the foci, and C the center, take any point in one of the branches, as P, and through it draw the tangent line THI'. From the foci let fall on the tangent the perpendic-

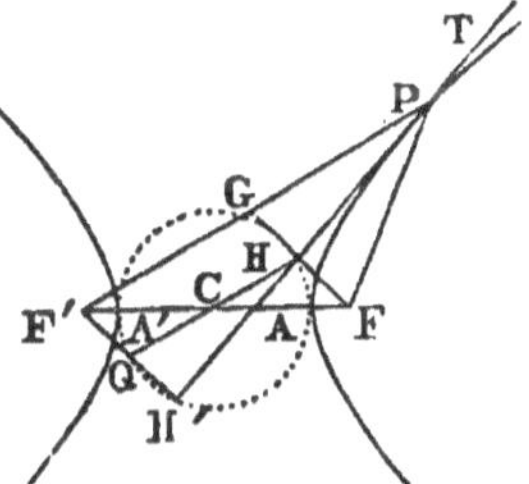

ulars FH, $F'H'$, draw PF and PF', and produce FH to intersect PF' in G. We are now to prove that H and H' are in the circumference of a circle of which AA' is the diameter.

Draw CH, producing it to meet $F'H'$ in Q. Then, because PH is a tangent to the curve, it bisects the angle FPF'; therefore the right-angled triangles, FPH and

7

HPG, being mutually equiangular, and having the side
PH common, are equal.   Whence, $FH = HG$ and $PF =$
PG.   But, by the definition of the hyperbola, $F'P - PF$
$= AA'$; hence $F'P - PG = F'G = AA'$.

Since CH bisects the sides $F'F$ and FG of the triangle
FGF', we have

$$F'F : FC :: F'G : CH$$
but $F'F = 2FC$; therefore $F'G = 2CH = AA'$

If then with C as a center and CA as a radius, a circumference be described, it will pass through the point H.

Again; the triangles FHC and $F'CQ$ are in all respects
equal; hence $CQ = CH$, and Q is also a point in the circumference of the circle of which AA'·is the diameter.
Therefore, the right-angled triangle $QH'H$, having for
its hypotenuse a diameter HQ of this circle, must have
the vertex, H' of its right angle at some point in the circumference.

Hence the theorem; *the perpendiculars let fall, etc.*

PROPOSITION VII.—THEOREM.

*The product of the perpendiculars let fall from the foci of
an hyperbola upon a tangent to the curve at any point, is equal
to the square of the semi-minor axis.*

Resuming the figure of the preceding proposition; then, since
the semi-minor axis, which we will
represent by B, is a mean proportional between the distances from
either focus to the extremities of
the major axis, we are to prove
that

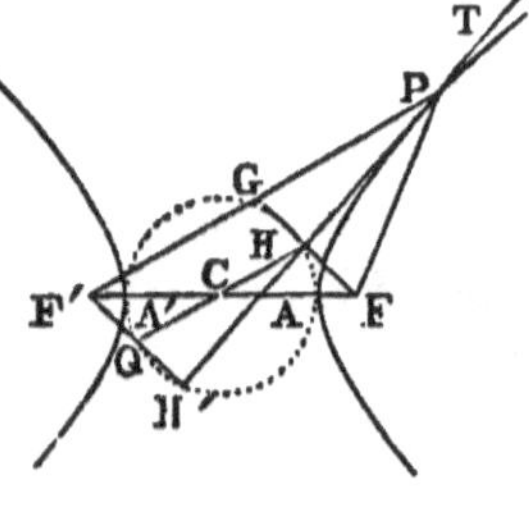

$$B^2 = FA \times FA' = FH \times F'H'$$

By the construction, the triangles FHC and CQF' are
equal; therefore $FH = F''Q$ (1)

Multiplying both members of eq. (1) by $F'H'$ it becomes

$$FH \cdot F'H' = F'Q \cdot F'H' \qquad (2)$$

Again, it was proved in the last proposition that the points H, H' and Q were in the circumference of the circle described on AA' as a diameter; therefore $F'H'$ and $F'A$ are secants to this circumference, and we have

$$F'Q : F'A' :: F'A : F'H' \quad \text{(Cor., Th. 18, B. III, Geom).}$$

Whence, $\qquad F'Q \cdot F'H' = F'A' \cdot F'A \qquad (3)$

But $F'A' = FA$, $F'A = FA'$, and $F'Q = FH$. Making these substitutions in eq. (3) it becomes

$$FH \cdot F'H' = FA \cdot FA' = B^2.$$

Hence the theorem: *the product of the perpendiculars, etc.*

Cor. 1. The triangles PFH, $PF'H'$ are similar; therefore, $\qquad PF : PF' :: FH : F'H'$

That is: *The distances from any point on the hyperbola to the foci, are, to each other, as the perpendiculars let fall from the foci upon the tangent at that point.*

Cor. 2. From the proportion in corrollary 1, we get

$$FH = \frac{PF \cdot F'H'}{PF'}; \text{ whence } \overline{FH}^2 = \frac{PF \cdot F'H' \cdot FH}{PF'}$$

But by the proposition, $F'H' \cdot FH = B^2$;

therefore, $\quad \overline{FH}^2 = \frac{B^2 \cdot PF}{PF'} = \frac{B^2 \cdot PF}{2CA + PF}$, because $F'G = AA' = 2CA$, and $PG = PF$.

In like manner it may be proved that

$$\overline{F'H'}^2 = \frac{B^2 \cdot PF'}{PF} = \frac{B^2(2CA + PF)}{PF}$$

PROPOSITION VIII.—THEOREM.

If a tangent be drawn to the hyperbola at any point, and also an ordinate to the major axis from the point of contact, then will the semi-major axis be a mean proportional between the distance from the center to the foot of the ordinate, and the distance from the center to the intersection of the tangent with this axis.

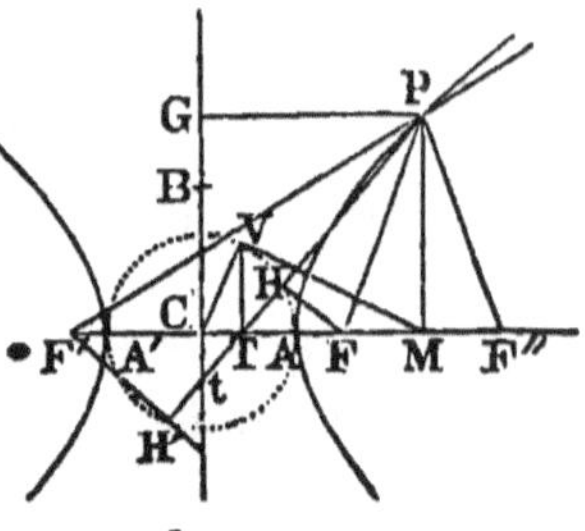

Let AA'' be the major axis, FF' the foci and C the center of the hyperbola. Through any point, as P, taken on one of the branches, draw the tangent PT intersecting the axis at T; also draw PF, PF' to the foci, and the ordinate PM to the axis. We are now to prove that

$$CT : CA :: CA : CM.$$

Because PT bisects the vertical angle of the triangle FPF' (Prop. 3), it divides the base into segments proportional to the adjacent sides (Th. 24, B. II, Geom.)

Therefore, $\qquad F'T : TF :: F'P : PF.$

Whence, $F'T-TF : F'T+TF :: F'P-PF : F'P+PF$

That is, $\quad 2CT : F'F :: AA'=2CA : F'P+PF$

Or, by inverting the means,

$$2CT : 2CA :: F'F : F'P+PF \qquad (1)$$

Now, making $MF''=MF$, and drawing PF'', we have, from the triangle $F'PF'''$,

$$F'F'' : F'P+PF'' :: F'P-PF'' : F'M-MF''$$

$$\text{(Prop 6, Pl. Trig.)}$$

But, because the triangle FPF'' is isosceles, and PM is a perpendicular from the vertical angle upon the base, $PF=PF''$, $F'F''=F'F+2FM=2CF+2FM=2CM;$

therefore the preceding proportion becomes

$$2CM : F'P+PF :: 2CA : F'F$$

or, $\qquad 2CM : 2CA :: F'P+PF : F'F \qquad (2)$

Multiplying proportions (1) and (2), term by term, observing that the terms of the second couplet of the resulting proportion are equal, we have

$$4CT \cdot CM : \overline{4CA}^2 :: 1 : 1$$

Whence, $\qquad CT \cdot CM = \overline{CA}^2;$

which, resolved into a proportion, becomes

$$CT : CA :: CA : CM.$$

Hence the theorem; *if a tangent be drawn, etc.*

SCHOLIUM.—The property of the hyperbola demonstrated in this proposition is not restricted to the major axis, but also holds true in reference to the minor axis.

The tangent intersects the minor axis at the point t, and PG is an ordinate to this axis from the point of contact. Now, the similar triangles tCT, THF, give the proportion

$$Ct : FH :: CT : TH \qquad (1)$$

and from the similar triangles $PMT, TF'H'$, we also have

$$PM : F'H' :: MT : H'T \qquad (2)$$

Multiplying proportions (1) and (2), term by term, we get

$$Ct \cdot PM : FH \cdot F'H' :: CT \cdot MT : TH \cdot H'T \qquad (3)$$

But $FH \cdot F'H' = B^2$ (Prop. 7). Moreover, drawing the ordinate TV, and the radius CV of the circle, and the line VM, we have by the proposition

$$CT : CA :: CA : CM$$

or, $\qquad CT : CV :: CV : CM$

Therefore, the triangles VCT and MCV, having the angle C common and the sides about this angle proportional, are similar (Cor. 2, Th. 17, B. II, Geom.); and because the first is right-angled, the second is also right-angled, the right angle being at V; hence

$$\overline{VT}^2 = CT \cdot MT \text{ (Th. 25, B. II, Geom).}$$

Also, AA' and HH' are two chords of a circle intersecting each other at T; hence

$$HT \cdot TH' = AT \cdot TA' = \overline{VT}^2 \text{ (Th. 17, B. III, Geom).}$$

Substituting for the terms of proportion (3) these several values, it becomes

$$Ct \cdot PM : B^2 :: \overline{VT}^2 : \overline{VT}^2 :: 1 : 1$$

Whence, $\qquad Ct \cdot PM = B^2$

Therefore, $\qquad Ct : B :: B : PM = CG$

7*

Cor. It has been proved that the triangle CVM is right-angled at V; therefore, VM is a tangent at the point V to the circumference on AA' as a diameter, and TM is its sub-tangent. But TM is also the sub-tangent on the major axis of the hyperbola answering to the tangent PT; hence

If a tangent be drawn to the hyperbola at any point, and through the point in which the tangent intersects the major axis an ordinate be drawn to the circle of which this axis is a diameter, the sub-tangent on the major axis corresponding to the tangent through the extremity of this ordinate will be the same as that of the tangent to the hyperbola.

PROPOSITION IX.—THEOREM.

In any hyperbola the square of the semi-major axis is to the square of the semi-minor axis, as the rectangle of the distances from the foot of any ordinate to the major axis, to the vertices of this axis, is to the square of the ordinate.

Resuming the figure to Proposition 8, the construction of which needs no further explanation, we are to prove that

$$\overline{CA}^2 : \overline{CB}^2 :: A'M \cdot AM : \overline{PM}^2,$$

assuming CB to represent the semi-minor axis.

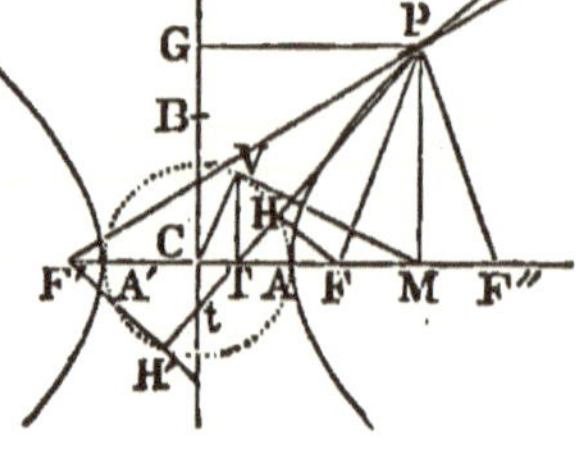

From the similar triangles PMT, THF and $TH'F'$, we derive the proportions

$$PM : FH :: MT : TH$$
$$PM : F'H' :: MT : TH'$$

Whence
$$\overline{PM}^2 : FH \cdot F'H' :: \overline{MT}^2 : TH \cdot TH' \qquad (1)$$

But $FH \cdot F'H'$ is equal to the square of the semi-minor axis (Prop. 7); and because the chords, HH' and AA', of the circle intersect each other at T, we have

$$TH \cdot TH' = AT \cdot TA' = \overline{VT}^2 \quad \text{(Th. 17, B. III, Geom.)}$$

These values of the consequents of proportion (1) being substituted, it becomes

$$\overline{PM}^2 : \overline{BC}^2 :: \overline{MT}^2 : \overline{VT}^2 \qquad (2)$$

The triangles CVT and TVM are similar, and give the proportion

$$\overline{MT}^2 : \overline{VT}^2 :: \overline{VM}^2 : \overline{CV}^2 = \overline{CA}^2 \qquad (3)$$

Comparing proportions (2) and (3), we find that

$$\overline{PM}^2 : \overline{BC}^2 :: \overline{VM}^2 : \overline{CA}^2 \qquad (4)$$

Because MV is a tangent and MA' a secant to the circle $AVA'H'$, we have

$$\overline{VM}^2 = A'M \cdot AM \quad \text{(Th. 18, B. III, Geom.)}$$

Placing this value of $\overline{VM}^2$ in proportion (4) and inverting the means of the resulting proportion, it becomes

$$\overline{PM}^2 : A'M \cdot AM :: \overline{BC}^2 : \overline{CA}^2$$

or, $$\overline{CA}^2 : \overline{BC}^2 :: A'M \cdot AM : \overline{PM}^2$$

Hence the theorem; *in any hyperbola the square of the, etc.*

Cor. Proportion (4) above may be put under the form

$$\overline{CA}^2 : \overline{BC}^2 :: \overline{VM}^2 : \overline{PM}^2 \qquad (a)$$

and from the right-angled triangle CVM we have

$$\overline{CV}^2 + \overline{VM}^2 = \overline{CM}^2$$

from which, because $CV = CA$, we get

$$\overline{VM}^2 = \overline{CM}^2 - \overline{CA}^2.$$

Also, the right-angled triangles CVM, VTM are similar; therefore, $$CM : VM :: VM : MT$$

Whence $$\overline{VM}^2 = CM \cdot MT.$$

Now, if in proportion (a) we place for $\overline{VM}^2$ these values, successively, we shall have the two proportions

$$\overline{CA}^2 : \overline{BC}^2 :: CM \cdot MT : \overline{PM}^2 \qquad (b)$$

and $$\overline{CA}^2 : \overline{BC}^2 :: \overline{CM}^2 - \overline{CA}^2 : \overline{PM}^2 \qquad (c)$$

Scholium 1.—Let us denote CA by a, CB by b, CM by x, and PM by y; then $A'M = x + a$ and $AM = x - a$. Because $\overline{CM}^2 - \overline{CA}^2 = (CM + CA)(CM - CA) = A'M \cdot AM$, proportion (c), by substitution, now becomes

$$a^2 : b^2 :: (x+a)(x-a) : y^2. \qquad (a')$$

Whence $\qquad a^2 y^2 = b^2 x^2 - a^2 b^2$

or, $\qquad a^2 y^2 - b^2 x^2 = -a^2 b^2.$

This equation is called, in analytical geometry, *the equation of the hyperbola referred to its center and axes*, in which x, the distance from the center to the foot of any ordinate to the major axis, is called the *abscissa*. The equation $a^2 y^2 - b^2 x^2 = -a^2 b^2$ therefore expresses the relation between the abscissa and ordinate of any point of the curve.

Scholium 2.—Let y' denote the ordinate and x' the abscissa of a second point of the hyperbola; then we shall have

$$a^2 : b^2 :: (x'+a)(x'-a) : y'^2$$

Comparing this proportion with proportion (a'), scholium 1, we find

$$y^2 : y'^2 :: (x+a)(x-a) : (x'+a)(x'-a)$$

That is: *In any hyperbola the squares of any two ordinates to the major axis are to each other, as the rectangles of the corresponding distances from the feet of these ordinates to the vertices of the axis.*

A similar property was proved for the ellipse and the parabola.

PROPOSITION X.—THEOREM.

The parameter of the major axis, or the latus-rectum, of the hyperbola is equal to the double ordinate to this axis through the focus.

Through the focus F of the hyperbola, of which AA' is the major and BB' the minor axis, draw the chord PP' at right angles to the major axis; then denoting the parameter by P, we are to prove that

$$AA' : BB' :: BB' : PP' = P \qquad \text{(Def. 11.)}$$

By definition 6, $\overline{BC}^2 = FA' \cdot FA$, and by proposition 9 we have

$$\overline{AC}^2 : \overline{BC}^2 :: FA' \cdot FA : \overline{PF}^2 = (\tfrac{1}{2}PP')^2 \text{ (Cor. Prop. 1.)}$$

Whence $\quad \overline{AC}^2 : \overline{BC}^2 :: \overline{BC}^2 : (\tfrac{1}{2}PP')^2$

Therefore $AC : BC :: BC : \tfrac{1}{2}PP'$ (Th. 10, B. II, Geom.)

Multiplying all the terms of this last proportion by 2, it becomes

$$2AC : 2BC :: 2BC : PP'$$

or, $\qquad AA' : BB' :: BB' : PP'$

Hence the theorem; *the parameter of the major axis, etc.*

PROPOSITION XI.—THEOREM.

If from the vertices of any two conjugate diameters of the hyperbola ordinates be drawn to either axis, the difference of the squares of these ordinates will be equal to the square of one half the other axis.

Let AA', BB' be the axes, and PP', QQ' any two conjugate diameters of the conjugate hyperbolas represented in the figure. Then, drawing the ordinates QV, PM, to the major axes, and the ordinates $PS = MC$, $QD = VC$, to the minor axis, it is to be proved that

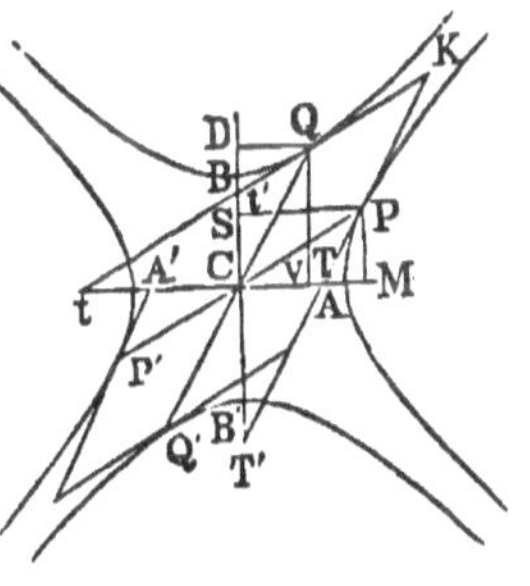

$$\overline{CA}^2 = \overline{MC}^2 - \overline{VC}^2$$

and that $\qquad \overline{CB}^2 = \overline{QV}^2 - \overline{PM}^2$

Draw the tangents PT and Qt, the first intersecting the major axis at T and the minor axis at T', and the second intersecting the minor axis at t' and the major axis at t.

Now, by proposition 8, we have, with reference to the tangent PT,

$$CT : CA :: CA : CM,$$

and by the scholium to the same proposition, we also have, with ference to the tangent Qt to the conjugate hyperbola,

$$Ct : CA' = CA :: CA : CV$$

The first proportion gives $\overline{CA}^2 = CT \cdot CM$, and the second $\overline{CA}^2 = Ct \cdot CV$,

Whence $CT \cdot CM = Ct \cdot CV$, which, in the form of a proportion, becomes

$$CM : CV :: Ct : CT \qquad (1)$$

From the similar triangles tCQ, CTP, we get

$$Ct : CT :: QC : PT \qquad (2)$$

and from the triangles CQV, TPM

$$QC : PT :: CV : TM \qquad (3)$$

Comparing proportions (1), (2) and (3), it is seen that

$$CM : CV :: CV : TM$$

Whence $\overline{CV}^2 = CM \cdot TM$; but $TM = CM - CT$;

Therefore $\overline{CV}^2 = \overline{CM}^2 - CT \cdot CM$.

And because $CT \cdot CM = \overline{CA}^2$ (Prop. 8), we have

$$\overline{CV}^2 = \overline{CM}^2 - \overline{CA}^2$$

or, $$\overline{CA}^2 = \overline{CM}^2 - \overline{CV}^2$$

Again we have

$$CT' : CB :: CB : PM \quad \text{(Scho., Prop. 8)}$$

and $Ct' : CB :: CB : CD = QV$ (Prop. 8)

Whence $CT' \cdot PM = Ct' \cdot QV$, which, resolved into a proportion, becomes

$$PM : QV :: Ct' : CT' \qquad (4)$$

From the similar triangles, $T'CP$, $Ct'Q$, we get

$$Ct' : CT' :: t'Q : CP \qquad (5)$$

And from the triangles $t'DQ$, CPM, we also get

$$t'Q : CP :: t'D : PM \qquad (6)$$

From proportions (4), (5) and (6) we deduce

$$PM : QV :: t'D : PM$$

Whence $\overline{PM}^2 = QV \cdot t'D$; but $t'D = C\!\!\!\!\!\! - Ct'$;

therefore, $\overline{PM}^2 = \overline{QV}^2 - Ct' \cdot QV = \overline{QV}^2 - Ct' \cdot CD.$

And because $Ct' \cdot CD = CB^2$ (Prop. 8) we have

$$\overline{PM}^2 = \overline{QV}^2 - \overline{CB}^2$$

or
$$\overline{CB}^2 = \overline{QV}^2 - \overline{PM}^2$$

Hence the theorem; *from the vertices of any two, etc.*

Cor. By corollary to proposition 9 we have

$$\overline{CA}^2 : \overline{CB}^2 :: \overline{CM}^2 - \overline{CA}^2 : \overline{PM}^2$$

In like manner, in reference to the conjugate hyperbola, we shall have

$$\overline{CB}^2 : \overline{CA}^2 :: \overline{CD}^2 - \overline{CB}^2 : \overline{QD}^2$$
$$:: \overline{QV}^2 - \overline{CB}^2 : \overline{CV}^2$$

or,
$$\overline{CB}^2 : \overline{QV}^2 - \overline{CB}^2 :: \overline{CA}^2 : \overline{CV}^2$$

By composition, $\overline{CB}^2 : \overline{QV}^2 :: \overline{CA}^2 : \overline{CA}^2 + \overline{CV}^2$

But by this proposition we have

$$\overline{CA}^2 = \overline{CM}^2 - \overline{CV}^2 ;\ \text{hence } \overline{CA}^2 + \overline{CV}^2 = \overline{CM}^2$$

therefore $\qquad \overline{CB}^2 : \overline{QV}^2 :: \overline{CA}^2 : \overline{CM}^2$

Whence $\qquad CB : QV :: CA : CM$

or, $\qquad CA : CB :: CM : QV$

PROPOSITION XII.—THEOREM.

The difference of the squares of any two conjugate diameters of an hyperbola is constantly equal to the difference of the squares of the axes.

In the figure, which is the same as that of the preceding proposition, PP' and QQ' are any two conjugate diameters (Def. 10). It is to be proved that

$$\overline{PP'}^2 - \overline{QQ'}^2 = \overline{AA'}^2 - \overline{BB'}^2$$

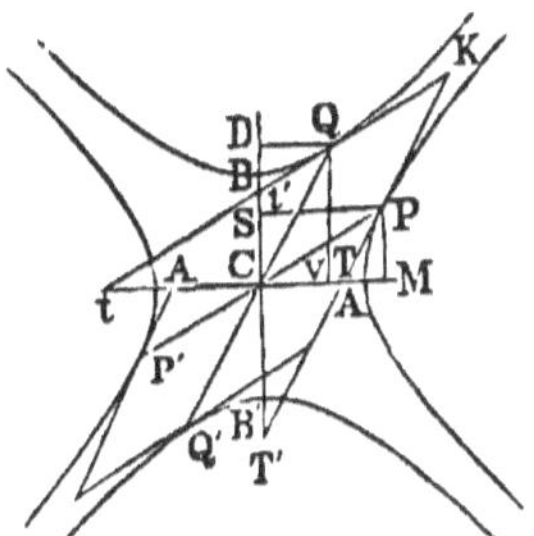

By proposition 11 we have

$$\overline{CA}^2 = \overline{CM}^2 - \overline{CV}^2$$

and

$$\overline{CB}^2 = \overline{QV}^2 - \overline{PM}^2$$

therefore $\quad \overline{CA}^2 - \overline{CB}^2 = \overline{CM}^2 + \overline{PM}^2 - (\overline{CV}^2 + \overline{QV}^2)$

or, $\qquad\qquad \overline{CA}^2 - \overline{CB}^2 = \overline{CP}^2 - \overline{CQ}^2$

Multiplying each member of this equation by 4, observing that $4\overline{CA}^2 = \overline{AA'}^2$ &c., it becomes

$$\overline{AA'}^2 - \overline{BB'}^2 = \overline{PP'}^2 - \overline{QQ'}^2$$

Hence the theorem; *the difference of the squares, etc.*

PROPOSITION XIII.—THEOREM.

The parallelogram formed by drawing tangent lines through the vertices of any two conjugate diameters of the hyperbola is equivalent to the rectangle contained by the axes.

Let $LMNO$ be a parallelogram formed by drawing tangent lines through the vertices of the two conjugate diameters PP', QQ' of the conjugate hyperbolas represented in the figure. It is to be proved that area $LMNO = AA' \times BB'$.

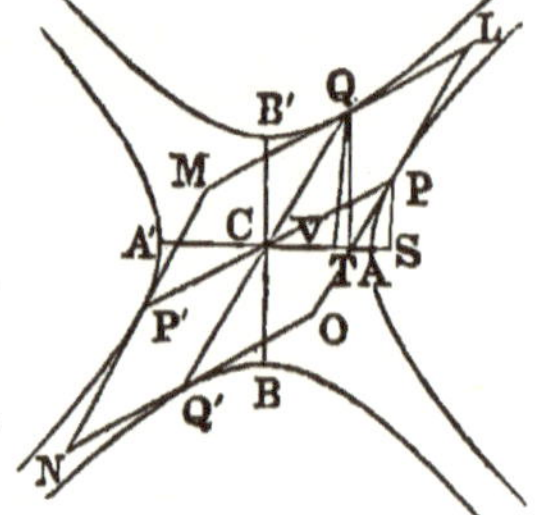

We have $\quad CA : CB :: CS : QV \quad$ (1) (Cor. Prop 11.)

Also, $\qquad CT : CA :: CA : CS$ (2) $\qquad$ (Prop. 8.)

Multiplying proportions (1) and (2), term by term, omitting in the first couplet of the result the common factor CA, and in the second the common factor CS, we find

$$CT : CB :: CA : QV$$

Whence $\qquad CT \cdot QV = CA \cdot CB$

But $CT \cdot QV$ measures twice the area of the triangle CQT, and this triangle is equivalent to the half of the parallelogram $QCPL$, because they have the common base QC and are between the same parallels QC, LT (Th. 30, B. I, Geom.)

Now the parallelogram $QCPL$ is one-fourth of the parallelogram $LMNO$, and $CA \cdot CB$ measures one fourth of the rectangle contained by the axes; therefore the parallelogram and rectangle are equivalent.

Hence the theorem; *the parallelogram formed, etc.*

PROPOSITION XIV.—THEOREM.

If a tangent to the hyperbola be drawn through the vertex of the transverse axis, and an ordinate to any diameter be drawn from the same point, the semi-diameter will be a mean proportional between the distances, on the diameter, from the center to the tangent, and from the center to the ordinate.

Let CA be the semi-major axis and CP any semi-diameter of the hyperbola. Draw the tangents At, PT, the ordinate AH to the diameter, and the ordinate PM to the major axis. It is now to be proved that $\overline{CP}^2 = Ct \cdot CH$.

We have $CT : CA :: CA : CM,$ \qquad (Prop. 8)

also $CA : Ct :: CM : CP$ from the similar $\triangle$'s CAt, CMP

Multiplying these proportions term by term, omitting in the result the common factor in the first couplet, and also that in the second, we find

$$CT : Ct :: CA : CP \qquad (1)$$

Again we have

$CP : CT :: CH : CA$ from the similar $\triangle$'s CPT, CHA.

Proceeding with these last proportions as with those above, we find

$$CP : Ct :: CH : CP$$

Whence, \qquad $\overline{CP}^2 = Ct \cdot CH.$

Hence the theorem; *if a tangent to the hyperbola, etc.*

Cor. 1. From proportion (1) we get $CT \cdot CP = Ct \cdot CA$; but the triangles CTP, CAt, having a common angle, C, are

8

to each other as the rectangles of the sides about this angle (Th. 23, B. II, Geom.) Therefore $\triangle CTP = \triangle CtA$.

Cor. 2. If from the equivalent areas $\triangle CTP$, $\triangle CtA$ we take the common area $CTVt$ there will remain $\triangle TAV = \triangle tVP$.

Cor. 3. If we add to each of the triangles TAV, tVP, the trapezoid $VAMP$, we shall have area $\triangle TMP =$ area $tAMP$.

PROPOSITION XV.—THEOREM.

If through any point of an hyperbola there be drawn a tangent, and an ordinate to any diameter, the semi-diameter will be a mean proportional between the distances on the diameter from the center to the tangent, and from the center to the ordinate.

Take any point as D on the hyperbola of which CA is the semi-major axis, and through this point draw the tangent DT and the semi-diameter CD, also take any other point, as P, on the curve, and draw the tangent Pt, the ordinate PH to 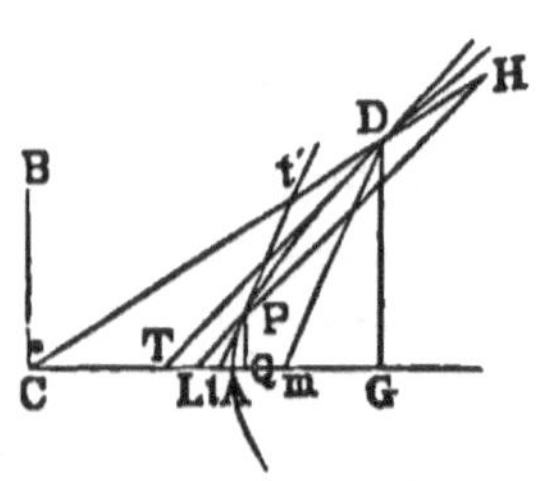
the diameter through D, and the ordinates PQ and DG to the axis. The semi-diameter CD and the tangent Pt intersect each other at t'. We will now prove that
$$\overline{CD}^2 = Ct' \cdot CH$$

Let CB represent the semi-conjugate axis, then by corollary to proposition 9 (proportion (b)) we have

$$\overline{CA}^2 : \overline{CB}^2 :: CG \cdot TG : \overline{DG}^2$$
and
$$\overline{CA}^2 : \overline{CB}^2 :: CQ \cdot tQ : \overline{PQ}^2$$

Whence $CG \cdot TG : CQ \cdot tQ : \overline{DG}^2 : \overline{PQ}^2$

but $\overline{DG}^2 : \overline{PQ}^2 :: \overline{TG}^2 : \overline{LQ}^2$, from the similar $\triangle$'s TGD, LQP;

therefore $\qquad CG \cdot TG : CQ \cdot tQ : : \overline{TG}^2 : \overline{LQ}^2 \qquad$ (1)

Drawing Dm parallel to Pt we have the similar $\triangle$'s mGD, tQP which give the proportion

$$DG : PQ : : Gm : Qt. \qquad (2)$$

The $\triangle$'s TGD, LQP also give

$$DG : PQ : : TG : LQ \qquad (3)$$

From proportions (2) and (3) we get

$$TG : LQ : : Gm : Qt \qquad (4)$$

Multiplying proportions (1) and (4) term by term, there results,

$$CG \cdot \overline{TG}^2 : CQ \cdot tQ \cdot LQ : : \overline{TG}^2 \cdot Gm : \overline{LQ}^2 \cdot Qt$$

Dividing the first and third terms of this proportion by $\overline{TG}^2$ and the second and fourth terms by $Qt \cdot LQ$ it becomes

$$CG : CQ : : Gm : LQ$$

or $\qquad CG : Gm : : CQ : LQ \qquad$ (5)

Whence $\qquad CG : CG{-}Gm : : CQ : CQ{-}LQ$

That is $\qquad CG : Cm : : CQ : CL \qquad$ (6)

Again $\qquad CT \cdot CG = \overline{CA}^2 = CQ \cdot Ct,$ (Prop. 8.)

therefore $\qquad CG : Ct : : CQ : CT$

The antecedents in this last proportion and in proportion (6) are the same, the consequents are therefore proportional, and we have

$$Ct : CT : : Cm : CL$$

We have also, $Cm : CD : : Ct : Ct'$ from the similar $\triangle$'s CmD, Ctt'

And $CT : CD : : CL : CH$ from the similar $\triangle$'s CTD CLH

By the multiplication of the last three proportions term by term we find

$$Ct \cdot Cm \cdot CT : \overline{CD}^2 \cdot CT : : Cm \cdot Ct \cdot CL : CL \cdot Ct' \cdot CH$$

Whence $\qquad CT : \overline{CD}^2 \cdot CT : : CL : CL \cdot Ct' \cdot CH$

or $\qquad 1 : \overline{CD}^2 : : 1 : Ct' \cdot CH$

therefore $\qquad \overline{CD}^2 = Ct' \cdot CH$

Hence the theorem; *if through any point of an, etc.*

REMARK.—The property of the hyperbola just established is the generalization of that demonstrated in the preceding proposition.

PROPOSITION XVI.—THEOREM.

The square of any semi-diameter of the hyperbola is to the square of its semi-conjugate as the rectangle of the distances from the foot of any ordinate to the first diameter, to the vertices of that diameter, is to the square of the ordinate.

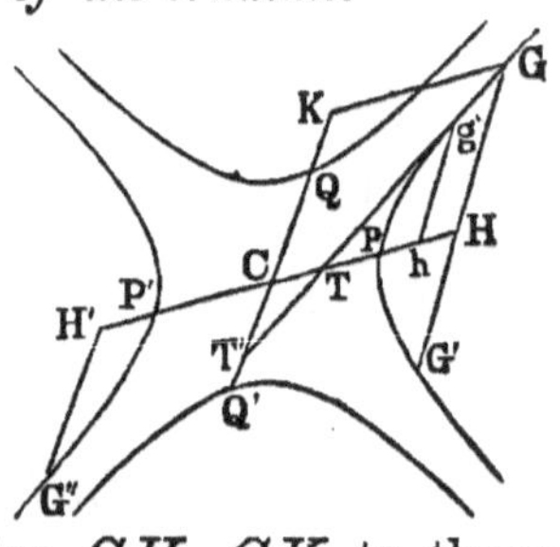

Let PP' and QQ' be any two conjugate diameters of the conjugate hyperbolas represented in the figure. Through any point as G draw the tangent GT intersecting the first diameter at T and the second at T', and from the same point draw the ordinates GH, GK, to these diameters.

We will now prove that,

$$\overline{CP}^2 : \overline{CQ}^2 :: PH \cdot P'H : \overline{GH}^2$$

By the preceding proposition we have $\overline{CP}^2 = CT \cdot CH$ and multiplying each member of this equation by CH it becomes $\overline{CP}^2 \cdot CH = CT \cdot \overline{CH}^2$

Whence $\overline{CP}^2 : \overline{CH}^2 :: CT : CH$ from which by division we get $\overline{CP}^2 : \overline{CH}^2 - \overline{CP}^2 :: CT : CH - CT = TH,$ (1)

Again we have $\overline{CQ}^2 = CT' \cdot CK$ (Prop. 15) and multiplying each member of this equation by CK it becomes $\overline{CQ}^2 \cdot CK = CT' \cdot \overline{CK}^2$

Whence $\overline{CQ}^2 : \overline{CK}^2 :: CT' : CK = GH$ (2)

The similar $\triangle$'s TCT', THG give the proportion
$$CT' : GH :: CT : TH$$ (3)

Comparing proportions (2) and (3) we obtain
$$\overline{CQ}^2 : \overline{CK}^2 :: CT : TH$$ (4)

And by comparing proportions (1) and (4) we obtain

$$\overline{CQ}^2 : \overline{CK}^2 : \overline{CP}^2 : \overline{CH}^2 - \overline{CP}^2$$

or $\quad \overline{CP}^2 : \overline{CQ}^2 : \overline{CH}^2 - \overline{CP}^2 : \overline{CK}^2 = \overline{GH}^2$

But because $CF = CP'$ and $\overline{CH}^2 - \overline{CP}^2 = (CH - CP)$ $(CH + CP) = PH \cdot (CH + CP)$ the last proportion above becomes $\quad \overline{CP}^2 : \overline{CQ}^2 :: PH \cdot P'H : \overline{GH}^2$

Hence the theorem; *The square of any semi-diameter, etc.*

REMARK.—The property of the hyperbola with reference to any two conjugate diameters just demonstrated is the same as that with reference to the axes established in proposition 9.

Cor. If the ordinate GH be produced to intersect the curve at G' and the above construction and demonstration be supposed made for the point G' instead of G, we should finally get the same proportion as before, except the fourth term, which would be $\overline{G'H}^2$; therefore, $G'H = GH$. Hence we conclude that

Any diameter of the hyperbola bisects all the chords drawn parallel to a tangent line through the vertex of that diameter.

PROPOSITION XVII.—THEOREM.

The squares of the ordinates to any diameter of the hyperbola are to one another as the rectangles of the corresponding distances from the feet of these ordinates to the vertices of the diameter.

Resuming the figure to the proposition which precedes and drawing any other ordinate gh to the diameter PP', it is to be proved that

$$\overline{GH}^2 : \overline{gh}^2 :: PH \cdot P'H : Ph \cdot P'h$$

By the foregoing proposition we have two proportions following, viz:

$$\overline{CP}^2 : \overline{CQ}^2 :: PH \cdot P'H : \overline{GH}^2$$
$$\overline{CP}^2 : \overline{CQ}^2 :: Ph \cdot P'h : \overline{gh}^2$$

8*

Since the ratio $\overline{CP}^2 : \overline{CQ}^2$ is common to these proportions the remaining terms are proportional.

That is $\qquad \overline{GH}^2 : \overline{gh}^2 :: PH \cdot P'H : Ph \cdot P'h$

Hence the theorem—*The squares of the ordinates, etc.*

PROPOSITION XVIII.—THEOREM.

If a cone be cut by a plane making an angle with its base greater than that made by an element of the cone, the section will be an hyperbola.

Let the △'s MVN, BVR be the sections of two opposite cones by a plane through the common axis, and BH a line in this section not passing through the vertex, and making with MN the $\llcorner BHN >$ the $\llcorner BMN$. Through this line pass a plane at right angles to the first plane, making in the lower cone the section 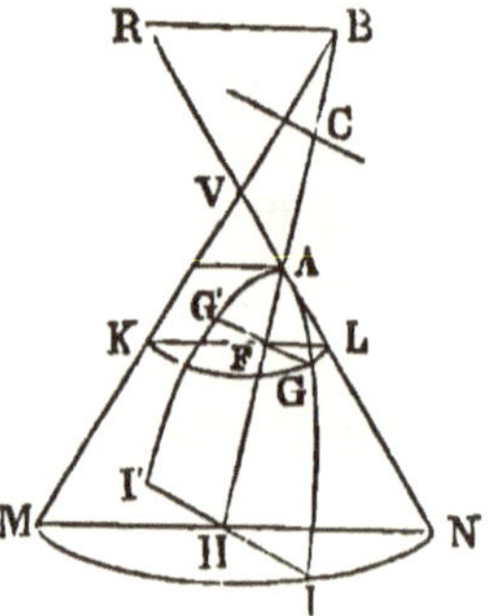 $IGAG'I'$; then will this section be one of the branches of an hyperbola.

Let KL and MN be the diameters of two circular sections made by planes at right angles to the axis of the cone, and at F and H, the intersections of these lines with BH, erect the perpendiculars FG, HI to the plane MVN. FG is the intersection of the plane of the section $IGAG'I'$ with the plane of the circle of which KL is the diameter and is a common ordinate of the section and of the circle; so likewise is HI a common ordinate of the section and of the circle of which MN is the diameter.

Now by the similar △'s AFL, AHN, and BFK, BHM we have

$$AF : AH :: FL : HN \qquad (1)$$

and $\qquad BF : BH :: FK : HM \qquad (2)$

Multiplying proportions (1) and (2), term by term, we get

$$AF\cdot BF : AH\cdot BH :: FL\cdot FK : HN\cdot HM \quad (3)$$

But because LGK and NIM are semi-circles, $\overline{FG}^2 = FL\cdot FK$ and $\overline{HI}^2 = HN\cdot HM$. Substituting these values for the terms of the last couplet of proportion (3) it becomes

$$AF\cdot BF : AH\cdot BH :: \overline{FG}^2 : \overline{HI}^2$$

If we denote any two ordinates of the corresponding section of the opposite cone by fg and hi we should have in like manner

$$Af\cdot Bf : Ah\cdot Bh :: (fg)^2 : (hi)^2$$

If, therefore, AB be taken as a diameter of the curves cut out of the opposite cones by a plane through AH, at right angles to the plane VMN, we have proved that these curves possess the property which was demonstrated in the preceding proposition to belong to the hyperbola.

Hence the theorem; *if a curve be cut by a plane, etc.*

ASYMPTOTES.

DEFINITION.—An **Asymptote** to a curve is a straight line which continually approaches the curve without ever meeting it, or, which meets it only at an infinite distance.

We shall for the present assume, what will be afterwards proved, that the diagonals of the rectangle constructed by drawing tangent lines through the vertices of the axis of the hyperbola possess the property of asymptotes, and they are therefore called *the asymptotes* of the hyperbola.

PROPOSITION XIX.—THEOREM.

If an ordinate to the transverse axis of an hyperbola be produced to meet the asymptotes, the rectangle of the segments into which it is divided by either of its intersections with the curve will be equivalent to the square of the semi-conjugate axis.

Let *CA*, *CB* be the semi-axes and *Ct*, *Ct'* the asymptotes of an hyperbola.— Through any point, as *P*, of the curve, draw the ordinate *PQ* to the major axis and produce it to meet the asymptotes at *n* and *n'*. By the enunciation we are required to prove that $\overline{CB}^2 = Pn \cdot Pn'$

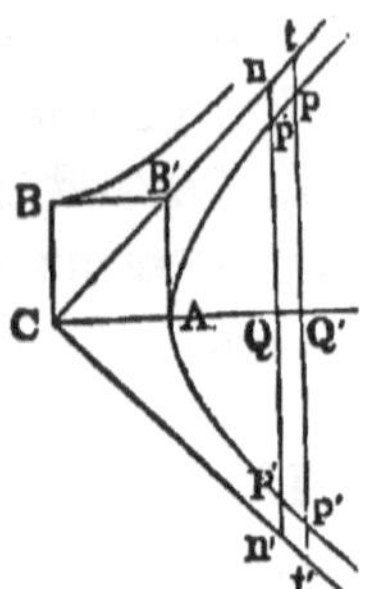

By Cor. proposition 9 we have

$$\overline{CA}^2 : \overline{CB}^2 :: \overline{CQ}^2 - \overline{CA}^2 : \overline{PQ}^2 \qquad (1)$$

And from the similar triangles *CAB'*, *CQn*

$$\overline{CA}^2 : \overline{AB'}^2 = \overline{CB}^2 :: \overline{CQ}^2 : \overline{Qn}^2 \qquad (2)$$

Comparing proportions (1) and (2) we find

$$\overline{CQ}^2 : \overline{CQ}^2 - \overline{CA}^2 : \overline{Qn,}^2 : \overline{PQ}^2 \text{ which gives by}$$

division $\overline{CA}^2 : \overline{CQ}^2 :: \overline{Qn}^2 - \overline{PQ}^2 : \overline{Qn}^2$

or $\qquad \overline{CA}^2 : \overline{Qn}^2 - \overline{PQ}^2 :: \overline{CQ}^2 : \overline{Qn}^2 \qquad (3)$

From proportions (2) and (3) we get

$$\overline{CA}^2 : \overline{CB}^2 :: \overline{CA}^2 : \overline{Qn}^2 - \overline{PQ}^2$$

In this proportion the antecedents are the same the consequents are therefore equal; that is

$$\overline{CB}^2 = \overline{Qn}^2 - \overline{PQ}^2 = (Qn + PQ)(Qn - PQ) = Pn \cdot Pn'$$

Hence the theorem ; *if an ordinate to the major axis, etc.*

Cor. Let us take another point *p* in the curve and from it draw the ordinate *pQ'* to the major axis ; then, as before, we shall have $\overline{CB}^2 = pt \cdot pt'$; *t* and *t'* being the intersections of the ordinate, produced, with the asymptotes.

Whence $Pn \cdot Pn' = pt \cdot pt'$, which in the form of a proportion becomes $Pn : Pt :: pt' : Pn'$

PROPOSITION XX.—THEOREM.

The parallelograms formed by drawing through the different points of the hyperbola lines parallel to and meeting the asymptotes are equivalent one to another, and any one is equivalent to one half of the rectangle contained by the semi-axes.

Let CA, CB be the semi-axes and Cn, Cn' the asymptotes of an hyperbola. From any point, as P, of the curve draw the ordinate PQ to the transverse axis, producing it to meet the asymptotes at n, n', and through P and the vertex A draw parallels to the asymptotes, forming the parallelograms $PmCt$, $AECD$. This last is a rhombus 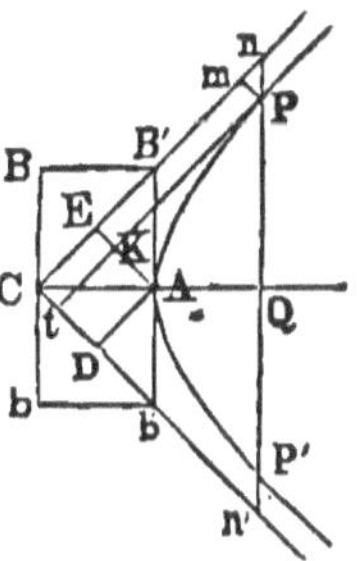 because its adjacent sides CE, CD are equal, being the semi-diagonals of equal rectangles.

It will now be proved that

Area $PmCt$ = area $AECD=\frac{1}{2}$ Rect. $AB'BC$.

By the proposition which precedes we have

$$\overline{CB}^2 = Pn \cdot Pn' \tag{1}$$

And from the similar triangles $AB'E$, Pnm, and the similar triangles ADb', Ptn' we also have

$$AE : AB' = CB : : mP : Pn$$
$$AD : Ab' = CB : : Pt : Pn'$$

Multiplying these proportions, term by term, we find

$$AE \cdot AD : \overline{CB}^2 : : mP \cdot Pt : Pn \cdot Pn'$$

By equation (1) the consequents of this proportion are equal, therefore the antecedents are also equal.

That is, $\qquad AE \cdot AD = mP \cdot Pt$

If the first member of this equation be multiplied by sin. $\lfloor DAE$, and the second member by the sine of the equal $\lfloor mPt$ it becomes

$$AE \cdot AD \cdot \text{sin. } DAE = mP \cdot Pt \cdot \text{sin } mPt$$

But $AE \cdot AD \cdot \sin DAE$ measures the area of the rhombus $AECD$ and $mP \cdot Pt$ sin. mPt measures the area of the parallelogram $PmCt$; therefore the parallelogram and the rhombus are equivalent. Moreover, because the $\triangle$'s AEC, ADC are equal, and the $\triangle$'s AEC, AEB' are equivalent, it follows that the rhombus $AECD$ is equivalent to the $\triangle AB'C$, or, to one half of the rectangle contained by the semi-axes.

Hence the theorem; *the parallelograms formed, etc.*

Cor. 1. If from the rhombus $AECD$ and the parallelogram $PmCt$ the common part be taken, there will remain the parallelogram $AKtD$, equivalent to the parallelogram $PmEK$, and if to each of these the curvilinear area AKP be added, we shall have

$$\text{Area } APmE = \text{ area } APtD.$$

Had we proceeded in the same way with the parallelogram $PmCt$ and any parallelogram other than $AECD$ we should have had a like result; therefore

If from any two points in the hyperbola parallels be drawn to each asymptote, the area bounded by the parallels to one asymptote, the other asymptote, and the curve will be equivalent to the other area like bounded.

Scholium.—If the product $AE \cdot AD$, which is a constant quantity be denoted by a, the distance Cm by x, and the distance $mp = Ct$ by y, then, by this proposition, we shall have the equation $xy = a$, which, in analytical geometry, is called the equation of the hyperbola referred to its center and asymptotes.

Cor. 2. In the equation $xy = a$, y expresses the distance of any point of the curve from the asymptote on which x is estimated. From this equation we get $y = \dfrac{a}{x}$. Now it is evident that as x increases y decreases, and finally when x becomes infinite, y becomes zero. That is, the asymptote continually approaches the hyperbola without ever meeting it, or without meeting it within a finite distance. We were, therefore, justified in assuming that the diagonals of the rectangle formed by the tangents through the vertices of the axes were asymptotes to the hyperbola.

ANALYTICAL GEOMETRY.

ANALYTICAL GEOMETRY.

GENERAL DEFINITIONS AND REMARKS.

Analytical Geometry, as the terms imply, proposes to investigate geometrical truths by means of analysis. In it the magnitudes under consideration are represent by simbols, such as letters, terms, simple or combined, and equations; and problems are then solved and the properties and relations of magnitude established by processes purely algebraic.

A *single letter*, without an exponent, will always be understood as denoting the length of a line; and in general, any *expression of the first degree* denotes the *length* of a line and is, for this reason, said to be *linear;* so likewise, an equation all of whose terms are of the first degree is called a *linear equation.*

An expression of the second degree will represent the *measure* of a surface, and *an expression of the third degree* will represent the *measure* of a volume.

When a term is of a higher degree than the third, a sufficient number of its literal factors, to reduce it to this degree, must be regarded as *numerical* or *abstract.*

The subject of Analytical Geometry naturally resolves itself into two parts.

First. That which relates to the solution of *determinate problems;* that is, problems in which it is required to determine certain unknown magnitudes from the relations which they bear to others that are known. In this case we must be able to express the relations between the known and unknown magnitudes by independent equations equal in number to the required magnitudes.

(96)

After having obtained, by a solution of the equations of the problem, the algebraic expressions for the quantities sought, it may be necessary, or, at least desirable, to construct their values, by which we mean, to draw a geometrical figure in which the parts represent the given and determined magnitudes, and have to each other the relations imposed by the conditions of the problem. This is called *the construction of the expression.*

This branch of analytical geometry, which may be termed *Determinate Geometry*, being of the least importance, relatively, will be omitted, after this reference, in the present treatise, and we shall pass at once to division.

Second. That which has for its object to discover and discuss the general properties of geometrical magnitudes. In this the magnitudes are represented by equations expressing relations between *constant* quantities, and, either two or three *indeterminate* or *variable* quantities, and for this reason it is sometimes called *Indeterminate Geometry.*

GENERAL PROPERTIES

OF

GEOMETRICAL MAGNITUDES.

CHAPTER I.

OF POSITIONS AND STRAIGHT LINES IN A PLANE, AND THE TRANSFORMATION OF CO-ORDINATES.

DEFINITIONS.

1. Co-ordinate Axes are two straight lines drawn in a plane through any assumed point and making with each other any given angle. One of these lines is the axis of *abscissas* or the axis of X; the other is the axis of *ordinates*, or the axis of Y, and their intersection is the *origin of co-ordinates.*

2. Abscissas are distances estimated from the axis of Y on lines parallel to the axis of X; *ordinates* are distances
9

estimated from the axis of X on lines parallel to the axis of Y.

3. The abscissa and ordinate of a point together are called the *co-ordinates* of the point.

4. The co-ordinate axes are said to be *rectangular* when they are at right angles to each other, otherwise they are *oblique*.

5. The two different directions in which distances may be estimated from either axis, on lines parallel to the other, are distinguished by the signs *plus* and *minus*.

6. Abscissas are designated by the letter x and ordinates by the letter y, and when unaccented they are called *general* co-ordinates, because they refer to no particular one of the points under consideration. When particular points are to be considered the co-ordinates of one are denoted by x' and y'; of another by x'' and y'', etc., which are read x prime, y prime, x second, y second, etc.

ILLUSTRATIONS.—Through any point A draw the lines XX', YY' making with each other any given angle. Call XX' the axis of abscissas and YY' the axis of ordinates. A is the origin of co-ordinates, or zero point. The four angular spaces into which the plane is divided are named, respectively, *first, second, third,* and *fourth* angles. YAX is the first angle, YAX' is the second angle, $Y'AX'$ is the third angle, and $Y'AX$ is the fourth angle.

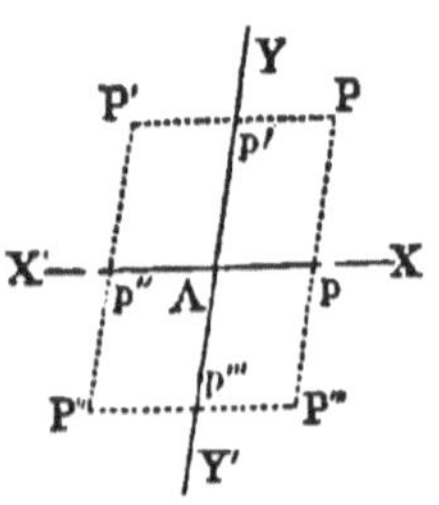

Take any point, as P, in the first angle, and from it draw Pp parallel to the axis of Y and Pp' parallel to the axis of X, the first meeting the axis of X at p, and the second the axis of Y at p'; then $p'P=Ap$ is the abscissa, and $pP=Ap'$ is the ordinate of the point P.

Now produce Pp' to P' making $p'P'=p'P$, and from P' draw a parallel to the axis of Y meeting the axis of X at p''; then the point P' is in the second angle, and $p'P'$

$=Ap''$ is its abscissa, and $p''P'=Ap'$ is the ordinate. By like constructions we determine the position of the point P'' in the third angle, and that of the point P''' in the fourth angle.

It is evident that the abscissas of these four points are numerically equal, as are likewise their ordinates; but if we have reference to the algebraic signs of the co-ordinates, each point will be assigned to its appropriate angle and will be completely distinguished from the others. Abscissas estimated to the right of the axis of Y are *positive* and those estimated to the left are *negative*. Ordinates estimated from the axis of X upwards are *positive*, those estimated downwards are *negative*.

We shall therefore have for points

> In the 1st angle, x positive, y positive.
> " " 2d " x negative, y positive.
> " " 3d " x negative y negative.
> " " 4th " x positive y negative.

From what precedes we see that the position of a point in the plane of the co-ordinate axis is fully determined by its co-ordinates. To construct this position we lay off on the axis of X the given abscissa, to the right, or to the left of the origin, according to the sign; also lay off on the axis of Y the given ordinate, upwards from the origin if the sign be plus, downwards if it be minus. The lines drawn through the points thus found, parallel to the co-ordinate axes, will intersect at the required point and fix its position.

As rectangular co-ordinates are more readily apprehended than oblique, and as discussions and algebraic expressions are generally less complicated where references are made to the former, than when made to the latter, rectangular co-ordinates will be habitually employed in the following pages. When we have occasion to use others it will be so stated.

PROPOSITION I.

To find the equation of a straight line,

Let XX', YY' be two rectangular co-ordinate axes. A being the
origin draw any line as $L'L$ through
this point, and designate the natural tangent of the angle LAX by a.

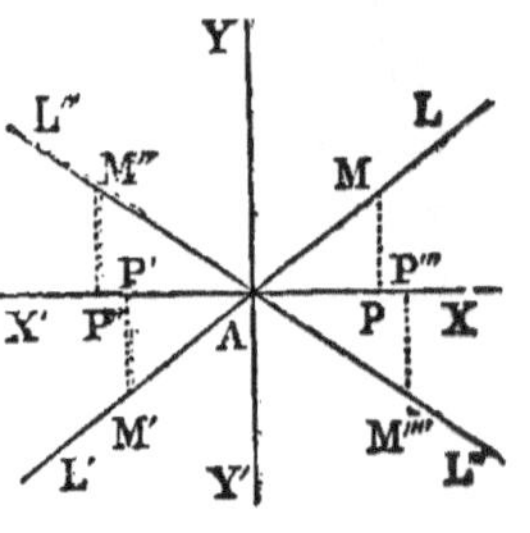

Then take any distance on AX
as AP, and represent it by x, and
the perpendicular distance PM y.

Then by trigonometry we have

$$\text{Rad} : \text{tan. } MAP :: AP : PM$$

or $$1 : a :: x : y$$

Whence $$y = ax \qquad (1)$$

Now this equation is general ; that is, it applies to any
point M on the line AL, because we can make x greater
or less, and PM will be greater or less in like proportion
and M will move along on the line AL as we move P on
the line AX. Because the point M will continue on the
line AL through all changes of x and y, we say that $y = ax$
is the equation of the line AL.

Now let us diminish x to 0, and the equation reduces
to $y = 0$ at the same time, which brings M to the point A.

Let x pass the line YY', then AP' becomes $-x$, and
the corresponding value of y will be $P'M'$, and, being below the line $X'X$, will, therefore, be *minus*.

Therefore $$y = ax.$$

is the general equation of the line LL', extending indefinitely in either direction.

If the tangent a becomes less, the line will incline more
towards the line $X'X$. When $a = 0$ the line will coincide
with XX'.

Now let AP''' be $+x$, and a become $-a$, then $P'''M'''$
will correspond to y, and becomes *minus* y, because it is

below the axis $X'X'$. Or, algebraically $y=-ax$, indicating some point M''' below the horizontal axis.

It is, therefore, obvious that $y=ax$ may represent any line, as LL', passing through A *from the* 1st *into the* 3d *quadrant*, and that $y=-ax$ may be made to represent any line, as $L''L'''$, passing through A from the 2d into the 4th quadrant.

Therefore $$y=\pm ax$$

may be made to represent any straight line passing through the zero point.

In case we have $-a$ and $-x$, that is, both a and x minus at the same time, their product will be $+ax$, showing that y must be *plus* by the rules of algebra.

As an exercise, let the learner examine these lines and see whether they correspond to the equation.

When we have $-a$ we must draw the line from A to the right and *below* AX; then XAL''' is the angle whose natural tangent is $-a$. But the opposite angle $X'AL''$ is the same in value.

When we have $-x$ we must take the distance as AP'' to the left of the axis YY', and the corresponding line $P''M''$ is above XX', and therefore *plus*, as it ought to be.

But the equation of a straight line passing through the **zero point** is not sufficiently general for practical application; we will therefore suppose a line to pass in any direction across the axis YY', cutting it at the distance AB or AD ($\pm b$) or b distance above or below the zero point A, and find its equation.

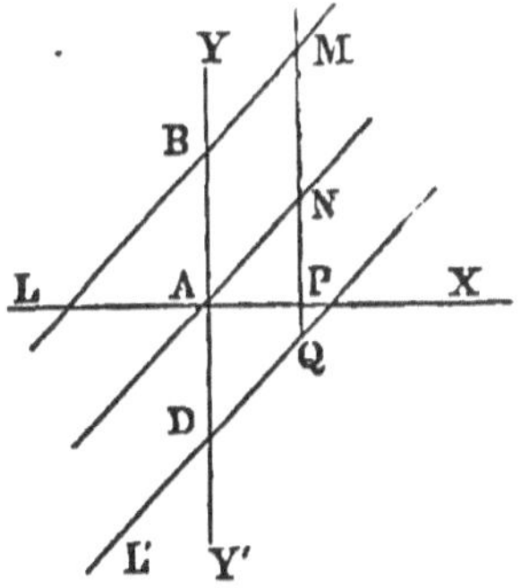

Through the zero point A draw a line, AN, parallel to ML.

Take any point on the line AX and through P draw

9*

PM parallel to AY, then $ABMN$ will be a parallelogram.

Put $AP=x$. $PM=y$. The tangent of the angle $NAP=a$. Then will $NP=ax$.

To each of these equals add $NM=b$, then we shall have

$$y=ax+b$$

for the relation between the values of x and y corresponding to the point M, and as M is any variable point on the line ML corresponding to the variations of x, this equation is said to be *the equation of the line ML.*

When b is *minus* the line is then QL', and cuts the axis YY' in D, a point as far below A as B is above A.

Hence we perceive that the equation

$$y=\pm ax \pm b$$

may represent the equation of any line in the plane YAX.

If we give to a, x, and b, their proper signs, in each case of application we may write

$$y=ax+b$$

for the equation of any straight line in a plane.

Cor. Since the equation $y=ax+b$ truly expresses the relation between the co-ordinates of any point of the line, it follows that if the co-ordinates x' and y' of any particular point of the line be substituted for the variables x and y the equation must hold true; but if the co-ordinates x'' and y'', of any point out of the line be substituted for the variables, the equation cannot be true.

What appears in the particular case of a straight line are general principles which we thus enunciate, viz:

1st. *If the co-ordinates of a particular point, in any line whatever, be substituted for the variables in the equation of the line, the equation must be satisfied; but if the co-ordinates of a point out the line, be substituted for the variables in its equation, the equation cannot be satisfied.*

2d. *If the co-ordinates of any point be substituted for the variables in the equation of a line, and the equation be satisfied, the point must be on the line; but if the equation be not satisfied by the substitution, the point cannot be on the line.*

These are principles of the highest importance in analytical geometry, and should be thoroughly committed and fully understood by the student.

SCHOLIUM.—Instead of rectangular, let us assume the oblique co-ordinate axes AX and AY, making with each other an angle denoted by m. Through the origin draw the line AP making with the axis of x the angle $PAD=n$; then the angle $PAD'=m-n$. Take any point as P in the line and from it draw PD' and PD parallel, respectively, to the axes of X and Y.

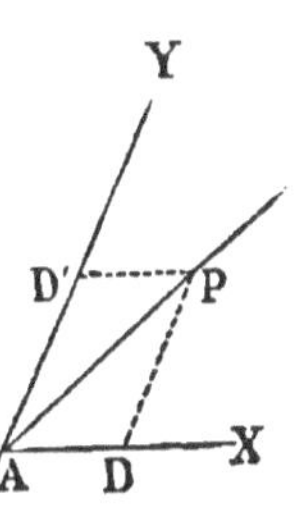

From the triangle APD we have (Prop. 4, Sec. 1, Plane Trig.)

$$PD : AD :: \text{Sin. } PAD = \text{Sin. } PAD'$$

or $$y : x :: \text{Sin. } n : \text{Sin. } (m-n.)$$

Whence $$y = \frac{\text{sin. } n}{\text{sin. } m-n} x$$

But $\dfrac{\text{sin } n}{\text{sin. } (m-n}$ is constant for the same line and may be represented by a.

Therefore, for any straight line passing through the origin of a system of oblique co-ordinate axes we have, as before, the equation

$$y = ax.$$

And if we denote by b the distance from the origin to the point at which a parallel line cuts the axis of Y above or below the origin we shall also have for the equation of this line

$$y = ax + b,$$

in which it must be remembered that a denotes the sine of the angle that the line makes with axis of x divided by the sine of the angle it makes with the axis of Y.

To fix in the minds of learners a complete comprehension of the equation of a straight line, we give the following practical

EXAMPLES.

1. Draw the line whose equation is $\quad y = 2x + 3.\qquad$ (1)

Then draw the line represented by $\quad y = -x + 2\qquad$ (2)

and determine where these two lines intersect.

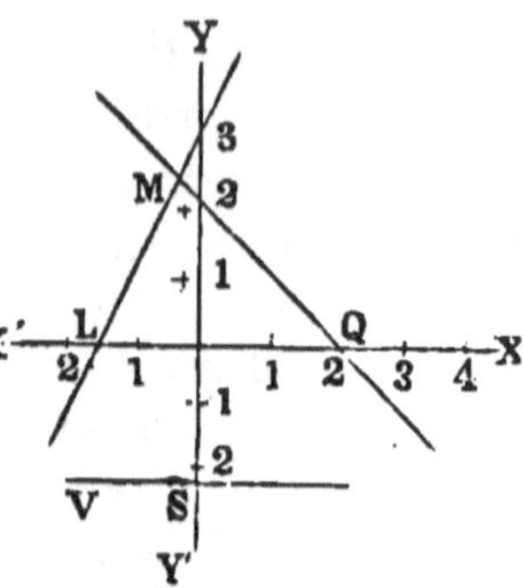

Draw YY' and XX' at right angles, and taking any convenient unit of measure lay it off on each of the axes from the origin in both positive and negative directions a sufficient number of times.

Equation (1) is true for all values of x and y. It is true then when $x=0$. But when $x=0$ the point on the line must be on the axis YY'.

When $x=0$. $y=3$.

This shows that the line sought for must cut YY' at the point $+3$.

The equation is equally true when $y=0$. But when $y=0$, the corresponding point on the line sought must be on the axis XX', and on the same supposition the equation becomes

$$0=2x+3, \quad \text{Or } x=-1\tfrac{1}{2}.$$

That is, midway between -1 and -2 is another point in the line which is represented by $y=2x+3$, but two points in any right line must define the line; therefore ML is the line sought.

Taking equation (2) and making $x=0$ will give $y=2$, and making $y=0$ will give $x=2$; therefore MQ must be the line whose equation is $y=-x+2$, and these two lines with the axis XX' form the triangle LMQ, whose base is $3\tfrac{1}{2}$ and altitude *about* $2\tfrac{1}{2}$.

But let the equations decide, (*not about,*) but exactly the position of the point M of intersection.

This point being in both lines, the co-ordinates x and y corresponding to this point are the same, therefore we may subtract one equation from the other, and the result will be a true equation, giving

$$3x+1=0. \quad \text{Or } x=-\tfrac{1}{3}.$$

Eliminating x from the two equations we find $y=2\tfrac{1}{3}$.

2. *For another example we require the projection of the line represented by the equation*

$$y=-\frac{x}{420}-2.$$

Making $x=0$, then $y=-2$. Making $y=0$, then $x=-840$.

Using the last figure, we perceive that the line sought for must pass through S two units below the zero point, and it must take such a direction SV as to meet the axis XX' at the distance of 840 units to the left of zero. Hence its *absolute* projection is practically impossible.

3. *Construct the line whose equation is* $\qquad$ $y=2x+5.$
4. *Construct the line whose equation is* $\qquad$ $y=-3x-3.$
5. *Construct the line represented by* $\qquad$ $2y=3x+5.$
6. *Construct the line represented by* $\qquad$ $y=4x-3.$

The lines represented by equations 4 and 6 will intersect the axis of Y at the same point. Why?

7. *Construct the line whose equation is* $\qquad$ $y=2x+3.$
8. *Construct the line whose equation is* $\qquad$ $y=-2x-3.$

The last two lines intercept a portion of the axis of Y which is the base of an isosceles triangle of which the two lines are the sides. What are the base and perpendicular, and where the vertex of the triangle?

ANS. The base is 6, the perpendicular $1\frac{1}{2}$, vertex on the axis of X.

Construct the lines represented by the following equations.

9. $\qquad$ $3x+5y-15=0$
10. $\qquad$ $2x-6y+7=0$
11. $\qquad$ $x+y+2=0$
12. $\qquad$ $-x+y+3=0$
13. $\qquad$ $2x-y+4=0$

PROPOSITION II

To find the distance between two given points in the plane of the co-ordinate axis. Also, to find the angle made by the line joining the two given points, and the axis of X.

Let the two given points be P and Q, and because the point P is said to be given, we know the two distances

$$AN=x', \quad NP=y'.$$

And because the point Q is given we know the two distances.

$$AM=x'' \text{ and } MQ=y''.$$

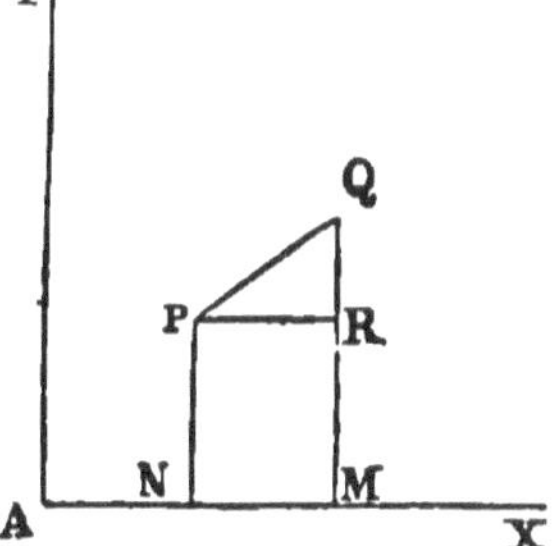

Then, $\qquad AM-AN=NM=PR=x''-x'$;

and $\qquad MQ-MR=QR=y''-y'$.

In the right angled triangle PRQ we have

$$(PR)^2+(RQ)^2=(PQ)^2. \quad \text{But } D=PQ.$$

That is $\quad D^2=(x''-x')^2+(y''-y')^2$,

Or $\qquad D=\sqrt{(x''-x')^2+(y''-y')^2}$

Thus we discover that the distance between any two given points is *equal to the square root of the sum of the squares of the differences of their abscissas and ordinates.*

If one of these points be the origin or zero point, then $x'=0$ and $y'=0$, and we have

$$D=\sqrt{(x'')^2+(y'')^2},$$

a result obviously true.

To find the angle between PQ and AX.

PR is drawn parallel to AX, therefore the angle sought is the same in value as the angle QPR.

Designate the tangent of this angle by a, then by trigonometry we have

$$PR : RQ : : \text{radius} : \text{tan. } QPR.$$

That is, $\quad x''-x' : y''-y' : : 1 : a.$

Whence $\qquad a=\dfrac{y''-y'}{x''-x'}$

In case $y''=y'$, PQ will coincide with PR, and be parallel to AX, and the tangent of the angle will then be 0, and this is shown by the equation, for then

$$a=\frac{0}{x''-x'}=0$$

In case $x''=x'$, then PQ will coincide with RQ and be parallel to AY, and tangent a will be infinite, and this too the equation shows, for if we make $x''=x'$ or $x''=x'=0$, the equation will become

$$a=\frac{y''-y'}{0}=\infty$$

PROPOSITION III.

To find the equation of a line drawn through any given point.

Let P be the given point: The equation must be in the form

$$y = ax + b \qquad (1)$$

Because the line must pass through the given point whose co-ordinates are x' and y', we must have

$$y' = ax' + b. \qquad (2)$$

Subtracting equation (2) from equation (1) member from member, we have

$$y - y' = a(x - x') \qquad (3)$$

for the equation sought.

In this equation a is indeterminate, as it ought to be, because an infinite number of straight lines can be drawn through the point P.

We may give to y' and x' their numerical values, and give any value whatever to a, then we can construct a particular line that will run through the given point P.

Suppose $x' = 2$, $y' = 3$, and make $a = 4$.

Then the equation will become

$$y - 3 = 4(x - 2).$$

Or $\qquad\qquad y = 4x - 5.$

This equation is obviously that of a straight line, hence equation (3) is of the required form.

PROPOSITION IV.

To find the equation of a line which passes through two given points.

Let AX and AY be the co-ordinate axes, and P and Q the given points. Denote the co-ordinates of P by x', y' and of Q by x'', y''.

The required equation must be of the form

$$y = ax + b \qquad (1)$$

We will now determine such values for a and b as will cause the line represented by this equation to pass through the given points.

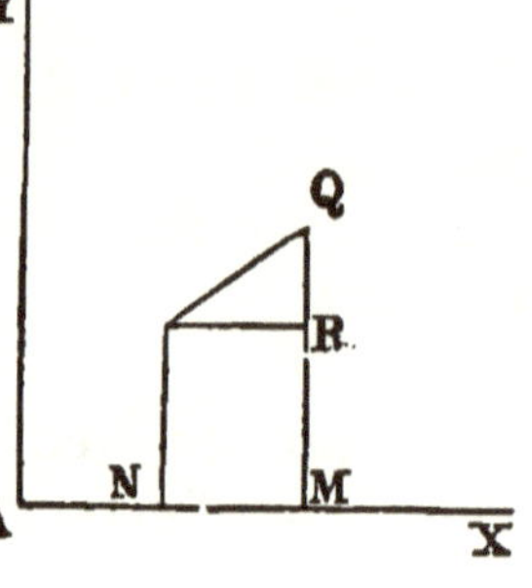

As the line is to pass through the point P, the co-ordinates x', y' of this point when substituted for the variables x, y must satisfy the equation, and we shall have

$$y'=ax'+b \qquad (2)$$

And because the line is to pass through the point Q, whose co-ordinates are x'', y'' we will also have

$$y''=ax''+b \qquad (3)$$

Subtracting eq. (2) from eq. (3) member from member, we get

$$y''-y'=a\,(x''-x')$$

Whence

$$a=\frac{y''-y'}{x''-x'} \qquad (4)$$

From eqs. (1) and (2) we obtain in like manner

$$y-y'=a(x-x') \qquad (5)$$

Substituting for a in eq. (5) its value in eq. (4) we find

$$y-y'=\frac{y''-y'}{x''-x'}(x-x') \qquad (6)$$

for the equation sought.

If we subtract eq. (3) from eq. (1) member from member, and substitute for a in the resulting equation its value in eq. (4) we find

$$y-y''=\frac{y''-y'}{x''-x'}(x-x'') \qquad (7)$$

for the required equation.

By simply clearing eqs. (6) and (7) of fractions and reducing, it may be shown that they are in fact but different forms of the same equation.

To prove that either of these equations is that of a line passing through the points P and Q, we have but to substitute in it, for x and y, the co-ordinates of these points. It will be found that when these substitutions are made for either point, the equation will be satisfied.

We will illustrate the use of these equations by the following

EXAMPLES.

1. The co-ordinates of P are $x'=3$, $y'=4$, and of Q, $x''=-1$, $y''=3$.

What is the equation of the line that passes through these points ?

Here
$$a=\frac{y''-y'}{x''-x'}=\frac{3-4}{-1-3}=\tfrac{1}{4}$$

And the equation $y-y'=\dfrac{y''-y'}{x''-x'}(x-x')$ becomes

$$y-4=\tfrac{1}{4}(x-3) \text{ or } y=\tfrac{1}{4}x+3\tfrac{1}{4}$$

By substituting in the equation $y-y''=\dfrac{y''-y'}{x''-x'}(x-x'')$

we get $y-3=\tfrac{1}{4}(x+1)$ or $y=\tfrac{1}{4}x+3\tfrac{1}{4}$, the same as that above.

2. Find the equation of the straight line that is determined by the points whose co-ordinates are $x'=-4$, $y'=-1$ and $x''=4\tfrac{1}{2}$, $y''=-1\tfrac{0}{8}$

Ans. $y=-\tfrac{4}{51}x-1\tfrac{16}{51}$.

3. The co-ordinates of one point are $x'=6$, $y'=5$, and of another they are $x''=-3$, $y''=3$. What is the equation of the straight line that passes through these points ?

Ans. $y=\tfrac{2}{9}x+3\tfrac{2}{3}$.

PROPOSITION V.

To find the equation of a straight line which shall pass through a given point and make, with a given line, a given angle.

The equation of the given line must be in the form
$$y=ax+b. \tag{1}$$

Because the other line must pass through a given point its equation must be (Prop. III.)

$$y - y' = a'(x - x'). \qquad (2)$$

We have now to determine the value of a'.

When a and a' are equal, the two lines must be parallel, and the inclination of the two lines will be greater or less according to the relative values of a and a'.

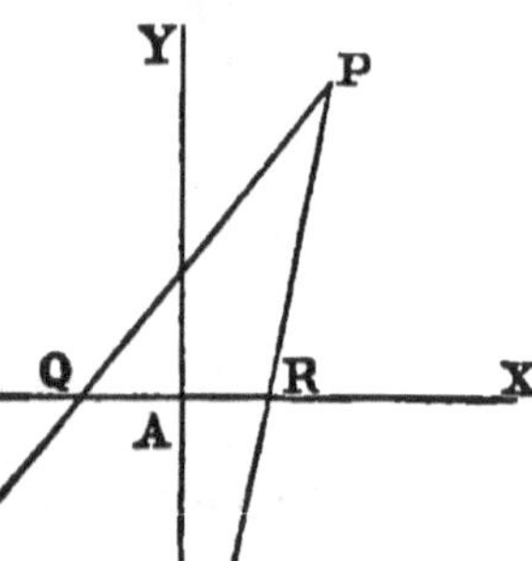

Let PQ be the given line, making with the axis of X an angle whose tangent is a and PR the other line which shall pass through the given point P and make with PQ, a given angle QPR. The tangent of the angle PRX is equal to a'.

Because $PRX = PQR + QPR$.

$$QPR = PRX - PQR$$

$$\text{Tan. } QPR = \text{tan. } (PRX - PQR.)$$

As the angle QPR is supposed to be known or given, we may designate its tangent by m, and m is a known quantity.

Now by trigonometry we have

$$m = \text{tan. } (PRX - PQR) = \frac{a' - a}{1 + aa'}. \qquad (3)$$

Whence $\qquad a' = \dfrac{a + m}{1 - ma}.$

This value of a' put in eq. (2) gives

$$y - y' = \left(\frac{a + m}{1 - ma} \right)(x - x') \qquad (4)$$

for the equation sought.

Cor. 1. When the given inclination is 90°, m its tangent is infinite, and then $a' = -\dfrac{1}{a}$. We decide this in the following manner.

An infinite quantity cannot be increased or diminished

relatively, by the addition or subtraction of finite quantities, therefore, on that supposition,

$$\frac{a+m}{1-ma} \text{ becomes} \frac{m}{-ma} \text{ or } -\frac{1}{a}.$$

APPLICATION.—To make sure that we comprehend this proposition and its resulting equation, we give the following example :

The equation of a given line is $y=2x+6$.

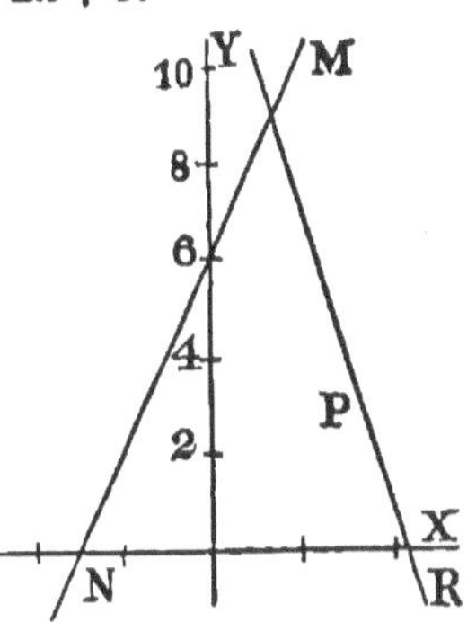

Draw another line that will intersect this at an angle of 45° and pass through a given point P, whose co-ordinates are

$$x'=3\tfrac{1}{2},\ y'=2.$$

Draw the line MN corresponding to the equation $y=2x+6$. Locate the point P from its given co-ordinates.

Because the angle of intersection is to be 45°, $m=1$, and $a=2$.

Substituting these values in eq. (4) we have

$$y-2=-3(x-3\tfrac{1}{2}).$$

Or $\qquad\qquad y=-3x+12\tfrac{1}{2}.$

Constructing the line MR corresponding to this equation, we perceive it must pass through P and make the angle NMR 45°, as was required.

The teacher can propose any number of like examples.

Cor. Equation (3) gives the tangent of the angle of the inclination of any two lines which make with the axis of X angles whose tangents are a and a'. That is, we have in general terms

$$m=\frac{a'-a}{1+aa'}.$$

In case the two lines are parallel $m=0$. Whence $a'=a$, an obvious result.

In case the two lines are perpendicular to each other, m must be infinite, and therefore we must put

$$1+aa'=0$$

to correspond with this hypothesis, and this gives

$$a'=-\frac{1}{a}$$

a result found in Cor. 1.

To show the practical value of this equation we require the angle of inclination of the two lines represented by the equations $y=3x-6$, and $y=-x+2$.

Here $a=3$ and $a'=-1$. Whence

$$m=\frac{-4}{1-3}=2.$$

This is the natural tangent of the angle sought, and if we have not a table of natural tangents at hand, we will take the log. of the number and add 10 to the index, then we shall have in the present example 10.301030 for the log. tangent which corresponds to $63° 26' 6''$ nearly.

The sign of the tangent determines the direction in which the angles are estimated.

2. What is the inclination of the two lines whose equation are

$$2y=5x+8$$
and
$$3y=-2x+6 \;?$$

Ans. The tangent of their inclination is $4\frac{3}{4}$

Log. 4.75 plus 10 = 10.676694.

The inclination of the lines is therefore $78° 6' 5''$.

3. Find the equation of a line which will make an angle of $56°$ with the line whose equation is

$$2y=5x+4.$$

As the required line is to pass through no particular point, but is merely to make a given angle with the known line, we may assume it to pass through the origin of co-ordinates. Its equation will then be of the form

$y=a'x$. We must now determine such a value for a' that the two lines will make with each other an angle of 56°.

Represent the tangent of the given angle by t; then by corollary (2)

$$t=\frac{a'-\frac{5}{2}}{1+\frac{5}{2}a'}$$

In the tables we find that log. tangent of 56° to be 10. 171013, from which subtracting 10 to reduce it to the log. of the natural tangent and we have 0.171013 for the log. of t. The number corresponding to this is 1.483.

Whence $$\frac{a'-\frac{5}{2}}{1+\frac{5}{2}a'}=1.483$$

From which we find $a'=-1,473$ nearly and the equation of the line making with the given line, an angle of 56° is therefore

$$y=-1.473x.$$

PROPOSITION VI.

To find the co-ordinates which will locate the point of intersection of two straight lines given by their equations.

We have already done this in a particular example in Prop. I, and now we propose to deduce *general expressions* for the same thing.

Let $\qquad y=ax+b$ be the first line.

And $\qquad y=a'x+b'$ be the second line.

For their point of intersection y and x in one equation will become the same as in the other.

Therefore we may subtract one equation from the other, and the result will be a true equation.

For the sake of perspicuity, let x_1 and y_1 represent the co-ordinates of the common point, then by subtraction

$$(a-a')x_1+b-b'=0$$

Whence $x_1=-\dfrac{(b-b')}{(a-a')}$ and $y_1=\dfrac{a'b-ab'}{a'-a}$.

10* H

EXAMPLE.

At what point will the lines represented by the two equations

$$y=-2x+1$$

and $\qquad y=5x+10$ intersect each other.

Here $a=-2$, $a'=5$, $b=1$, $b'=10$. Whence $x=-\tfrac{9}{7}$, $y=-3\tfrac{4}{7}$.

If we take another line *not parallel* to either of these, the three will form a triangle.

Then if we *locate* the three points of intersection and join them, we shall have the triangle.

PROPOSITION VII.

To draw a perpendicular from a given point to a given straight line and to find its length.

Let $y=ax+b$ be the equation of the given straight line, and x', y' the co-ordinates of the given point.

The equation of the line which passes through the given point must take the form

$$y-y'=a'(x-x'). \quad \text{(Prop. 3.)}$$

And as this must be perpendicular to the given line, we must have $a'=-\dfrac{1}{a}$. Therefore the equations for the two lines must be

$$y=ax+b \text{ for the given line;} \qquad (1)$$

and $\qquad y-y'=-\dfrac{1}{a}(x-x');$

Or $\qquad y=-\dfrac{1}{a}x+\left(\dfrac{x'}{a}+y'\right)$ for the perpendicular line $\quad(2)$

Let x_1 and y_1 represent the co-ordinates of the point of intersection of these two lines. Then by Prop. 6,

$$x_1 = -\left[\frac{b-\dfrac{x'}{a}-y'}{a+\dfrac{1}{a}}\right] \qquad \text{and } y_1 = \frac{\dfrac{b}{a}+a\left(\dfrac{x'}{a}+y'\right)}{\dfrac{1}{a}+a}$$

Or $x_1 = -\left(\dfrac{ab-x'-ay'}{a^2+1}\right)$, and $y_1 = \dfrac{b+ax'+a^2y'}{a^2+1}$

Or we may conceive x and y to represent the co-ordinates of the point of intersection, and eliminating y from eqs. (1) and (2) we shall find x as above, and afterwards we can eliminate y.

Now the length of the perpendicular is represented by

$$\sqrt{(x_1-x')^2+(y_1-y')^2}=D. \qquad (\textit{Prop. II.})$$

Whence $\sqrt{\left(\dfrac{-ab+ay'-a^2x'}{a^2+1}\right)^2+\left(\dfrac{b+ax'-y'}{a^2+1}\right)^2}=$ the perpendicular.

If we put $u=b+ax'-y'$, the quantities under the radical will become

$$\sqrt{\frac{a^2u^2}{(a^2+1)^2}+\frac{u^2}{(a^2+1)^2}}=\sqrt{\frac{(a^2+1)u^2}{(a^2+1)^2}}=\pm\frac{u}{\sqrt{a^2+1}}.$$

Whence the perpendicular $=\pm\dfrac{b+ax'-y'}{\sqrt{a^2+1}}.$

EXAMPLES.

1. The equation of a given line is $y=3x-10$, and the co-ordinates of a given point are $x'=4$ and $y'=5$.

What is the length of the perpendicular from this given point to the given straight line? Ans. $\frac{1}{10}\sqrt{90}.$

2. The equation of a line is $y=-5x-15$, and the co-ordinates of a given point are $x'=4$ and $y'=5$.

What is the length of the perpendicular from the given point to the straight line? Ans. $7.84+.$

PROPOSITION VIII.

To find the equation of a straight line which will bisect the angle contained by two other straight lines.

Let $$y=ax+b \qquad (1)$$
and $$y=a'x+b' \qquad (2)$$

be the equations of two straight lines which intersect; the co-ordinates of the point of intersection are

$$x_1=-\left(\frac{b-b'}{a-a'}\right) \quad y_1=\frac{a'b-ab'}{a'-a} \quad \text{(Prop. VI.}$$

We now require a third line which shall pass through the same point of intersection and form such an angle with the axis of X (the tangent of which may be represented by m) that this line will bisect the angle included between the other two lines. Whence by (Prop. V.) tho equation of the line sought must be

$$y-y_1=m(x-x_1) \qquad (3)$$

in which we are to find the value of m.

Let PN represent the line corresponding to equation (1) PM the line whose equation is (2), and PR the line required.

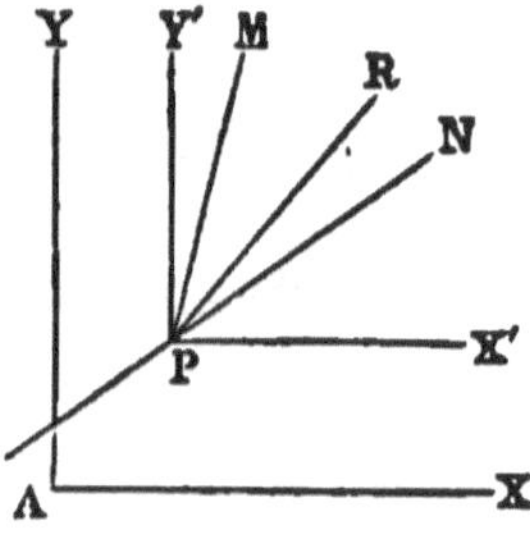

Now the position or inclination of PN to AX depends entirely on the value of a, and the inclination of PM depends on a' and both are independent of the position of the point P.

Now $RPN=RPX'-NPX'$ and $MPR=MPX'-RPX'$.

Whence by the application of a well known equation in plane trigonometry, (Equation (29), p. 253 Plane Trig.) we have

$$\tan. RPN=\tan. (RPX'-NPX')=\frac{m-a}{1+am}$$

And $$\tan. MPR=\tan. (MPX'-RPX')=\frac{a'-m}{1+a'm}$$

But by hypothesis these two angles RPN and MPR are to be equal to each other. Therefore

$$\frac{m-a}{1+am}=\frac{a'-m}{1+a'm}.$$

Whence
$$m^2+\frac{2(1-aa')}{a'+a}m=1. \qquad (4)$$

This equation will give two values of m; one will correspond to the line PR, and the other to a line at right angles to PR.

If the proper value m be taken from this equation and put in eq. (3), we shall have the equation required.

Practically we had better let the equations stand as they are, and substitute the values of a, a' x, and y, corresponding to any particular case.

To illustrate the foregoing proposition we propose the following

EXAMPLES.

Two lines intersect each other:

$$y=-2x+5 \text{ is the equation of one line.} \qquad (1)$$
$$y=4x+6 \text{ is that of the other line.} \qquad (2)$$

Find the equation of the line which bisects the angle contained by these two lines :

Here $\qquad a=-2,\ a'=4,\ b=5,\ b'=6.$

Whence $\qquad x_1=-\tfrac{1}{6},$ and $y_1=\tfrac{16}{3}.$

Thus (3) becomes

$$y-\tfrac{16}{3}=m(x+\tfrac{1}{6}).$$

And eq. (4) becomes

$$m^2+9m=1.$$

Whence $\qquad m=0.1097$ or $m=-9.1097.$

$$y-\tfrac{16}{3}=0.1097(x+\tfrac{1}{6}).$$

Or $\qquad y-\tfrac{16}{3}=-9.1097(x+\tfrac{1}{6}).$

Equation (4) is that of the line required; (3) that of the line at right angles to the line required.   All will be obvious if we construct the lines represented by the eqs. (1), (2), (3), and (4).

For another example, find the equation of a line which bisects the angle contained by the two lines whose equations are

$$y=x+12, \qquad y=-20x+2.$$

Here $a=1$, $a'=-20$.   Whence (4) becomes

$$m^2-\tfrac{42}{19}m=1.$$

Therefore $m=-0.385$, or $+2.6$.

Note.—Two straight lines whose equations are

$$y=ax+b \quad \text{and} \quad y'=a+b'$$

will always intersect at a point (unless $a=a'$) and with the axis of Y form a triangle.   The area of such triangle is expressed by

$$-\left(\frac{b-b'}{a-a'}\right) \times \left(\frac{b\infty b'}{2}\right)$$

From the given equations we find the co-ordinates of the intersection of the lines to be

$$x_1=-\tfrac{10}{21},\ y_1=\tfrac{242}{21}$$

For the line bisecting the angle included between the given lines we have either

$$y-\tfrac{242}{21}=-0.385(x+\tfrac{10}{21}) \qquad (1)$$

or, $\qquad y-\tfrac{242}{21}=2.6(x+\tfrac{10}{21}) \qquad (2)$

By transposition and reduction (1) becomes

$$y=-0.385x+11.75 \qquad (3)$$

and (2) becomes $\qquad y=2.6x+12.76 \qquad (4)$

The lines represented by eqs. (3) and (4) are at right angles to each other.   The latter line bisects the angle included between the given lines, and the former the adjacent or supplemental angle.

3. From the intersection of two lines whose equations are

$$3y+5x=4 \qquad (1)$$

and
$$2y=3x+4 \qquad (2)$$

A third line is drawn making, with the axis of X, an angle of 30°. Find the intersection of the given lines and the equation of the third line.

$Ans.$ $\begin{cases} \text{The co-ordinates of the points of intersection} \\ \text{are } x_1=-\tfrac{4}{19},\ y_1=\tfrac{32}{19}, \text{ and the required equation} \\ \text{is } y-\tfrac{32}{19}=0{,}5773(x+\tfrac{4}{19}). \end{cases}$

4. Two lines are represented by the equations

$$2y-3x=-1$$

and
$$2y+3x=3$$

What kind of a triangle do these lines form with the intercepted portion of the axis of Y, and what are its sides and its area?

$Ans.$ $\begin{cases} \text{The triangle is isosceles; its base on the axis} \\ \text{of } Y \text{ is 2, the other sides are each } 1.201+, \quad \text{and} \\ \text{its area } 0.66+. \end{cases}$

5. Two lines are given by the equations

$$-2\tfrac{1}{5}y+3\tfrac{1}{2}x=-2\tfrac{1}{4}$$

and
$$2\tfrac{2}{5}y-\tfrac{2}{3}x=4$$

Required the equation of the line drawn from the point whose co-ordinates are $x''=3$, $y''_1=0$ to the intersection of the given lines and the distance between these two points.

$Ans.$ $\begin{cases} \text{The equation sought is } y=-0.717x+2.1523 \text{ and} \\ \text{the distance is } \sqrt{(1.8)^2+(2.52)^2}. \end{cases}$

TRANSFORMATION OF CO-ORDINATES.

It is often desirable to change the reference of points from one system of co-ordinate axes to another differing from the first either in respect to the origin or the direction of the axes, or both. The operation by which this is done is called the *transformation of co-ordinates*. The system of co-ordinate axes from which we pass is the *primitive* system and that to which we pass is the *new* system.

Let AX and AY be the primitive axes. Take any point, as A', the co-ordinates of which referred to AX and AY are $x=a$, $y=b$ and through it draw the new axes $A'X'$, and $A'Y'$ parallel to the primative axes. Then denoting the co-ordinates of any point, as

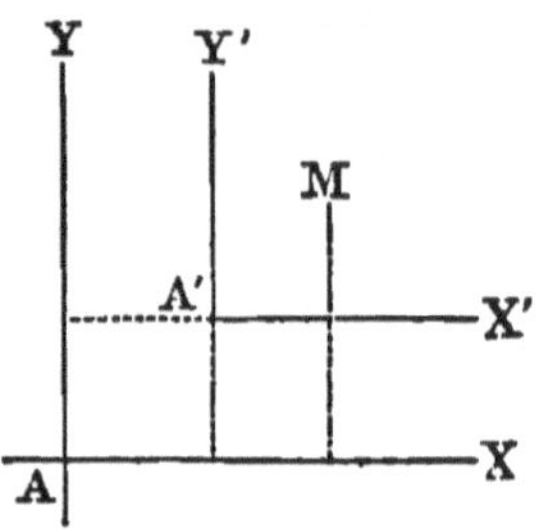

M, referred to the primitive axes by x and y, and the co-ordinates of the same point referred to the new axes by x' and y', it is apparent that

$$x=a+x'$$
$$y=b+y'$$

By giving to a and b suitable signs and values we may place the new origin at any point in the plane of the primitive axes and the above formulas are those for passing from one system of axes to a system of parallel axes having a different origin.

The formulas for the transformation of co-ordinates must express the values of the primitive co-ordinates of points in terms of the new co-ordinates and those quantities which fix the position of the new in respect to the primitive axes.

PROPOSITION IX.

To find the formulas for passing from a system of rectangular to a system of oblique co-ordinates from a different origin.

Let AX, AY be the primitive axes and $A'X'$, $A'Y'$ the new axes. Through any point as M draw MP' parallel to $A'Y'$ and MP perpendicular to $A'X$. Then $A'P'$ is the new abscissa, $P'M$ the new ordinate of the point M, and AP and PM are respectively the primitive abscissa and ordinate of the same point.

Let $AB=a$, $BA'=b$, $AP=x$, $PM=y$, $A'P'=x'$, $P'M=y'$ the angle $X'A'X''=m$, and the angle $Y'A'X'''=n$. Now by trigonometry we have

$A'K=x'\cos.m, KP'=LH=x'\sin.m$

$\qquad P'H=KL=y'\cos.n.$

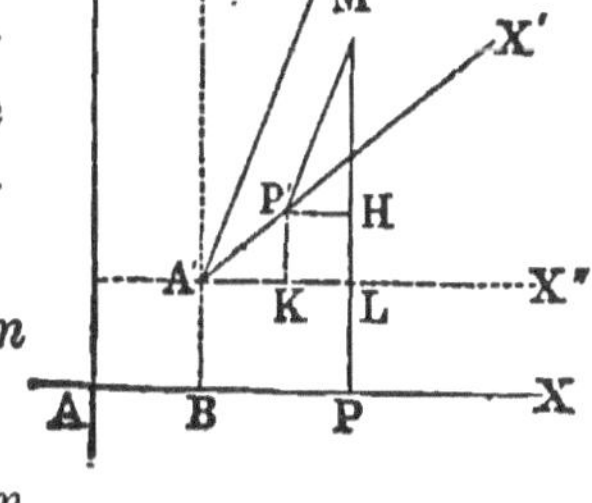

And $\qquad\qquad MH=y'\sin.n.$

Whence $x=a+x'\cos.m+y'\cos.n$, $y=b+x'\sin.m+y'\sin.n$, the formulas required.

Scholium.—In case the two systems have the same origin, we merely suppress a and b, and then the formulas required are

$$x=x'\cos.m+y'\cos.n.\quad y=x'\sin.m+y'\sin.n.$$

PROPOSITION X.

To find the formulas for passing from a system of oblique co-ordinates to a system of rectangular co-ordinates, the origin being the same.

Take the formulas of the last problem

$$x=x'\cos.m+y'\cos.n,\quad y=x'\sin.m+y'\sin.n.$$

We now regard the oblique as the primitive axes, and require the corresponding values on the rectangular axes. That is, we require the values of x' and y'. If we multiply the first by sin. n, and the second by cos. n, and subtract their products, y' will be eliminated, and if x' be eliminated in a similar manner, we shall obtain

$$x'=\frac{x\sin.n-y\cos.n}{\sin.(n-m)}\qquad y'=\frac{y\cos.m-x\sin m}{\sin.(n-m)}$$

Scholium.—If the zero point be changed at the same time in reference to the oblique system, we shall have

$$x'=a+\frac{x\sin.n-y\cos.n}{\sin.(n-m)}\qquad y'=\ b+=\frac{y\cos.m-x\sin.m}{\sin.(n-m)}$$

We will close this subject by the following

11

The equation of a line referred to rectangular co-ordinates is

$$y=a'x+b'.$$

Change it to a system of oblique co-ordinates having the same zero point.

Substituting for x and y their values as above, we have

$$x'\sin. m+y'\sin. n=a'(x\cos. m+y'\cos. n)+b'.$$

Reducing

$$y'=\frac{(a'\cos. m-\sin. m)x'}{\sin. n-a'\cos. m}+\frac{b'}{\sin. n-a'\cos. m}$$

POLAR CO-ORDINATES.

There are other methods by which the relative positions of points in a plane may be analytically established than that of referring them to two rectilinear axes intersecting each other under a given angle.

For example, suppose the line AB to revolve in a plane about the point A. If the angle that this line makes with a fixed line passing through A be known, and also the length of AB, it is evident that the extremity B of this line will be determined, and that there is no point whatever in the plane the position of which may not be assigned by giving to AB and the angle which it makes with the fixed line appropriate values.

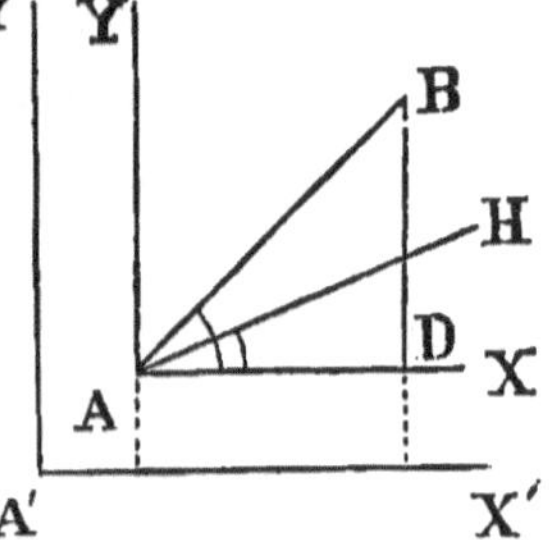

The variable distance AB is called the *radius vector*, the angle that it makes with the fixed line the *variable angle* and the point A about which the radius vector turns, the *pole*. The radius vector and the variable angle together constitute a system of *polar co-ordinates*.

Denote variable angle BAD by v, the radius vector by r and by x and y, the co-ordinates of B referred to the rectangular axes AX, AY; then by trigonometry we have

$$x = r \cos. v \text{ and } y = r \sin. v.$$

Now from the first of these we have $r = \dfrac{x}{\cos. v}$ (v may revolve all the way round the pole), and as x and cos. v are both positive and both negative at the same time, that is, both positive in the first and fourth quadrants, and both negative in the second and third quadrants, therefore r *will always be positive.*

Consequently, should a negative radius appear in any equation, we *must infer* that some incompatible conditions have been admitted into the equation.

PROPOSITION XI.

To find the formulas for changing the reference of points from a system of rectangular co-ordinate axes to a system of polar co-ordinates.

Let $A'X$, $A'Y$ be the co-ordinate axes, A the pole AB the radius vector of any point, and AD parallel to $A'X$ the fixed line from which the variable angle is estimated. Denote the co-ordinates $A'E$, AE of the pole by a and b and the radius vector AB by r.

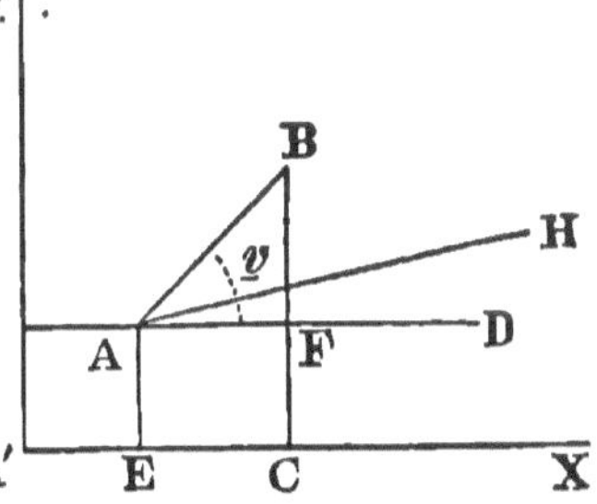

Draw BC perpendicular to $A'X$; then is $A'C = x$ the abscissa, and $BC = y$ the ordinate of the point B. From the figure we have

$$A'C = A'E + EC = A'E + AF = A'E + AB \cos. v$$

$$\text{and } BC = BF + FC = BF + AE = AE + AB \sin. v$$

Whence
$$x = a + r \cos. v$$
$$y = b + r \sin. v.$$

SCHOLIUM.—If instead of estimating the variable angle from the line AD, which is parallel to the axis $A'X$, we estimate it from the line AH which makes with the axis the given angle $HAD = m$ we shall have

$$x = a + r \cos. (v + m)$$
$$y = b + r \sin. (x + m)$$

CHAPTER II.

THE CIRCLE.

LINES OF THE SECOND ORDER.

Straight lines can be represented by equations of the first degree, and they are therefore called lines of the first order. The circumference of a circle, and all the conic sections, are lines of the second order, because the equations which represent them are of the second degree.

PROPOSITION I.

To find the equation of a circle.

Let the origin be the center of the circle. Draw AM to any point in the circumference, and let fall MP perpendicular to the axis of X. Put $AP = x$, $PM = y$ and $AM = R$.

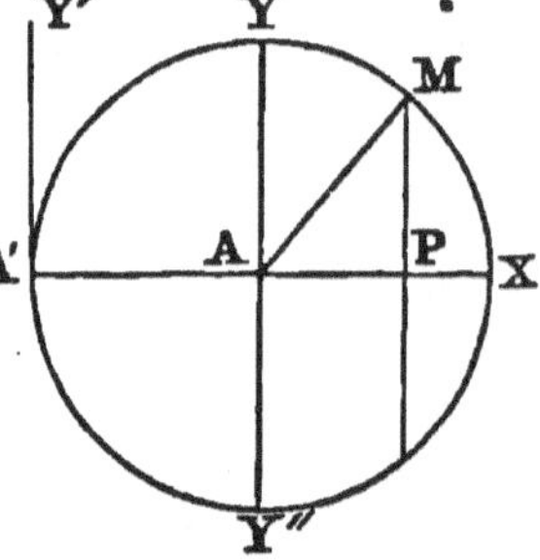

Then the right angled triangle APM gives

$$x^2 + y^2 = R^2 \qquad (1)$$

and this is the equation of the circle when the zero point is the center.

When $y=0$, $x^2=R^2$, or $\pm x=R$, that is, P is at X or A'. When $x=0$, $y^2=R^2$, or $\pm y=R$, showing that M on the circumference is then at Y or Y''.

When x is positive, then P is on the right of the axis of Y, and when negative, P is on the left of that axis, or between A and A'.

When we make *radius unity*, as we often do in trigonometry, then $x^2+y^2=1$, and then giving to x or y any value *plus* or *minus* within the limit of unity, the equation will give us the corresponding value of the other letter.

In trigonometry y *is called the sine of the arc* XM, *and* x *its cosine.*

Hence in trigonometry we have $\sin.^2+\cos.^2=1$.

Now if we remove the origin to A' and call the distance $A'P=x$, then $AP=x-R$, and the triangle APM gives
$$(x--R)^2+y^2=R^2.$$
Whence $\qquad\qquad y^2=2Rx-x^2.$

This is the equation of the circle, when the origin is on the circumference.

When $x=0, y=0$ at the same time. When x is greater than $2R$, y *becomes imaginary,* showing that *such an hypothesis is inconsistent with the existence of a point in the circumference of the circle.*

There is still a more general equation of the circle when the zero point is neither at the center nor in the circumference.

The figure will fully illustrate.

Let $AB=c$, $BC=b$. Put $AP=x$, or $AP'=x$, and PM or $P'M''=y$, CM, CM', &c. each$=R$.

In the circle we observe *four* equal right angled· triangles. The *numerical* expression is the same for each. Signs only *indicate positions.*

11*

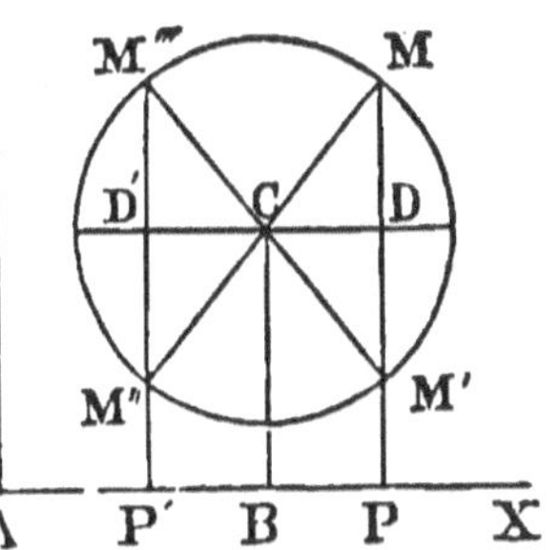

Now in case CDM is the triangle we *fix upon*,

We put $AP=x$, then $BP=CD=(x-c)$,

$\qquad PM=y$, $MD=y-CB=(y-b)$.

Whence $\qquad (x-c)^2+(y-b)^2=R^2 \qquad\qquad$ (1)

In case CDM' is the triangle, we put $AP=x$ and PM' $=y$.

Then $\qquad\qquad (x-c)^2+(b-y)^2=R^2 \qquad\qquad$ (2)

In case $CD'M'''$ is the triangle, we put $AP'=x$, $P'M'''$ $=y$.

Then $\qquad\qquad (c-x)^2+(y-b)^2=R^2 \qquad\qquad$ (3)

If $CD'M''$ is the triangle, we put $P'M''=y$.

Then $\qquad\qquad (c-x)^2+b-y)^2=R^2 \qquad\qquad$ (4)

Equations (1), (2), (3), and (4), are in all respects numerically the same, for $(c-x)^2=(x-c)^2$, and $(b-y)^2=(y-b)^2$. Hence we may take equation (1) to represent the general equation of the circle referred to rectangular co-ordinates.

The equation $\qquad (x-c)^2+(y-b)^2=R^2 \qquad\qquad$ (1)

includes all the others by attributing proper values and signs to c and b.

If we suppose both c and b equal 0, it transfers the zero point to the center of the circle, and the equation becomes

$$x^2+y^2=R^2$$

To find where the circle cuts the axis of X we must make $y=0$. This reduces the general equation (1) to

$$(x-c)^2+b^2=R^2.$$

Or $\qquad\qquad (x-c)^2=R^2-b^2.$

Now if b is numerically greater than R, the first member being *a square*, (and therefore positive,) must be equal to a negative quantity, which is impossible,—showing that in that case the circle does not meet or cut the axis of X, *and this is obvious from the figure.*

In case $b=R$, then $(x-c)^2=0$, or $x=c$, showing that the

circle would then *touch* the axis of X. If we make $x=0$, eq. (1) becomes

$$c^2+(y-b)^2=R^2.$$

Or $$(y-b)^2=R^2-c^2.$$

This equation shows that if c is greater than R, the circle does not cut the axis of Y, and this is also obvious from the figure.

If c be less than R, the second member is positive in value, and

$$y=b\pm\sqrt{R^2-c^2},$$

showing that if the circumference cut the axis at all, it must be in two points, as at M'', M'''.

PROPOSITION II.

The supplementary chords in the circle are perpendicular to each other.

Definition.—Two lines drawn, one through each extremity of any diameter of a curve, and which intersect the curve in the same point, are called *supplementary chords*.

That is, the chord of an arc, and the chord of its supplement.

In common geometry this proposition is enunciated thus:

All angles in a semi-circle are right angles.

The equation of a straight line which will pass through the given point B, must be of the form (Prop. III. Chap. L)

$$y-y'=a(x-x'). \qquad (1)$$

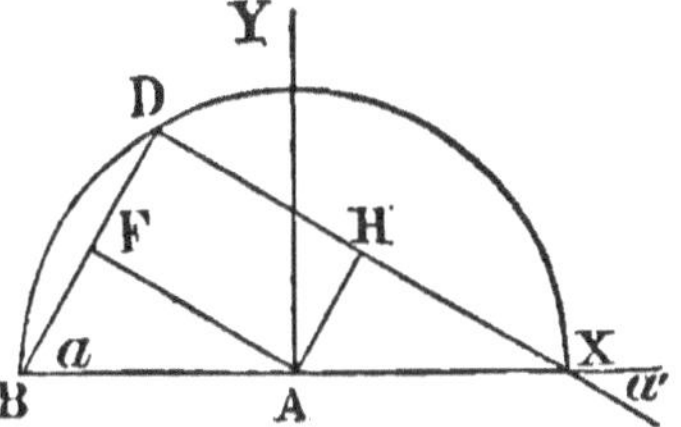

The equation of a straight line which will pass through the given point X, must be of the form $$y-y'=a'(x-x'). \qquad (2)$$

At the point B, $y'=0$, and $x'=-R$, or $-x'=R$. Therefore eq. (1) becomes

$$y=a(x+R). \qquad (3)$$

And for like reason eq. (2) becomes

$$y=a'(x-R). \qquad (4)$$

For the point in which these lines intersect x and y in eq. (3) are the same as x and y in eq. (4); hence, these equations may be multiplied together under this supposition, and the result will be a true equation. That is, .

$$y^2=aa'(x^2-R^2). \qquad (5)$$

But as the point of intersection must be on the *curve*, *by hypothesis*, therefore, x and y must conform to the following equation:

$$y^2+x^2=R^2. \quad \text{Or } y^2=-1(x^2-R^2). \qquad (6)$$

Whence $\qquad aa'=-1$, or $aa'+1+0.$

This last equation shows that the two lines are perpendicular to each other, as proved by (Cor. 2, Prop. 5., Chap. 1.)

Because a and a' are indeterminate, we conclude that an infinite number of supplemental chords may be drawn in the semi-circle, which is obviously true.

PROPOSTION III.

To find the equation of a line tangent to the circumference of a circle at a given point.

Let C be the center of the circle, P the point of tangency, and Q a point assumed at pleasure in the circumference.

Denote the co-ordinates of P by x', y', and those of Q, by x'', y'',

The equation of a line passing through two points whose co-or-

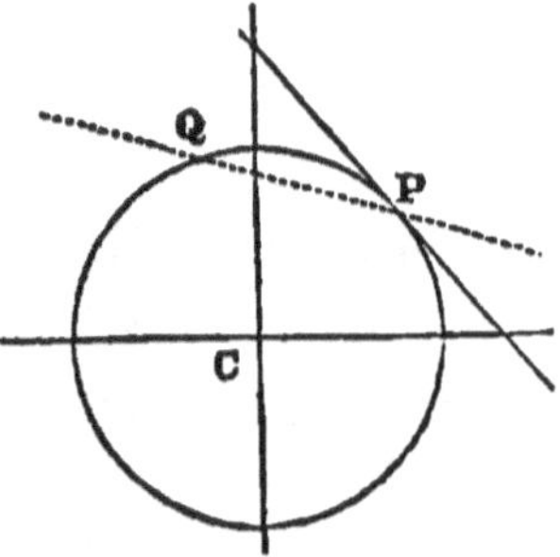

dinates are x', y' and x'', y'' is of the form (Prop. 4, Chap. 1).

$$y - y' = \frac{y' - y''}{x' - x''}(x - x'').\qquad (1)$$

We are to introduce in this equation, first, the condition that the points P and Q are in the circumference of the circle, which will make the line a secant line, and then the further condition that the point Q shall coincide with the point P, which will cause the secant line to become the required tangent line.

Because the points P and Q are in the circumference of the circle, we have

$$x'^2 + y'^2 = R^2$$

and
$$x''^2 + y''^2 = R^2$$

Whence by subtraction and factoring,

$$(x' + x'')\,(x' - x'') + (y' + y'')\,(y' - y'') = 0 \qquad (2)$$

from which we find

$$\frac{y' - y''}{x' - x''} = -\frac{x' + x''}{y' + y''}$$

This value of $\dfrac{y' - y''}{x' - x''}$ substituted in equation (1) gives us for the equation of the secant line,

$$y - y' = -\frac{x' + x''}{y' + y''}(x - x') \qquad (3)$$

Now, if we suppose this line to turn about the point P until Q unites with P, we shall have $x'' = x'$ and $y'' = y'$, and the secant line will become a tangent to the circumference at the point P.

Under this supposition eq. (3) becomes

$$y - y' = -\frac{x'}{y'}(x - x'), \qquad (4)$$

in which $\dfrac{x'}{y'}$ is the value of the tangent of the angle which the tangent line makes with axis of X.

By clearing this equation of fractions, and substituting for $x'^2+y'^2$ its value, R^2, we have finally for the equation of the tangent line,

$$yy'+xx'=R^2. \qquad (5)$$

This is the general equation of a tangent line; x',y', are the co-ordinates of the tangent point, and x, y, the co-ordinates of any other point in the line.

SCHOLIUM 1.—For the point in which the tangent line cuts the axis of X, we make $y'=0$, then

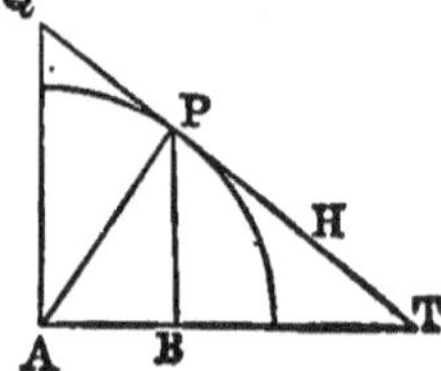

$$x=\frac{R^2}{x'}=AT.$$

For the point in which it meets the axis of Y, we make $x'=0$, and

$$y=\frac{R^2}{y'}=AQ.$$

SCHOLIUM 2.—A line is said to be *normal* to a curve when it is perpendicular to the tangent line at the point of contact.

Join A, P, and if APT is a right angle, then AP is a *normal*, and AB, a portion of the axis of X *under it*, is called the *subnormal*. The line BT *under* the tangent is called the *subtangent*.

Let us now discover whether APT is or is not a right angle.

Put $a'=$ the tangent of the angle PAT, then by trigonometry

$$a'=\frac{y'}{x'}.$$

But
$$a=-\frac{x'}{y'}. \qquad \text{Eq. (6)}$$

Whence
$$aa'=-1. \quad \text{Or} \quad a'=-\frac{1}{a}.$$

Therefore AP is at right angles to PT. (Prop. 5. Chap. 1.) That is, *a tangent line to the circumference of a circle at any point is perpendicular to the radius drawn to that point.*

SCHOLIUM 3.—Admitting the principle, which is a well-known truth of elementary geometry, demonstrated in the preceding scholium, we would not, in getting the equation of a tangent line to the circle, draw a line cutting the curve in *two* points, but would draw the tangent line PT at once, and *admit* that the angle APT was a right angle. Then it is clear that the angle $APB=$ the angle PTB.

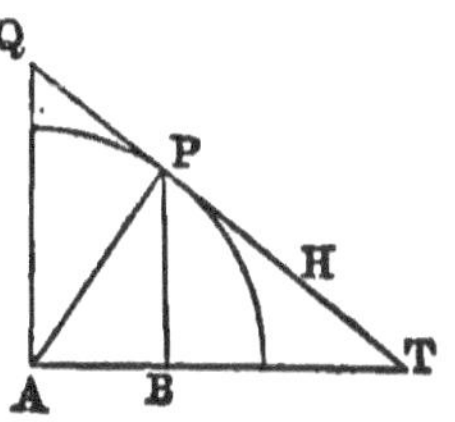

Now to find the equation of the line, we let x' and y' represent the co-ordinates of the point P, and x and y the general co-ordinates of the line, and a the tangent of its angle with the axis of X, then (by Prop III, Chap. I,) we have

$$y-y'=a(x-x').$$

Now the triangle APB gives us the following expression for the tangent of the angle APB, or its equal PTB,

$$a=-\frac{x'}{y'}$$

This value of a put in the preceding equation, will give us

$$y'-y=-\frac{x'}{y'}(x'-x).$$

Or $\qquad y'^2-yy'=-x'^2+xx'.$

Whence $\qquad yy'+xx'=R^2,$ the same as before.

PROPOSITION IV.

To find the equation of a line tangent to the circumference of a circle, which shall pass through a given point without the circle.

Let H (see last figure to the preceding proposition) be the given point, and x'' and y'' its co-ordinates, and x' and y' the co-ordinates of the point of tangency P.

The equation of the line passing through the two points H and P must be of the form

$$y-y''=a(x-x'') \qquad (1)$$

in which $\qquad a=\dfrac{y'-y''}{x'-x''}.$

Since PH is supposed to be tangent at the point P,

and x' and y' are the co-ordinates of this point, equation (6) Prop. 3, gives us

$$a=-\frac{x'}{y'}.$$

Placing this value of a in equation (1) we have

$$y-y''=-\frac{x'}{y'}(x-x'')$$

for the equation sought.

This equation combined with

$$x'^2+y'^2=R^2,$$

which fixes the point P on the circumference will determine the values of x' and y', and as there will be two real values for each, it shows that two tangents can be drawn from H, or from any point without the circle, which is obviously true.

SCHOLIUM. We can find the value of the tangent PT by means of the similar triangles ABP, PBT, which give

$$x':R::y':PT.$$

$$PT=R\frac{y'}{x'}.$$

More general and elegant formulas, applicable to all the conic sections, will be found in the calculus for the *normals, subnormals, tangents* and *subtangents*

OF THE POLAR EQUATION OF THE CIRCLE.

The polar equation of a curve is the equation of the curve expressed in terms of polar co-ordinates. The variable distance from the pole to any point in the curve is called the *radius vector*, and the angle which the radius vector makes with a given straight line is called the *variable angle*.

PROPOSITION V.

To find the polar equation of the circle.

When the center is the pole or the fixed point, the equation is

$$r^2 = x^2 + y^2 = R^2 \qquad (1)$$

and the radius vector R is then *constant*.

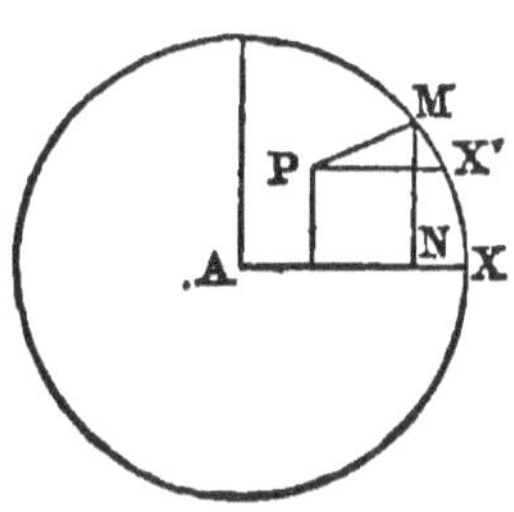

Now let P be the pole, and the co-ordinates of that point referred to the center and rectangular axes be a and b. Make $PM = r$, and $MPX' = v$ the variable angle; $AN = x$ and $NM = y$. Then (Prop. 11, Chap. 1.) we have

$$x = a + r \cos. v, \text{ and } y = b + r \sin. v.$$

These values of x and y substituted in eq. (1), (observing that $\cos.^2 v + \sin.^2 v = 1$,) will give

$$r^2 + 2(a \cos. v + b \sin. v)r + a^2 + b^2 - R^2 = 0$$

which is the polar equation sought.

Scholium 1.—P may be at any point on the plane. Suppose it at B'. Then $a = -R$ and $b = 0$. Substituting these values in the equation, and it reduces to

$$r^2 - 2Rr \cos. v = 0.$$

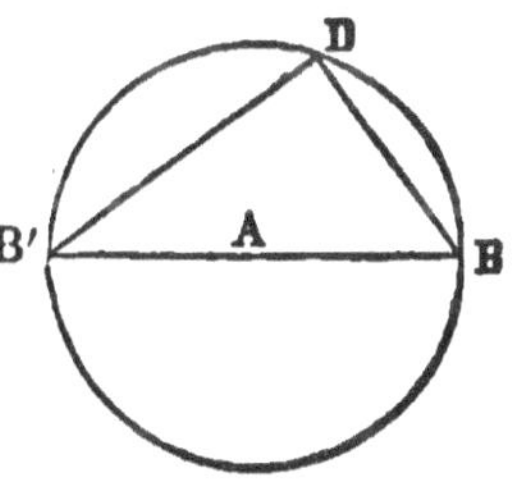

As there is no absolute term, $r = 0$ will satisfy the equation and correspond to one point in the curve, and this is true, as P is supposed to be in the curve. Dividing by r, and

$$r = 2R \cos. v.$$

This value of r will be positive when $\cos. v.$ is positive, and negative when $\cos. v$ is negative; but r being a radius vector can never be negative, and the figure shows this, as r never passes to *the left* of B, but runs into zero at that point.

When $v = 0$, $\cos. v = 1$, then $r = BB'$. When $v = 90$, $\cos. v = 0$, and r becomes 0 at B', and the variations of v from 0 to 90, determine all the points in the semi-circumference BDB'.

12

134 ANALYTICAL GEOMETRY.

Scholium 2.—If the pole be placed at B, then $a=+R$ and $b=0$, which reduces the general equation to

$$r=-2R \cos. v.$$

Here it is necessary that cos. v should be negative to make r positive, therefore v must commence at $90°$ and vary to $270°$; that is, be on the left of the axis of Y drawn through B, and this corresponds with the figure.

Application. The polar equation of the circle in its most general form is

$$r^2+2(a \cos. v+b \sin v)r+a^2+b^2=R^2. \qquad (1)$$

If we make $b=0$, it puts the polar point somewhere on the axis of X, and reduces the equation to

$$r^2+2a \cos. v.r+a^2=R^2. \qquad (2)$$

Now if we make $v=0$, then will cos. $v=1$, and the lines represented by $\pm r$ would refer to the points X, X', in the circle.

This hypothesis reduces the last equation to

$$r^2+2ar=(R^2-a^2) \qquad (3)$$

and this equation is the same in form *as the common quadratic in algebra*, or in the same form as

$$x^2 \pm px=q.$$

Whence $\qquad x=r, \qquad 2a=\pm p, \qquad$ and $\qquad R^2-a^2=q$

$$a=\pm \tfrac{1}{2}p, \quad R=\sqrt{q+a^2}=\sqrt{q+\tfrac{1}{4}p^2}.$$

These results show us that if we describe a circle with the radius $\sqrt{q+\tfrac{1}{4}p^2}$, and place P on the axis of X at a distance from the center equal to to $\tfrac{1}{2}p$, then PX represents one value of x, and PX' the other. That is,

$$x=-\tfrac{1}{2}p +\sqrt{q+\tfrac{1}{4}p^2}=PX.$$

Or $\qquad x=-\tfrac{1}{2}p-\sqrt{q+\tfrac{1}{4}p^2}=PX',$

and this is the common solution.

When p is negative, the polar point is laid off to the left from the center at P'.

The operation refers to the right angled triangle APM.

$$AP=\tfrac{1}{2}p, \quad PM=\sqrt{q}, \quad \text{and} \quad AM=\sqrt{q+\tfrac{1}{4}p^2}.$$

Let the form of the quadratic be

$$x^2\pm px=-q.$$

Then comparing this with the polar equation of the circle, we have

$$2a=\pm p. \quad R^2-a^2=-q.$$
$$a=\pm\tfrac{1}{2}p. \quad R=\pm\sqrt{\tfrac{1}{4}p^2-q}.$$

Take $AX=R$ and describe a semicircle. Take $AP=\tfrac{1}{2}p$ and $AP'=-\tfrac{1}{2}p$. From P and P' draw the lines PM, and $P'M'$ to touch the circle; and draw AM, AM'.

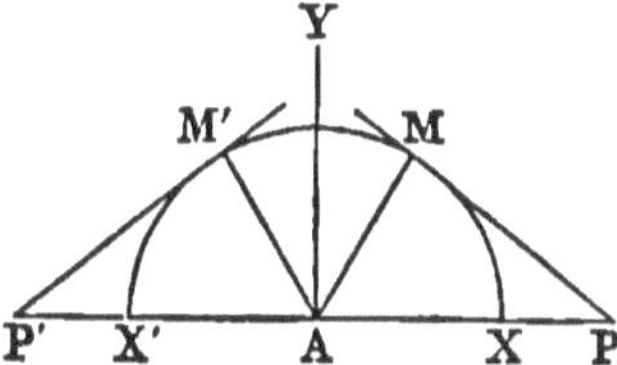

Here AP is the hypotenuse of a right angled triangle. In the first case AP was a side.

In this figure as in the other, $PM=\sqrt{q}$; but here it is inclined to the axis of X; in the first figure it was perpendicular to it.

The figure thus drawn, we have PX for one value of x, and PX' is the other, which may be *determined geometrically*.

If
$$x^2+px=-q$$
$$x=-\tfrac{1}{2}p+\sqrt{\tfrac{1}{4}p^2-q}=PX, \quad \text{or} \quad x=-\tfrac{1}{2}p-\sqrt{\tfrac{1}{4}p^2-q}=PX'.$$

Observe that the first part of the value of x, is *minus, corresponding to a position from P to the left.*

If
$$x^2-px=-q,$$
we take P' for one extremity of the line x.

$$x=\tfrac{1}{2}p+\sqrt{\tfrac{1}{4}p^2-q}=P'X, \quad \text{or} \quad x=\tfrac{1}{2}p-\sqrt{\tfrac{1}{4}p^2-q}=P'X'.$$

Here the first part of the value of x, $(\tfrac{1}{2}p)$, is *plus, because it is laid off to the right of the point P'.*

Because $R=\sqrt{\tfrac{1}{4}p^2-q}$ R or AM becomes less and less as the numerical value of q approaches the value of $\tfrac{1}{4}p^2$. When these two are equal, $R=0$, and the circle becomes a point. When q is greater than $\tfrac{1}{4}p^2$, the circle has *more than vanished,* giving no real existence to any of these lines, and the values of x are said to be *imaginary.*

We have found another method of *geometrizing* quadratic equations, which we consider well worthy of notice, although it is of but little practical utility.

It will be remembered that the equation of a straight line passing through the origin of co-ordinates is

$$y=ax, \qquad (1)$$

and that the general equation of the circle is

$$(x\mp c)^2+(y\mp b)^2=R^2. \qquad (2)$$

If we make $b=0$, the center of the circle must be somewhere on the axis of X.

Let AM represent a line, the equation of which is $y=ax$, and if we take $a=1$, AM will incline 45° from either axis, as represented in the figure. Hence $y=x$, and making $b=0$, if these two values be substituted in eq. (2) and that equation reduced, we shall find

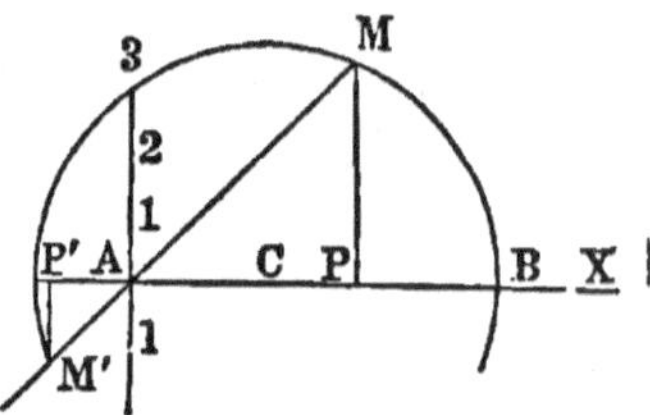

$$y^2\mp cy=\frac{R^2-c^2}{2}. \qquad (3)$$

This equation has the common *quadratic form.*

Equation (1) responds to any point in the straight line $M'M$. Equation (2) responds to any point in the circumference BMM'.

Therefore equation (3) which results from the combination of eqs. (1) and 2), must respond to the points M and M', the points in which the circle cuts the line.

That is, PM and $P'M'$ are the two roots of equation (3), and when one is above the axis of X, as in this figure, it is the *positive* root, and $P'M'$ being below the axis of X, it is the *negative* root.

When both roots of equation (3) are positive, the circle will cut the line in two points above the axis of X. When the two roots are *minus*, the circle will cut the line in two points below the axis of X.

When the two roots of any equation in the form of eq. (3) are equal and positive, the circle will *touch* the line above the axis of X. If the roots are *equal* and *negative*, the circle will touch the line below the axis of X. In case the roots of eq. (3) are *imaginary*, the circle will not meet the line.

We give the following examples for illustration:

$$y^2-2y=5.$$

To determine the values of y by a geometrical construction of this kind, we must make

$$c=-2, \quad \text{and} \quad \frac{R^2-c^2}{2}=5.$$

Whence $R=3.74$, the radius of the circle. Take any distance on the axes for the unit of measure, and set off the distance c on the axis of X from the origin, for the center of the circle; to the right, if c is *negative*, and to the left, if c is positive.

Then from the center, with a radius equal to $R=\sqrt{2q+c^2}$, describe a circumference cutting the line drawn midway between the two axes, as in the figure.

In this example the center of the circle is at C, the distance of two units from the origin A, to the right. Then, with the radius 3.74 we described the circumference, cutting the line in M and M', and we find by measure (when the construction is accurate) that $MP=4.44$, the positive root, and $M'P'=-1.44$, the negative root.

For another example we require the roots of the following equation by construction:

$$y^2+6y=27.$$

N. B. When the numerals are too large in any equation for convenience, we can always reduce them in the following manner:

Put $\qquad y=nz$, then the equation becomes

$$n^2z^2+6nz=27.$$

Or $\qquad\qquad z^2+\frac{6}{n}z=\frac{27}{n^2}.$

12*

Now let $n=$ *any number what-ever.* If $n=3$, then

$$z^2+2z=3.$$

Here $c=2$. $\dfrac{R^2-c^2}{2}=3.$

Whence $R=\sqrt{10}=3.16.$

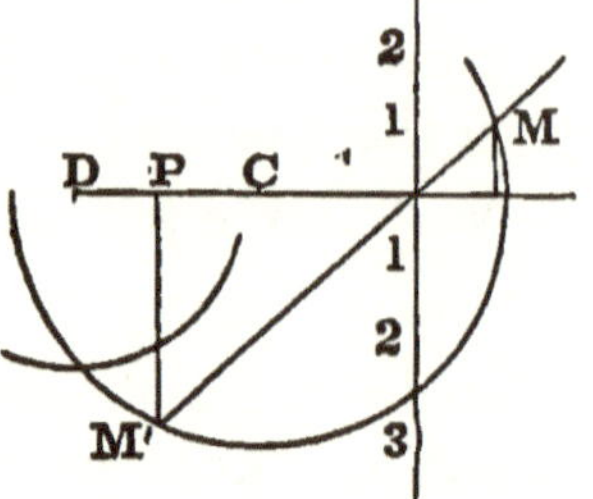

At the distance of two units to the left of the origin, is the center of the circle. We see by the figure that 1 is the positive root, and —3 the negative root.

But $\qquad y=nz, \quad n=3, \quad z=1, \quad y=3$ or —9.

We give one more example.

Construct the equation

$$y^2+4y=-6.$$

Here $c=4$, and $\dfrac{R^2-c^2}{2}=-6.$ $\qquad$ Whence $R=2.$

Using the same figure as before, the center of the circle to this example is at D, and as the radius is only 2, the circumference does not cut the line $M'M$, showing that the equation has no *real roots.*

We have said that this method of finding the roots of a quadratic was of little practical value. The reason of this conclusion is based on the fact that it requires more labor to obtain the value of the radius of the circle than it does to find the roots themselves.

Nevertheless this method is an interesting and instructive application of geometry in the solution of equations.

When we find the polar equation of the parabola, we shall then have another method of constructing the roots of quad-ratics which will not require the extraction of the square root.

To facilitate the geometrical solution of quadratic equations which we have thus indicated, the operator should provide himself with an accurately constructed scale, which is represented in the following figure. It consists of two lines, or axes, at right angles to each other, and another line drawn through their intersection and making with them an angle of 45°. On the axes, any convenient unit, as the inch, the half, or the fourth of an inch, etc., is laid off a sufficient number of times, to the right

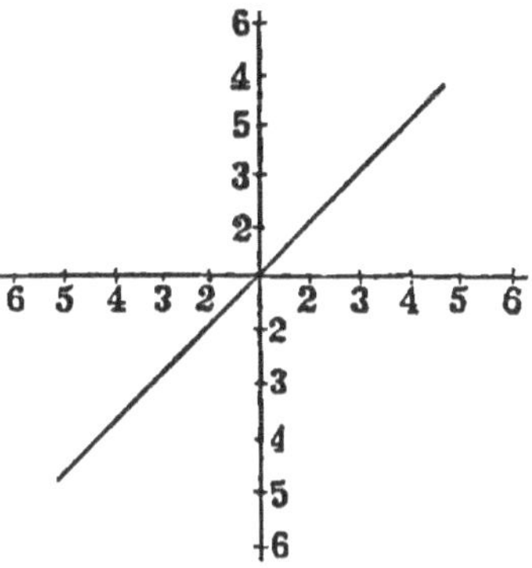

and the left, above and below the origin, from which the divisions are numbered 1, 2, 3, etc., or 10, 20, 30, etc., or .1, .2, .3, etc. To use this scale, a piece of thin, transparent paper, through which the numbers may be distinctly seen, is fastened over it, and with the proper center and radius the circumference of a circle is described. The distances from the axis of X of the intersections of this circumference, with the inclined line through the origin, will be the roots of the equation, and their numerical values may be determined by the scale.

By removing one piece of paper from the scale and substituting another, we are prepared for the solution of another equation, and so on.

EXAMPLES.

1. Given $x^2+11x=80$, to find x.   *Ans.* $x=5$, or—16.
2. Given $x^2-3x=28$, to find x.   *Ans.* $x=7$, or—4.
3. Given $x^2-x=2$, to find x.   *Ans.* $x=2$, or—1.
4. Given $x^2-12x=-32$, to find x.   *Ans.* $x=4$, or 8.
5. Given $x^2-12x=-36$, to find x.   *Ans.* Each value is 6.
6. Given $x^2-12x=-38$, to find x.   Both values imaginary.
7. Given $x^2+6x=-10$, to find x.   Both values imaginary.
8. Given $x^2=81$, to find x.   *Ans.* $x=9$, or—9.

For example 8, $c=o$ and $\dfrac{R^2-c^2}{2}=81$;

Whence, $R=9\sqrt{2}$.

This method may therefore be used for extracting the square root of numbers.   In such cases, the center of the circle is at the zero point.

CHAPTER III.

THE ELLIPSE.

We have already developed the properties of the *Ellipse, Parabola* and *Hyperbola* by geometrical processes, and it is now proposed to re-examine these curves, and develop their properties by analysis.

As he proceeds, the student cannot fail to perceive the superior beauty and simplicity of the analytical methods of investigation; and, even if a knowledge of the conic sections were not, as it is, of the highest practical value, the mental discipline to be acquired by this study would, of itself, be a sufficient compensation for the time and labor given to it.

As all needful definitions relating to these curves have been given in the CONIC SECTIONS, we shall not repeat them here, but will refer those to whom such reference may be necessary to the appropriate heads in that division of the work.

PROPOSITION I.

To find the equation of the ellipse referred to its axes as the axes of co-ordinates, the major axis and the distance from the center to the focus being given.

Let AA' be the major axis, F, F' the foci, and C the center of an ellipse. Make $CF=c$ $CA=A$. Take any point on the curve, and from it let fall the perpendicular Pt on the major axis; then, by our conventional notation, is $Ct=x$, $tP=y$.

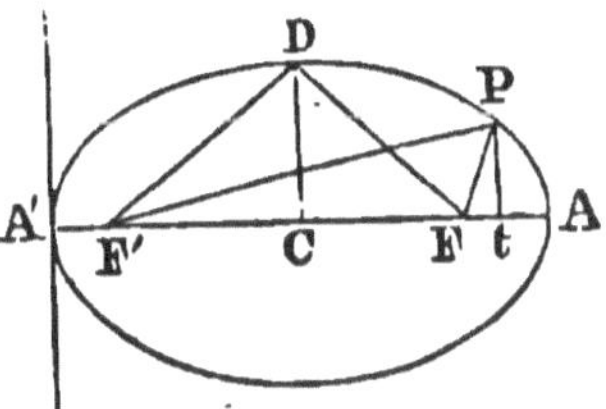

As $F'P+PF=2A$, we may put $F'P=A+z$, and $PF=A-z$. Then the two right angled triangles $F'Pt$, FPt, give us

$$(c+x)^2+y^2=(A+z)^2 \qquad (1)$$
$$(x-c)^2+y^2=(A-z)^2 \qquad (2)$$

For the points in the curve which cause t to fall between C and F, we would have

$$(c-x)^2+y^2=(A-z)^2 \qquad (3)$$

But when expanded, there is no difference between eqs. (2) and (3), and by giving proper values and signs to x and y, eqs. (1) and (2) will respond to *any point* in the curve as well as to the point P.

Subtracting eq. (2) from eq. (1), member from member, and dividing the resulting equation by 4, we find

$$cx=Az, \text{ or } z=\frac{cx}{A} \qquad (4)$$

This last equation shows that $F'P$, the radius vector, varies as the abscissa x.

Add eqs. (1) and (2), member to member, and divide the result by 2, and we have

$$c^2+x^2+y^2=A^2+z^2$$

Substituting the value of z^2 from eq. (4), and clearing of fractions, we have

$$c^2A^2+A^2x^2+A^2y^2=A^4+c^2x^2.$$
$$\text{Or,} \qquad A^2y^2+(A^2-c^2)x^2=A^2(A^2-c^2). \qquad (5)$$

Now conceive the point P to move along describing the curve, and when it comes to the point D, so that DC makes a right angle with the axis of X, the two triangles DCF and DCF' are right angled and equal. DF and

DF' each is equal to A, and as CF, CF', each is equal to c, we have

$$\overline{DC^2}=A^2-c^2.$$

It is customary to denote DC half the *minor* axis of the ellipse by B, as well as half the *major* axis by A, and adhering to this notation

$$B^2=A^2-c^2. \tag{6}$$

Substituting this in eq. (5), we have for the equation of the ellipse

$$A^2y^2+B^2x^2=A^2B^2,$$

referred to its center for the origin of co-ordinates.

If we wish to transfer the origin of co-ordinates from the center of the ellipse to the extremity A' of its major axis, we must put

$$x=-A+x', \quad \text{and} \quad y=y'.$$

Substituting these values of x and y in the last equation, and reducing, we have

$$y'^2=\frac{B^2}{A^2}(2Ax'-x'^2).$$

Or without the primes, we have

$$y^2=\frac{B^2}{A^2}(2Ax-x^2),$$

for the equation of the ellipse when the origin is at the extremity of the major axis.

Cor. 1. If it were possible for B to be equal to A, then c must be equal to 0, as shown by eq. (6). Or, while c has a value, it is impossible for B to equal A.

If $B=A$, then $c=0$, and the equation becomes

$$A^2y^2+A^2x^2=A^2A^2.$$

Or
$$y^2+x^2=A^2,$$

the equation of the circle. Therefore the circle may be called an ellipse, whose *eccentricity is zero*, or whose eccentricity is *infinitely small*.

Cor. 2. To find where the curve cuts the **axis of** X, make $y=0$ in the equation, then

$$x=\pm A,$$

showing that it extends to equal distances from the center.

To find where the curve cuts the axis of Y, make $x=0$, and then

$$y=\pm B.$$

Plus B refers to the point D, $-B$ indicates the point directly opposite to D, on the lower side of the axis of X.

Finally, let x have any value whatever, less than A, then

$$y=\pm\frac{B}{A}(A^2-x^2)^{\frac{1}{2}}.$$

an equation showing two values of y, numerically equal, indicating that the curve is symmetrical in respect to the axis of X.

If we give to y any value less than B, the general equation gives

$$x=\pm\frac{A}{B}(B^2-y)^{\frac{1}{2}}.$$

Showing that the curve is symmetrical in respect to the axis of Y.

Scholium.—The ordinate which passes through one of the foci, corresponds to $x=c$. But $A^2-B^2=c^2$. Hence A^2-c^2 or $A^2-x^2=B^2$. Or $(A^2-x^2)^{\frac{1}{2}}=B$, and this value substituted in the last equation, gives $y=\pm\dfrac{B^2}{A}$. Whence $\dfrac{2B^2}{A}$ is the measure of the parameter of any ellipse.

PROPOSITION II.

Every diameter of the ellipse is bisected in the center.

Through the center draw the line DD'. Let x, and y, denote the co-ordinates of the point D, and x', y', the co-ordinates of the point D'.

The equation of the curve is

$$A^2y^2 + B^2x^2 = A^2B^2.$$

The equation of a line passing through the center, must be of the form

$$y = ax.$$

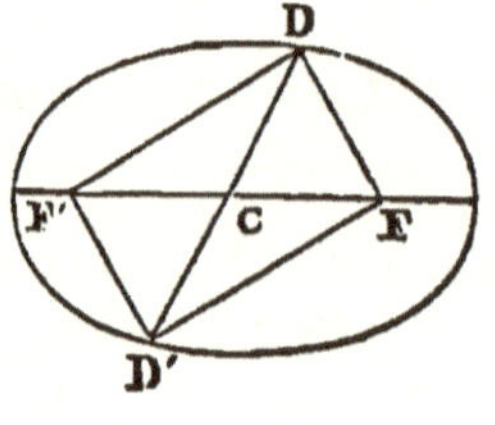

This equation combined with the equation of the curve, gives

$$x = \frac{AB}{\sqrt{a^2A^2 + B^2}}, \qquad y = \frac{aAB}{\sqrt{a^2A^2 + B^2}}.$$

$$x' = -\frac{AB}{\sqrt{a^2A^2 + B^2}}, \qquad y' = -\frac{aAB}{\sqrt{a^2A^2 + B^2}}.$$

These equations show that the co-ordinates of the point D, are the same as those of the point D', except opposite in signs. Hence *DD' is bisected at the center.*

PROPOSITION III.

The squares of the ordinates to either axis of an ellipse are to one another as the rectangles of their corresponding abscissas.

Let y be any ordinate, and x its corresponding abscissa. Then, by the first proposition, we shall have

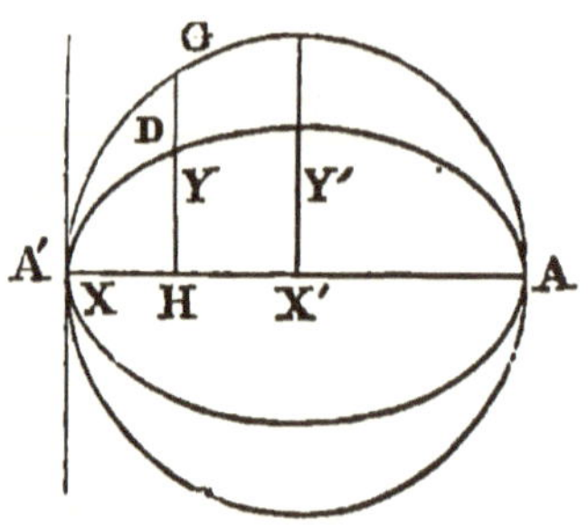

$$y^2 = \frac{B^2}{A^2}(2A - x)x.$$

Let y' be any other ordinate, and x' its corresponding abscissa, and by the same proposition we must have

$$y'^2 = \frac{B^2}{A^2}(2A - x')x'.$$

Dividing one of these equations by the other, omitting common factors in the numerator and denominator of the second member of the new equation, we shall have

$$\frac{y^2}{y'^2}=\frac{(2A-x)x}{(2A-x')x'}.$$

Hence, $\qquad y^2 : y'^2=(2A-x)x : (2A-x')x'.$ $\qquad$ (1)

By simply inspecting the figure, we cannot fail to perceive that $(2A-x)$, and x, are the abscissas corresponding to the ordinate y, and $(2A-x')$ and x' are those corresponding to y'.

If we transfer the origin to the lower extremity of the conjugate axis, the equation of the ellipse may be put under the form $\qquad x^2=\frac{A^2}{B^2}(2B-y)y,$

and by a process in all respects similar to the above, we prove that $\qquad x^2 : x'^2 : : (2B-y)y : (2B-y')y'.$

Therefore, *the squares of the ordinates, etc.*

SCHOLIUM.—Suppose one of these ordinates, as y' to represent half the *minor axis*, that is, $y'=B$. Then the corresponding value of x' will be A and $(2A-x',)$ will be A, also. Whence proportion (1) will become

$$y^2 : B^2=(2A-x)x : A^2.$$

In respect to the third term we perceive that if $A'H$ is represented by x, AH will be $(2A-x)$, and if G is a point in the circle, whose diameter is $A'A$, and GH the ordinate, then

$$(2A-x)x=\overline{GH}^2,$$

and the proportion becomes

$$y^2 : B^2=\overline{GH}^2 : A^2.$$
$$\text{Or}\qquad y : GH=B : A.$$
$$\text{Or}\qquad A : B= GH : y=DH.$$

If a circumference be described on the conjugate axis as a diameter, and an ordinate of the circle to this diameter be denoted by X and the corresponding ordinate of the ellipse by x, it may be shown in like manner that

$$A : B::x : X.$$

13 $\qquad\qquad$ K

PROPOSITION IV.

The area of an ellipse is a mean proportional between the areas of two circles, the diameter of the one being the major axis, and of the other the minor axis.

On the major axis $A'A$ of the ellipse as a diameter describe a circle, and in the semicircle $A'D$ A inscribe a polygon of any number of sides. From the vertices of the angles of this polygon draw ordinates to the major axis, and join the points in which they 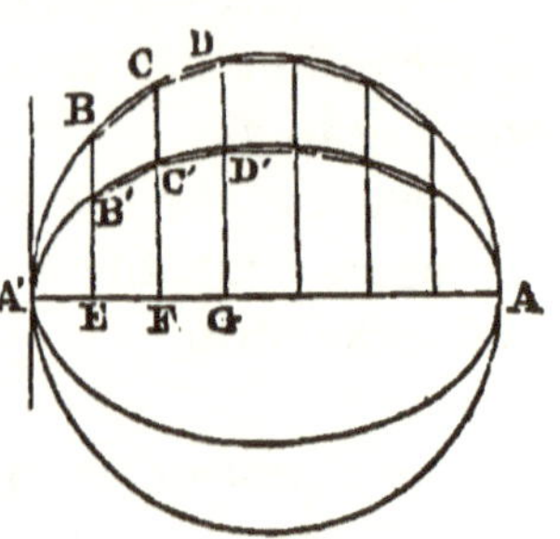
intersect the ellipse by straight lines, thus constructing a polygon of the same number of sides in the semi-ellipse $A'D'A$. Take the origin of co-ordinates at A', and denote the ordinates BE, CF, etc., of the circle by Y, Y', etc., the ordinates $B'E$, $C'F$, etc., of the ellipse by y, y', etc., and the corresponding abscissas, which are common to ellipse and circle, by x, x', etc.

Then by the scholium to Prop. 3, we have

$$Y : y :: A : B$$

and

$$Y' : y' :: A : B,$$

whence

$$Y : Y' :: y : y',$$

from which, by composition, we get

$$Y + Y' : y + y' : Y : y :: A : B$$

But the area of the trapezoid $BEFC$ is measured by

$$\left(\frac{Y+Y'}{2}\right)(x'-x) \quad \text{or} \quad (Y+Y')\left(\frac{x'-x}{2}\right),$$

and that of the trapezoid $B'EFC'$ by

$$\left(\frac{y+y'}{2}\right)(x'-x) \quad \text{or} \quad (y+y')\left(\frac{x'-x}{2}\right)$$

therefore,

$$\frac{\text{trapez. } BEFC}{\text{trapez. } B'EFC'} = \frac{Y+Y'}{y+y'} = \frac{A}{B}$$

That is, trapez. $BEFC$: trapez. $B'EFC'$: $A : B$;
or, in words, *any trapezoid of the semi-circle is to the corresponding trapezoid of the semi-ellipse as A is to B.*

From this we conclude that the sum of the trapezoids in the semi-circle is to the sum of the trapezoids in the semi-ellipse as A is to B. But by making these trapezoids indefinitely small, and their number, therefore, indefinitely great, the first sum will become the area of the semi-circle and the second, the area of the semi-ellipse.

Hence,

Area semi-circle : area semi-ellipse : : $A : B$

or,   area circle : area ellipse : : $A : B$

That is, πA^2 : area ellipse : : $A : B$

Whence,   area ellipse $= \dfrac{\pi A^2.B}{A} = \pi A.B$

But $\pi A.B$ is a mean proportional between πA^2 and πB^2.

Hence; *The area of an ellipse is a mean proportional, etc.*

SCHOLIUM.—Hence the common rule in mensuration to find the area of an ellipse.

RULE.—*Multiply the semi-major and semi-minor axes together, and multiply that product by* 3.1416.

PROPOSITION V.

To find the product of the tangents of the angles that two supplementary chords through the vertices of the transverse axis of an ellipse make with that axis, on the same side.

Let x, y, be the co-ordinates of any point, as P, and x', y', the co-ordinates of the point A'.

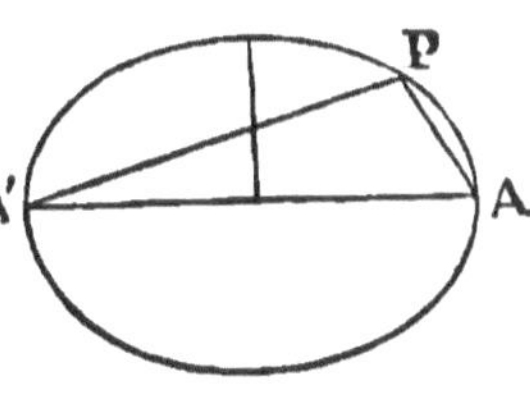

Then the equation of a line which passes through the two points A' and P, (Prop. 3, Chap. 1,) will be

$$y-y'=a(x-x'). \qquad (1)$$

The equation of the line which passes through the points A and P, will be of the form

$$y-y''=a'(x-x''). \qquad (2)$$

For the given point A', we have $y'=0$, and $x'=-A$. Whence eq. (1) becomes

$$y=a(x+A). \qquad (3)$$

For the given point A we have $y''=0$, and $x''=A$, which values substituted in eq. (2) give

$$y=a'(x-A). \qquad (4)$$

As y and x are the co-ordinates of the same point P in both lines, we may combine eqs. (3) and (4) in any manner we please. Multiplying them member by member, we have

$$y^2=aa'(x^2-A^2). \qquad (5)$$

Because F is a point in the ellipse, the equation of the curve gives

$$y^2=\frac{B^2}{A^2}(A^2-x^2)=-\frac{B^2}{A^2}(x^2-A^2). \qquad (6)$$

Comparing eqs. (5) and (6), we find

$$aa'=-\frac{B^2}{A^2}$$

for the equation sought.

SCHOLIUM 1.—In case the ellipse becomes a circle, that is, in case $A=B$, $aa'+1=0$, showing that the angle $A'PA$ would then be a right angle, as it ought to be, by (Prop. II, Chap. II.)

Because $\dfrac{B^2}{A^2}$ is less than *unity*, or aa' less than 1,* or *radius ;* the two angles $PA'A$ and PAA' are together less than $90°$; therefore, the angle at P is obtuse, or greater than $90°$.

SCHOLIUM 2.—Since aa' has a constant value, the sum of the two, $a+a'$, will be least when $a=a'$.

---

* In trigonometry we learn that tan. x cot. $x=R^2-1$. That is, the product of two tangents, the sum of whose arcs is $90°$, is equal to 1. When the sum is less than $90°$, the product will be a fraction.

Hence the angle at P will be greatest when P is at the vertex of the minor axis, and the supplementary chords equal; and the angle at P will become nearer a right angle as P approaches A or A'.

PROPOSITION VI.

To find the equation of a straight line which shall be tangent to an ellipse.

Assume any two points, as P and Q, on the ellipse, and denote the co-ordinates of the first by x', y', and of the second by x'', y''. Through these points draw a line, the equation of which (Prop. 4, Chap. 1,) is

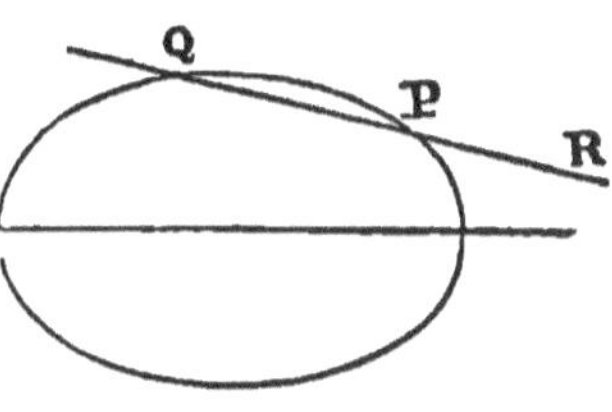

$$y—y'=a(x—x'), \qquad (1)$$

in which
$$a=\frac{y'—y''}{x'—x''}$$

We must now determine the value of a when this line becomes a tangent line to the ellipse.

Because the points P and Q are in the curve, the co-ordinates of those points must satisfy the following equations:

$$A^2y'^2+B^2x'^2=A^2B^2.$$
$$A^2y''^2+B^2x''^2=A^2B^2.$$

By subtraction $A^2(y'^2—y''^2)+B^2(x'^2—x''^2)=0.$

Or $\qquad A^2(y'+y'')(y'—y'')=—B^2(x'+x'')(x'—x'').$ (2)

Whence $\qquad a=\frac{y'—y''}{x'—x''}=—\frac{B^2(x'+x'')}{A^2(y'+y'')}.$

Now conceive the line to revolve on the point P until Q coincides with P, then PR will be tangent to the curve. But when Q coincides with P, we shall have

$$y'=y'' \text{ and } x'=x''.$$

13*

Under this supposition, we have

$$a = -\frac{B^2 x'}{A^2 y'}.$$

The value of a put in eq. (1), gives

$$y - y' = -\frac{B^2 x'}{A^2 y'}(x - x').$$

Reducing $\qquad A^2 yy' + B^2 xx' = A^2 y'^2 + B^2 x'^2.$

Or $\qquad A^2 yy' + B^2 xx' = A^2 B^2.$

This is the equation sought, x and y being the general co-ordinates of the line.

Scholium 1.—To find where the tangent meets the axis of X, we must make $y = 0$.

This gives $x = \dfrac{A^2}{x'} = CT.$

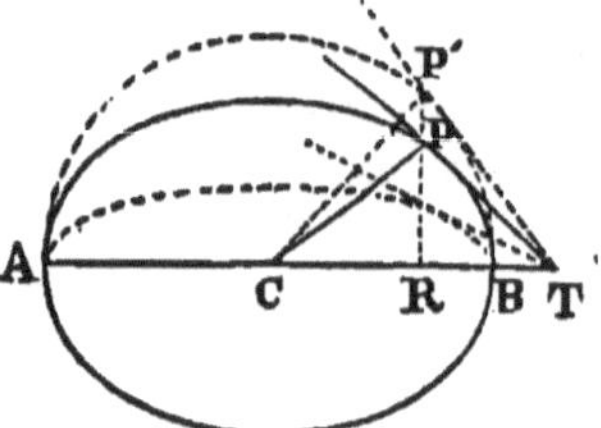

In case the ellipse becomes a circle, $B = A$, and then the equation will become $\qquad yy' + xx' = A^2,$

the equation for a tangent line to a circle; and to find where this tangent meets the axis of X, we make $y = 0$, and

$$x = \frac{A^2}{x'} = CT, \text{ as before.}$$

In short, as these results are both independent of B, the minor axis, it follows that the circle and all ellipses on the major axis AB have tangents terminating at the same point T on the axis of X, if drawn from the same ordinate, as shown in the figure.

Scholium 2.—To find the point in which the tangent to an ellipse meets the axis of Y, we make $x = 0$, then the equation for the tangent becomes

$$y = \frac{B^2}{y'}.$$

As this equation is independent of A, it shows that all ellipses having the same *minor axis*, have tangents terminating in the same point on the axis of Y, if drawn from the same abscissa.

Scholium 3. If from CT we subtract CR, we shall have RT, a common *subtangent* to a circle, and all ellipses which have $2A$ for a major diameter.   That is

$$RT = \frac{A^2}{x'} - x' = \frac{A^2 - x'^2}{x'}.$$

We can also find RT by the triangle PRT, as we have the tangent of the angle at T, $\left(-\frac{B^2 x'}{A^2 y'}\right)$ to the radius 1.

Whence we have the following proportion:

$$1 : -\frac{B^2 x'}{A^2 y'} = RT : y'$$

$$RT = -\frac{A^2 y'^2}{B^2 x'^2}.$$

The *minus* sign indicates that the measure from T is towards the left.

PROPOSITION VII.

To find the equation of a normal line to the ellipse.

Since the normal passes through the point of tangency, its equation will be in the form

$$y - y' = a'(x - x'). \qquad (1)$$

Because PN is at right angles to the tangent,

$$aa' + 1 = 0.$$

But by the last proposition

$$a = -\frac{B^2 x'}{A^2 y'}.$$

Whence $a' = \dfrac{A^2 y'}{B^2 x'}$, and this value of a' put in eq. (1) gives

$$y - y' = \frac{A^2 y'}{B^2 x'}(x - x'),$$

for the equation sought.

Scholium 1.—To find where the normal cuts the axis of X, we must make $y = 0$, then we shall have

$$x = \left(\frac{A^2 - B^2}{A^2}\right) x' = CN.$$

APPLICATION.—Meridians on the earth are ellipses; the semi-major axis through the equator is $A = 3963.$ miles, and the semi-minor axis from the center to the pole is $B = 3949.5$.

A plumb line is everywhere at right angles to the surface, and of course its prolongation would be a normal line like PN. *In latitude 42°, what is the deviation of a plumb line from the center of the earth ?* In other words, how far from the center of the earth would a plumb line meet the plane of the equator? Or, what would be the value of CN ?

As this ellipse differs but little from a circle, we may take CR for the cosine of 42°, which must be represented by x'. This being assumed, we have

$$x' = 2945. \quad \left(\frac{A^2 - B^2}{A^2}\right) 2945. = 20, + \text{miles} = CN. \quad Ans.$$

SCHOLIUM 2.—To find NR, the *subnormal*, we simply subtract CN from CR, whence

$$NR = x' - \left(\frac{A^2 - B^2}{A^2}\right) x' = \frac{B^2 x'}{A^2}.$$

We can also find the *subnormal* from the similar triangles PRT, PNR, thus :

$$TR : RP :: RP : RN.$$

$$-\frac{A\, y'^2}{B^2 x'} : y' :: y' : -NR. \quad \text{Whence } NR = \frac{B^2 x'}{A^2}.$$

PROPOSITION VIII.

Lines drawn from the foci to any point in the ellipse make equal angles with the tangent line drawn through the same point.

Let C be the center of the ellipse, PT the tangent line, and PF, PF', the two lines drawn to the foci.

Denote the distance $CF = \sqrt{A^2 - B^2}$ by c, CF'

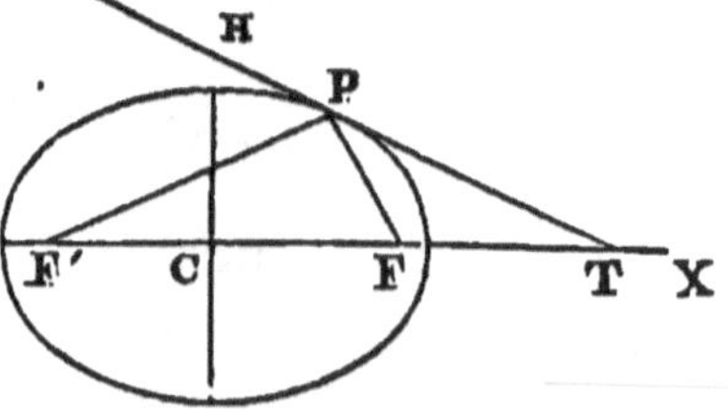

by $-c$, the angle FPT by V, and the tangents of the angles PTX, PFT, by a and a'.

Now $$FPT = PTX - PFT.$$

By trigonometry, (Eq. 29, p. 253, Robinson's Geometry), we have

$$\text{Tan. } FPT = \text{tan. } (PTX - \dot{P}FT).$$

That is, $$\text{tan. } V = \frac{a - a'}{1 + aa'}. \quad (1)$$

Prop. 6, gives us $a = -\dfrac{B^2 x'}{A^2 y'}$ x', y', being the co-ordinates of the point P.

Let x, y, be the co-ordinates of the point F, then from Prop. 4, Chap. 1, we have

$$a' = \frac{y' - y}{x' - x}.$$

But at the point F, $y = 0$ and $x = c$.

Whence $$a' = \frac{y'}{x' - c}.$$

These values of a and a' substituted in eq. (1) give

$$\text{Tan. } V = \frac{\dfrac{-B^2 x'}{A^2 y'} - \dfrac{y'}{x' - c}}{1 - \dfrac{B^2 x'}{A^2 (x' - c)}} = \frac{-B^2 x'^2 + B^2 c x' - A^2 y'^2}{A^2 y'(x' - c) - B^2 x' y'}.$$

$$\text{Tan. } V = \frac{B^2 c x' - A^2 B^2}{(A^2 - B^2) x' y' - A^2 c y'} = \frac{B^2 (c x' - A^2)}{c y'(c x' - A^2)} = \frac{B^2}{c y'}$$

observing that $A^2 y'^2 + B^2 x'^2 = A^2 B^2$, and $A^2 - B^2 = c^2$. The equation of the line PF will become the equation of the line PF' by simply changing $+c$ to $-c$, for then we shall have the co-ordinates of the other focus.

We now have

$$\text{tan. } FPT = \frac{B^2}{c y'}.$$

But if c is made $-c$, then

$$\text{tan. } F'PT = -\frac{B^2}{c y'}.$$

As these two tangents are *numerically* the same, differing only in signs, the lines are equally inclined to the straight lines from which the angles are measured, or the angles are supplements of each other.

Whence $\quad\quad\quad FPT + F'PT = 180.$

But $\quad\quad\quad F'PH + F'PT = 180.$

Therefore $\quad\quad FPT = F'PH.$

Cor. The normal being perpendicular to the tangent, it must bisect the angle made by the two lines drawn from the tangent point to the foci.

SCHOLIUM.—Any point in the curve may be considered as a point in a tangent to the curve at that point.

It is found by experiment that *light,* *heat* and *sound,* after they approach to, are reflected off, from any reflecting surface at equal angles; that is, for any ray, the angle of reflection is equal to the angle of incidence.

- Therefore, if a light be placed at one focus of an ellipsoidal reflecting surface, such as we may conceive to be generated by revolving an ellipse about its major axis, the reflected rays will be concentrated at the other focus. If the sides of a room be ellipsoidal, and a stove is placed at one focus, the heat will be concentrated at the other.

Whispering galleries are made on this principle, and all theaters and large assembly rooms should more or less approximate to this figure. The concentration of the rays of heat from one of these points to the other, is the reason why they are called the *foci*, or burning points.

PROPOSITION IX.

The product of the tangents of the angles that a tangent line to the ellipse and a diameter through the point of contact, make with the major axis on the same side, is equal to minus the square of the semi-minor divided by the square of the semi-major axis.

Let PT be the tangent line and PP' the diameter through the point of contact P, and denote the co-ordinates of P by x', y'. The equation of the diameter is

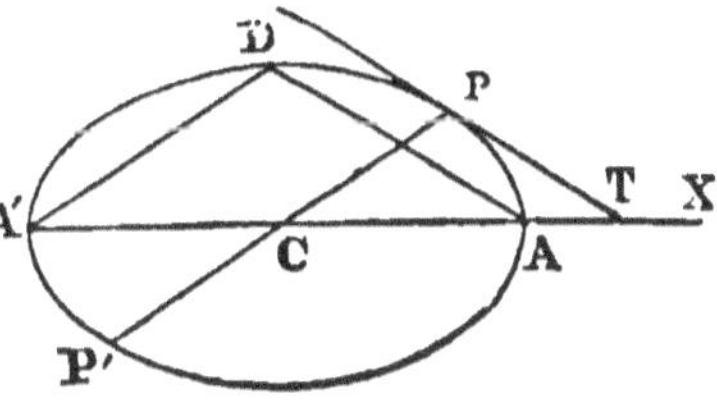

$$y = a'x,$$

in which a' is the tangent of the angle PCT.

Since this line passes through the point P, we must have

$$y' = a'x'$$

Whence

$$a' = \frac{y'}{x'} \qquad (1)$$

For the tangent of the angle PTX we have

$$a = -\frac{B^2 x'}{A^2 y'} \qquad (2)$$

Multiplying eqs. (1) and (2), member by member, we find

$$aa' = -\frac{B^2}{A^2}$$

SCHOLIUM.—The product of the tangents of the angles that a diameter and a tangent line through its vertex make with the major axis of an ellipse is the same (Prop. 5) as that of the tangents of the angles that supplementary chords drawn through the vertices of the major axis make with it.

Hence, if $a = a$, then $a' = a'$. That is, if the diameter is parallel to one of the chords, the tangent line will be parallel to the other chord, and conversely. This suggests an easy rule for drawing a tangent line to an ellipse at a given point, or parallel to a given line.

OF THE ELLIPSE REFERRED TO CONJUGATE DIAMETERS.

Two diameters of an ellipse are *conjugate* when either is parallel to the tangent lines drawn through the vertices of the other.

Since a diameter and the tangent line through its vertex make, with the major axis, angles whose tangents satisfy the equation

$$aa' = -\frac{B^2}{A^2}$$

it follows that the tangents of the angles which any two conjugate diameters make with the major axis must also satisfy the same equation.

Now let m be the angle whose tangent is a, and n be the angle whose tangent is a', then

$$a = \frac{\sin.\, m}{\cos.\, m}, \text{ and } a' = \frac{\sin.\, n}{\cos.\, n}.$$

Substituting these values in the last equation, and reducing, we obtain

$$A^2 \sin.\, m \sin.\, n + B^2 \cos.\, m \cos.\, n = 0,$$

which expresses the relation which must exist between A, B, m, and n, to fix the position of any two conjugate diameters in respect to the major axis, and this equation is called *the equation of condition for conjugate diameters.*

In this equation of condition, m and n are undetermined, showing that an infinite number of conjugate diameters might be drawn, but whenever any value is assigned to one of these angles, that value must be put in the equation, and then a deduction made for the value of the other angle.

PROPOSITION X.

To find the equation of the ellipse referred to its center and conjugate diameters.

The equation of the ellipse referred to its major and minor axes, is

$$A^2 y^2 + B^2 x^2 = A^2 B^2.$$

The formulas for changing rectangular co-ordinates

into oblique, the origin being the same, are (Prop. 9, Chap. 1,)

$$x = x' \cos. m + y' \cos. n. \qquad y = x' \sin. m + y' \sin. n.$$

Squaring these, and substituting the values of x^2 and y^2 in the equation of the ellipse above, we have

$$\left\{ \begin{array}{l} (A^2\sin^2 n + B^2\cos^2 n)y'^2 + (A^2\sin^2 m + B^2\cos^2 m)x'^2 \\ \quad + 2(A^2\sin. m \sin. n + B^2\cos. m \cos. n)y'x' \end{array} \right\} = A^2 B^2$$

But if we now assume the condition that the new axes shall be conjugate diameters, then

$$A^2\sin. m \sin. n + B^2\cos. m \cos. n = 0,$$

which reduces the preceding equation to $\qquad$ (F)

$$(A^2\sin.^2 n + B^2\cos.^2 n)y'^2 + (A^2\sin.^2 m + B^2\cos.^2 m)x'^2 = A^2 B^2,$$

which is the equation required. But it can be simplified as follows:

The equation refers to the two diameters $B''B'$ and $D''D'$ as co-ordinate axes. For the point B' we must make $y'=0$, then

$$x'^2 = \frac{A^2 B^2}{A^2\sin.^2 m + B^2\cos.^2 m} =$$

$$(CB')^2 = A'^2. \qquad\qquad\qquad\text{(P)}$$

Designating CB' by A', and CD' by B'.

For the point D' we must make $x'=0$. Then

$$y'^2 = \frac{A^2 B^2}{A^2\sin.^2 n + B^2\cos.^2 n} = (CD')^2 = B'^2. \quad\text{(Q)}$$

From (P) we have $(A^2\sin.^2 m + B^2\cos.^2 m) = \dfrac{A^2 B^2}{A'^2}.$

From (Q) $\qquad (A^2\sin.^2 n + B^2\cos.^2 n) = \dfrac{A^2 B^2}{B'^2}.$

These values put in (F) give

$$\frac{A^2 B^2}{B'^2}y'^2 + \frac{A^2 B^2}{A'^2}x'^2 = A^2 B^2.$$

Whence $\qquad A'^2 y'^2 + B'^2 x'^2 = A'^2 B'^2.$

14

We may omit the accents to x' and y', as they are general variables, and then we have

$$A'^2y^2 + B'^2x^2 = A'^2B'^2.$$

for the equation of the ellipse referred to its center and conjugate diameters.

SCHOLIUM.—In this equation, if we assign any value to x less than A', there will result two values of y, *numerically equal*, and to every assumed value of y less than B', there will result two corresponding values of x, numerically equal, differing only in signs, showing that the curve is symmetrical in respect to its conjugate diameters, *and that each diameter bisects all chords which are parallel to the other.*

OBSERVATION.—As this equation is of the same form as that of the general equation referred to rectangular co-ordinates on the *major* and *minor* axis, we may infer at once that we can find equations for ordinates, tangent lines, etc., referred to conjugate diameters, which will be in the same form as those already found, which refer to the axes. But as a general thing, it will not do to draw summary conclusions.

PROPOSITION XI.

As the square of any diameter of the ellipse is to the square of its conjugate, so is the rectangle of any two segments of the diameter to the square of the corresponding ordinate.

Let CD be represented by A', and CE by B', CH by x, and GH by y, then by the last proposition we have

$$A'^2y^2 + B'^2x^2 = A'^2B^2.$$

Which may be put under the form

$$A'^2y^2 = B'^2(A'^2 - x^2).$$

Whence $\quad A'^2 : B'^2 :: (A'^2 - x^2) : y^2.$

Or $\quad (2A')^2 : (2B')^2 :: (A'+x)(A'-x) : y^2.$

Now $2A'$ and $2B'$ represent the conjugate diameters $D'D$, $E'E$, and since CH represents x, $A'+x=D'H$, and

$A'-x=HD$. Also $y=GH$. Hence the above proportions correspond to

$$(D'D)^2 : (E'E)^2 :: D'H \times HD : (GH)^2.$$

SCHOLIUM.—As x is no particular distance from C, CF may represent x, then LF will represent y, and the proportion then becomes

$$(D'D)^2 : (E'E)^2 :: D'F \times FD : (LF)^2.$$

Comparing the two proportions, we perceive that

$$D'H \cdot HD : D'F \cdot FD :: \overline{GH^2} : \overline{LF^2}.$$

That is, *The rectangle of the abscissas are to one another as the squares of the corresponding ordinates.*

The same property as was demonstrated in respect to rectangular co-ordinates in Prop. 3.

In the same manner we may prove that

$$Eh \cdot hE' : Ef \cdot fE' :: (hg)^2 : (fe)^2$$

PROPOSITION XII.

To find the equation of a tangent line to an ellipse referred to its conjugate diameters.

Conceive a line to cut the curve in two points, whose co-ordinates are x', y', and x'', y'', x and y being the co-ordinates of any point on the line.

The equation of a line passing through two points is of the form

$$y-y'=a(x-x'), \tag{1}$$

an equation in which a is to be determined when the line touches the curve.

From the equation of the ellipse referred to its conjugate axes we have

$$A'^2 y'^2 + B'^2 x'^2 = A'^2 B'^2.$$
$$A'^2 y''^2 + B'^2 x''^2 = A'^2 B'^2.$$

Subtracting one of these equations from the other, and operating as in Prop. 6, we shall find

$$a = -\frac{B'^2 x'}{A'^2 y'}.$$

This value of a put in eq. (1) will give

$$y-y'=-\frac{B'^2x'}{A'^2y'}(x-x').$$

Reducing, and $A'^2y'y+B'^2x'x=A'^2B'^2,$

which is the equation sought, and it is in the same form as that in Prop. 6, agreeably to the observation made at the close of Prop. 10.

PROPOSITION XIII.

To transform the equation of the ellipse in reference to conjugate diameters to its equation in reference to the axes.

The equation of the ellipse in reference to its conjugate diameter is

$$A'^2y'^2+B'^2x'^2=A'^2B'^2. \qquad (1)$$

And the formulas for passing from oblique to rectangular axes are (Prop. 10, Chap. 1,)

$$x'=\frac{x\sin. n-y\cos. n}{\sin.(n-m)}, \qquad y'=\frac{y\cos. m-x\sin. m}{\sin. (n-m)}.$$

These values of x' and y' substituted in eq. (1) give

$$\left.\begin{array}{l}(A'^2\cos.^2 m+B'^2\cos.^2 n)y^2+(A'^2\sin.^2 m+B'^2\sin.^2 n)x^2 \\ -2(A'^2\sin. m\cos. m+B'^2\sin. n\cos. n)xy\end{array}\right\}=$$

$$A'^2B'^2\sin.^2(n-m).$$

This equation must be true for any point in the curve, x being measured on the major axis, and y the corresponding ordinate at right angles to it.

This being the case, such values of A', B', m, and n, must be taken as will reduce the preceding equation to the well known form

$$A^2y^2+B^2x^2=A^2B^2.$$

Therefore we must assume

$$A'^2\cos.^2 m+B'^2\cos.^2 n=A^2. \qquad (1)$$

$$A'^2\sin.^2 m+B'^2\sin.^2 n=B^2. \qquad (2)$$

$$A'^2\sin. m\cos. m+B'^2\sin. n\cos. n=0. \quad (3)$$

$$A'^2B'^2\sin.^2(n-m)=A^2B^2. \qquad (4)$$

The values of m and n must be taken so as to respond to the following equation, because the axes are in fact *conjugate diameters.*

$$A^2 \sin.m \sin.n + B^2 \cos.m \cos.n = 0. \quad (5)$$

These equations unfold two very interesting properties.

Scholium 1.—By adding eqs. (1) and (2) we find

$$A'^2 + B'^2 = A^2 + B^2.$$

Or $\qquad 4A'^2 + 4B'^2 = 4A^2 + 4B^2.$

That is, the sum of the squares of any two conjugate diameters is equal to the sum of the squares of the axes.

Scholium 2.—Equation eq. (3) or (5) will give us m when n is given ; or give us n when m is given.

Scholium 3.—The square root of eq. (4) gives

$$A'B' \sin.(n-m) = AB,$$

which shows the equality of two *surfaces,* one of which is obviously the rectangle of the two axes.

Let us examine the other.

Let n represent the angle NCB, and m the angle PCB. Then the angle NCP will be represented by $(n-m)$.

Since the angle MNK is the supplement of NCP, the two angles have the same sine and

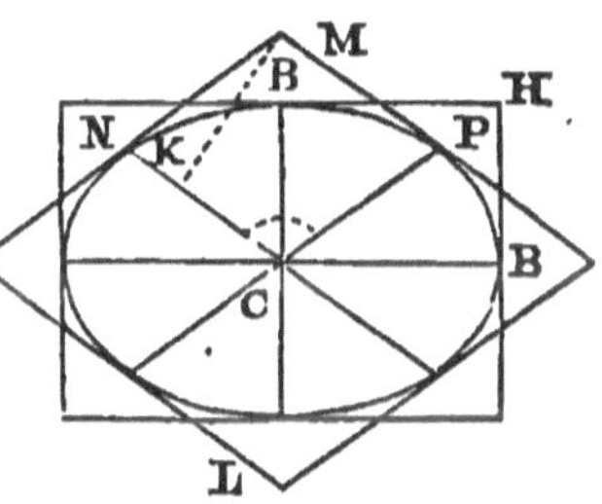

$$NM = A'.$$

In the right-angled triangle NKM, we have

$$1 : A' : : \sin.(n-m) : MK.$$
$$MK = A' \sin.(n-m).$$

But $\qquad\qquad NC = B'.$

Whence

$$MK \cdot NC = A'B' \sin.(n-m) = \text{the parallelogram } NCPM.$$

Four times this parallelogram is the parallelogram ML, and four times the parallelogram $DCBH$, which is measured by $A \times B$, is equal to the parallelogram HF. Hence eq. (4) reveals this general truth :

The rectangle which is formed by drawing tangent lines through

14* $\qquad\qquad$ L

the vertices of the axes of an ellipse is equivalent to any parallelogram which can be formed by drawing tangents through the vertices of conjugate diameters.

NOTE.—The student had better test his knowledge in respect to the truths embraced in scholiums 1 and 3, by an example:

Suppose the semi-major axis of an ellipse is 10, *and the semi-minor axis* 6, *and the inclination of one of the conjugate diameters to the axis of* X *is taken at* 30° *and designated by* m.

We are required to find A'^2 and B'^2, which together should equal A^2+B^2, or 136, and the area $NCPM$, which should equal AB, or 60, if the foregoing theory is true.

Equation (5) will give us the value of n as follows:

$$100 \cdot \tfrac{1}{2}\tan.n + 36 \cdot \tfrac{1}{2}\sqrt{3} = 0.$$

Or $$\tan.n = -\frac{36\sqrt{3}}{100}.$$

Log. $36 + \tfrac{1}{2}$ log. 3—log. 100 plus 10 added to the index to correspond with the tables, gives 9.794863 for the log. tangent of the angle n, which gives 31° 56' 42'', and the sign being negative, shows that 31° 56' 42'' must be taken below the axis of X, or we must take the supplement of it, NCB, for n, whence

$$n = 148° \; 3' \; 18'', \text{ and } (n-m) = 118° \; 3' \; 18''.$$

To find A'^2 and B'^2, we take the formulas from Prop. 10.

$$A'^2 = \frac{A^2 B^2}{A^2\sin.^2 30 + B^2\cos.^2 30} = \frac{100 \cdot 36}{100 \cdot \tfrac{1}{4} + 36 \cdot \tfrac{3}{4}} = \frac{3600}{52} = 69.23.$$

$$B'^2 = \frac{A^2 B^2}{A^2\sin.^2 31°56'42'' + B^2\cos.^2(31°56'42'')} = \frac{3600}{27 \cdot 99 + 25 \cdot 92} =$$

66·77. And their sum = 136.

This agrees with scholium 1.

As radius		10.000000
Is to	$A'\tfrac{1}{2}$(log.69.23)	0.920147
So is sine	$(n-m)$ 61° 56' 42''	9.945713
	log. $MK=$	0.865860
Log. $B'=\tfrac{1}{2}$ log. (66.77)............................		0.912290
$AB=60.$	log. 60 =	1.778150

PROPOSITION XIV.

To find the general polar equation of an ellipse.

If we designate the co-ordinates of the pole P, by a and b, and estimate the angles v from the line PX' parallel to the transverse axis, we shall have the following formulas :

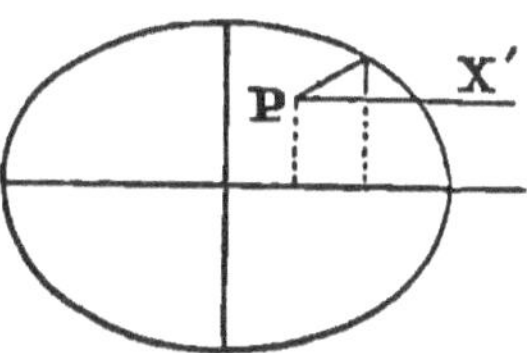

$$x=a+r \cos.v. \quad y=b+r \sin v.$$

These values of x and y substituted in the general equation $\quad A^2y^2+B^2x^2=A^2B^2,$
will produce

$$A^2 \sin.^2v \left| \begin{array}{l} r^2+2A^2b \sin.v \\ +2B^2a \cos.v \end{array} \right| r+A^2b^2+B^2a^2=A^2B^2,$$
$$B^2 \cos.^2v$$

for the general polar equation of the ellipse.

Scholium 1.—When P is at the center, $a=0$, and $b=0$, and then the general polar equation reduces to

$$r^2=\frac{A^2B^2}{A^2\sin.^2v+B^2\cos.^2v}$$

a result corresponding to equations (P) and (Q) in Prop. 10.

Scholium 2.—When P is on the curve $A^2b^2+B^2a^2=A^2B^2$, therefore

$$A^2\sin.^2v \left| \begin{array}{l} r^2+2A^2b \sin. v \\ +2B^2a \cos.v \end{array} \right| r=0.$$
$$B^2\cos.^2v$$

This equation will give two values of r, one of which is 0, as it should be. The other value will correspond to a chord, according to the values assigned to a, b, and v. Dividing the last equation by the equation $r=0$, and we have

$$A^2\sin.^2v \left| \begin{array}{l} r+2A^2b \sin.v \\ +2B^2a \cos.v \end{array} \right| =0.$$
$$B^2\cos.^2v$$

The value of r in this equation is the value of a chord.

When the chord becomes 0, the value of r in the last equation becomes 0 also, and then

$$A^2b \sin.v+B^2a \cos.v=0.$$

Or
$$\tan.v = -\frac{B^2 a}{A^2 b}$$

a result corresponding to Prop. 6, as it ought to do, because the *radius vector* then becomes tangent to the curve.

SCHOLIUM 3.—When P is placed at the extremity of the major axis on the right, and if $v=0$, then sin. $v=0$, and cos. $v=1$ $a=A$, and $b=0$; these values substituted in the general equation will reduce it to
$$B^2 r^2 + 2B^2 Ar = 0,$$
which gives $r=0$, and $r=-2A$, obviously true results.

When P is placed at either focus, then $a=\sqrt{A^2-B^2}=c$, and $b=0$. These values substituted, and we shall have

$$(A^2 \sin.^2 v + B^2 \cos.^2 v)r^2 + 2B^2 a \cos.v\cdot r = B^4.$$

It is difficult to deduce the values of r from this equation, therefore we adopt a more simple method.

Let F be the focus, and FP any radius, and put the angle $PFD=v$.

By Prop. 1, of the ellipse, we learn that

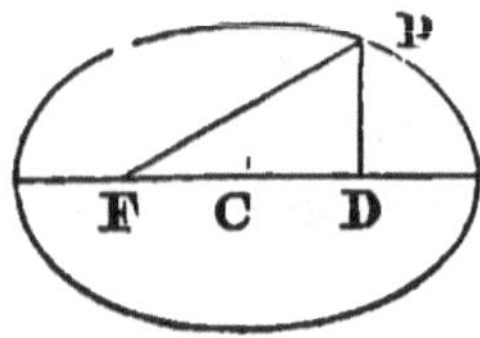

$$FP = r = A + \frac{cx}{A}, \qquad (1)$$

an equation in which $c=\sqrt{A^2-B^2}$, and x any variably distance CD.

Take the triangle PDF, and by trigonometry we have
$$1 : r :: \cos.v : c + x.$$
Whence
$$x = r \cos.v - c.$$
This value of x placed in (1), will give
$$r = A + \frac{cr.\cos.v - c^2}{A}$$
Whence
$$(A - c \cos.v)r = A^2 - c^2$$
Or
$$r = \frac{A^2 - c^2}{A - c \cos.v}.$$

This equation will correspond to all points in the curve by giving to cos.v all possible values from 1 to —1. Hence, the greatest value of r is $(A+c)$, and the least value $(A-c)$, obvious results when the polar point is at F.

The above equation may be simplified a little by introducing the *eccentricity*. The eccentricity of an ellipse is the distance from the center to either focus, when the semi-major axis is taken as unity. Designate the eccentricity by e, then

$$1 : e = A : c.$$

Whence
$$c = eA.$$

Substituting this value of c in the preceding equation, we have

$$r = \frac{A^2 - e^2 A^2}{A - eA \text{ cos. } v} = \frac{A(1 - e^2)}{1 - e \text{ cos. } v}$$

This equation is much used in astronomy.

PROPOSITION XV.—PROBLEM.

Given the relative values of three different radii, drawn from the focus of an ellipse, together with the angles between them, to find the relative major axis of the ellipse, the eccentricity, and the position of the major axis, or its angle from one of the given radii.

Let r, r', and r'', represent the three given radii, m the angle between r and r', and n that between r and r''. The angle between the radius r and the major axis is supposed to be unknown, and we therefore, call it x.

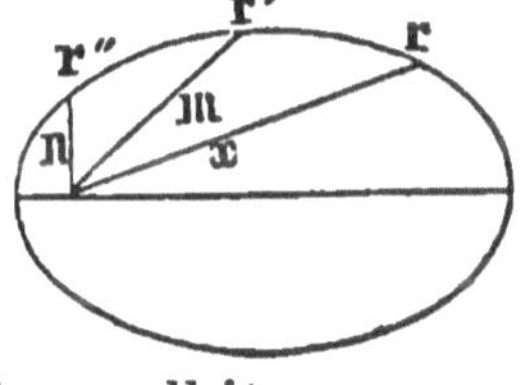

From the last proposition, we have

$$r = \frac{A(1 - e^2)}{1 - e \text{ cos. } x} \qquad (1)$$

$$r' = \frac{A(1 - e^2)}{1 - e \text{ cos. } (x + m)} \qquad (2)$$

$$r'' = \frac{A(1 - e^2)}{1 - e \text{ cos. } (x + n)} \qquad (3)$$

Equating the value of $A(1 - e^2)$ obtained from eqs. (1) and (2), and we have

$$r - re \text{ cos. } x = r' - r'e \text{ cos. } (x + m)$$

Or, $$e=\frac{r-r'}{r\,\cos.\,x-r'\,\cos.\,(x+m).} \qquad (4)$$

In like manner from eqs. (1) and (3), we have

$$r-re\,\cos.\,x=r''-r''e\,\cos.\,(x+n).$$

Or, $$e=\frac{r-r''}{r\,\cos.\,x-r''\,\cos.\,(x+n)} \qquad (5)$$

Equating the second members of eqs. (4) and (5), we have

$$\frac{r-r'}{r\,\cos.x-r'\,\cos.(x+m)}=\frac{r-r''}{r\,\cos.x-r''\,\cos.(x+n)}$$

Whence, $$\frac{r-r'}{r-r''}=\frac{r\,\cos.\,x-r'\,\cos.\,(x+m)}{r\,\cos.\,x-r''\,\cos.(x+n)}$$

$$=\frac{r\,\cos.\,x-r'\,\cos.\,x\,\cos.\,m+r'\,\sin.\,x\,\sin.\,m}{r\,\cos.\,x-r''\,\cos.\,x\,\cos.\,n+r''\,\sin.\,x\,\sin.\,n}$$

$$=\frac{r-r'\,\cos.\,m+r'\,\sin.\,m\,\tan.\,x}{r-r''\,\cos.\,n+r''\,\sin.\,n\,\tan.\,x}$$

For the sake of brevity, put $r-r'=d,$
$r-r''=d'$, the known quantity $r-r'\,\cos.\,m=a,$
and $r-r''\cos.n=b.$ Then the preceding equation becomes

$$\frac{d}{d'}=\frac{a+r'\sin.m\,\tan.x}{b+r''\sin.n\,\tan.x},$$

From which we get successively

$$db+dr''\,\sin.\,n\,\tan.\,x=ad'+d'r'\,\sin.\,m\,\tan.\,x$$

$$(dr''\,\sin.\,n-d'r'\,\sin.\,m)\,\tan.\,x=ad'-db,$$

$$\tan.\,x=\frac{ad'-db}{dr''\,\sin.n-d'r'\,\sin.m},$$

The value of x from this equation determines the position of the major axis with respect to that of r, which is supposed to be known, as it may be by observation.   -

Having x, eq. (4) or (5) will give e the eccentricity.   If the values of e found from these equations do not agree, the discrepancy is due to errors of observation, and in such cases the mean result is taken for the eccentricity.

Equations (1), (2) and (3) contain A, the semi-major axis, as a common factor in their second members. This factor, therefore, does not affect the relative values of r, r' and r'', and as it disappears in the subsequent part of the investigation, it shows that the angle x and the eccentricity are entirely independent of the magnitude of the ellipse. To apply the preceding formulas, we propose the following

EXAMPLE.

On the first day of August, 1846, an astronomer observed the sun's longitude to be 128° 47′ 31″, and by comparing this observation with observations made on the previous and subsequent days, he found its motion in longitude was then at the rate of 57′ 24″.9 per day. By like observations made on the first of September, he determined the sun's longitude to be 158° 37′ 46″, and its mean daily motion for that time 58′ 6″.6 ; and at a third time, on the 10th of October, the observed longitude was 196° 48′ 4″, and mean daily motion 59′ 22″.9. From these data are required the longitude of the solar apogee, and the eccentricity of the apparent solar orbit.

It is demonstrated in astronomy that the relative distances to the sun, when the earth is in different parts of its orbit, must be to each other inversely as the *square root* of the sun's apparent angular motion at the several points; therefore, $(r)^2$, $(r')^2$, and $(r'')^2$, must be in the proportion of

$$\frac{1}{57'\,24''\,9}, \quad \frac{1}{58'\,6''\,6}, \quad \text{and} \quad \frac{1}{59'\,22''\,9},$$

Or as the numbers

$$\frac{1}{3444.9}, \quad \frac{1}{3486.6}, \quad \text{and} \quad \frac{1}{3562.9}.$$

Multiply by 3562.9 and the proportion will not be changed, and we may put

$$r=\left(\frac{3562.9}{3444.9}\right)^{\frac{1}{2}}, \qquad r'=\left(\frac{3562.9}{3486.6}\right)^{\frac{1}{2}}, \quad \text{and } r''=1.$$

By the aid of logarithms we soon find

$r=1.016982$ $r'=1.010857$ and $r''=1.$

Hence $r-r'=d=0.006125,$ $r-r''=d'=0.016982.$

158° 37′ 46″	196° 48′ 4″
128 47 31	128 47 31
$m=$ 29 50 15	$n=$ 68 0 33

To substitute in our formulas, we must have the *natural sine* and cosine of m and n.

sin. $m=$ sin. 29° 50′15″$=0.497542$, cos.$=0.867440.$

sin. $n=$ sin. 68° 0′ 33″$=0.927238$, cos.$=0.374472.$

$r-r'$ cos. $m=a=0.140124.$

$r-r''$ cos. $n=b=0.642510.$

$ad'=0.0023695,$ $db=0.00393537.$

$d'r'$ sin. $m=0.008538616.$

dr'' sin. $n=0.005679332.$

These values substituted in the formula

$$\tan x=\frac{ad'-db}{dr''\sin n-d'r'\sin m}=\frac{db-ad'}{d'r'\sin m-dr''\sin n},$$

give

$$\tan x=\frac{.00156586}{.00285928}=\frac{15.6586}{28.5928}$$

Log. 15.6586 plus 10 to the index$=11.194746$

Log. 28.5928   1.456224

Log. tan. 28° 42′ 45″   9.738522

Long. of r 128° 47′ 31″

Long. apogee 100° 4′ 46″

According to observation, the longitude of the solar apogee on the 1st of January, 1800, was 99° 30′ 8″39, and it increases at the rate of 61″9 per annum. This would give, for the longitude of the apogee on the 1st of January, 1861, 100° 33′ 03″54.

To find e, the eccentricity, we employ eq. (5), which is

$$e = \frac{r-r''}{r\cos.x - r''\cos.(x+n)}.$$

Whence, by substituting the values of r, r'', cos.x, etc., we find

$$e = \frac{0.016982}{r\cos. 28°42'45'' - \cos. 96°43'18''} = \frac{.016982}{.891891 + .11694}$$

$$= \frac{.016982}{1.0088} = 0.016833$$

CHAPTER IV.

THE PARABOLA.

To describe a parabola.

Let CD be the directrix, and F the focus. Take a square, as DBG, and to one side of it, GB, attach a thread, and let the thread be of the *same length as the side* GB *of the square.*

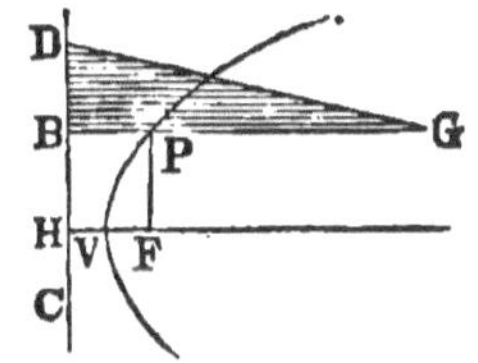

Fasten one end of the thread at the point G, the other end at F.

Put the other side of the square against CD, and with a pencil, P, in the thread, bring the thread up to the side of the square. Slide one end of the square along the line CD, and at the same time keep the thread close against the other side, permitting the thread to slide round the pencil P. As the side of the square, BD, is moved along the line CD, the pencil will describe the curve represented as passing through the points V and P.

$$GP + PF = \text{the thread.}$$
$$GP + PB = \text{the thread.}$$

By subtraction $PF - PB = 0$, or $PF = PB$.

This result is true at any and every position of the point P; that is, it is true for every point on the curve.

Hence, $$FV = VH.$$

If the square be turned over and moved in the opposite direction, the other part of the parabola, on the other side of the line FH may be described.

PROPOSITION I.

To find the equation of the parabola.

Take the axis of the parabola for the axis of abscissas and the line at right angles to it through the vertex for the axis of ordinates.

The perpendicular distance from the focus F to the directrix BH, is called p, a constant quantity, and when this constant is large, we have a parabola on a *large scale*, and when small, we have a parabola on a *small scale*.

By the definition of the curve, V is midway between F and the line BH, and $PF=PB$.

Put $VD=x$ and $PD=y$, and operate on the right angled triangle PDF.

$$FD=x-\tfrac{1}{2}p, \qquad PB=x+\tfrac{1}{2}p=PF.$$
$$(FD)^2+(PD)^2=(PF)^2.$$

That is, $\qquad (x-\tfrac{1}{2}p)^2+y^2=(x+\tfrac{1}{2}p)^2.$

Whence $\qquad y^2=2px$, the equation sought.

Cor. 1. If we make $x=0$, we have $y=0$ at the same time, showing that the curve passes through the point V, corresponding to the definition of the curve.

As $y=\pm\sqrt{2px}$, it follows that for every value of x there are two values of y, *numerically equal*, one $+$, the other $-$, which shows that the curve is symmetrical in respect to the axis of X.

Cor. 2. If we convert the equation $y^2=2px$ into a proportion, we shall have

$$x : y :: y : 2p,$$

a proportion showing that *the parameter of the axis is a third proportional to any abscissa and its corresponding ordinate.*

Cor. 3. If we substitute $\frac{1}{2}p$ for x in the equation $y^2 = 2px$ we get

$$y = p \text{ or } 2y = 2p.$$

That is *the parameter of the axis of the parabola is equal to the double ordinate through the focus, or, it is equal to four times the distance from the vertex to the directrix.*

PROPOSITION II.

The squares of ordinates to the axis of the parabola are to one another as their corresponding abscissas.

Let x, y, be the co-ordinates of any point P, and x', y', the co-ordinates of any other point in the curve.

Then by the equation of the curve we must have

$$y^2 = 2px. \tag{1}$$
$$y'^2 = 2px', \tag{2}$$

By division

$$\frac{y^2}{y'^2} = \frac{x}{x'}.$$

Whence

$$y^2 : y'^2 :: x : x'.$$

PROPOSITION III.

To find the equation of a tangent line to the parabola.

Draw the line SPQ intersecting the parabola in the two points P and Q. Denote the co-ordinates of the first point by x', y', and of the second, by x'', y''.

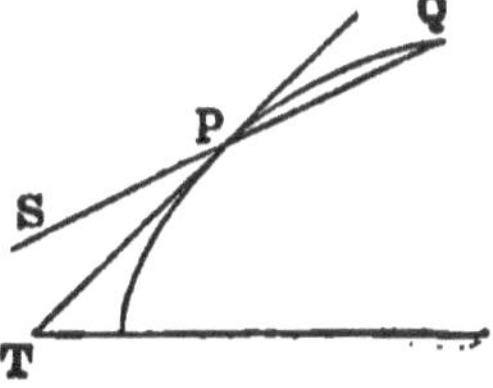

The equation of the straight line passing through these points is

$$y - y' = a(x - x') \tag{1}$$

in which a is equal to $\dfrac{y'-y''}{x'-x''}$

It is now required to find the value of a when the point Q unites with P, or, when the secant line becomes a tangent line at the point P.

Since P and Q are on the parabola we must have

$$y'^2=2px'$$

And
$$y''^2=2px''$$

Whence
$$y'^2-y''^2=2p(x'-x'')$$

Or
$$(y'-y'')(y'+y'')=2p(x'-x')$$

Therefore
$$a=\frac{y'-y''}{x'-x''}=\frac{2p-x}{y'+y''}$$

Substituting this value of a in eq. (1) we have for the equation of the secant line.

$$y-y'=\frac{2p}{y'+y''}(x-x') \qquad (2)$$

Now if this line be turned about P until Q coincides with P we shall have $y''=y'$ and the line becomes tangent to the curve at the point P.

Under this supposition the value of a becomes $\dfrac{p}{y'}$ and equation (2) reduces to

$$y-y'=\frac{p}{y'}(x-x')$$

Or
$$y\,y'-y'^2=px-px'$$

But $y'^2=2px'$; substituting this value y'^2 in the last equation, transposing and reducing, we have finally

$$y\,y'=p(x+x') \qquad (3)$$

for the equation of the tangent line.

Cor. To find the point in which the tangent meets the axis of X, we must make $y=0$, this makes

$$p(x+x')=0.$$

Or
$$x'=-x.$$

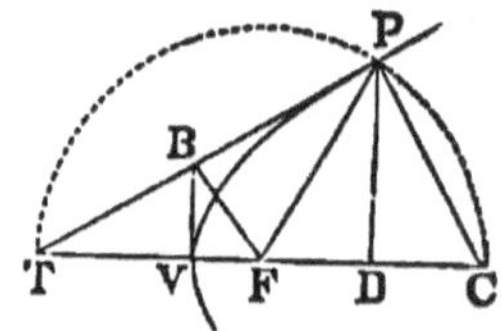

That is, *VD=VT, or the sub-tangent is bisected by the vertex.*

Hence, to draw a tangent line from any given point, as P, we draw the ordinate PD, then make $TV= VD$, and from the point T draw the line TP, and it will be tangent at P, as required.

PROPOSITION IV.

To find the equation of a normal line in the parabola.

The equation of a straight line passing through the point P is

$$y-y'=a(x-x'). \qquad (1) \ .$$

Let x_1, y_1, be the general co-ordinates of another line passing through the same point, and a' the tangent of the angle it makes with the axis of the parabola, its equation will then be

$$y_1-y'=a'(x_1-x'). \qquad (2)$$

But if these two lines are perpendicular to each other, we must have

$$aa'=-1. \qquad (3)$$

But since the first line is a tangent,

$$a=\frac{p}{y'}.$$

This value substituted in eq. (3) gives

$$a'=-\frac{y'}{p}.$$

And this value put in eq, (2) will give

$$y_1-y'=-\frac{y'}{p}(x_1-x')$$

for the equation required.

15*

Cor. 1. To find the point in which the normal meets the axis of X, we must make $y_1=0$. Then by a little reduction we shall have

$$p=x_1-x'.$$

But $VC=x_1$, and $VD=x'$. Therefore $DC=p$, that is,
The sub-normal is a constant quantity, double the distance between the vertex and focus.

Cor. 2. Since $TV=VD$, and $VF=\frac{1}{2}DC$, $TF=FC$. Therefore, if the point F be the center of a circle of which the radius is FC, the circumference of that circle will pass through the point P, because TPC is a right angle. Hence the triangle PFT is isosceles. Therefore, *If from the point of contact of a tangent line to the parabola a line be drawn to the focus it will make an angle with the tangent equal to that made by the tangent with the axis.*

Cor. 3. Now as V bisects TD and VB is, parallel to PD, the point B bisects TP. Draw FB, and that line bisects the base of an isosceles triangle, it is therefore perpendicular to the base. Hence, we have this general truth :

If from the focus of a parabola a perpendicular be drawn to any tangent to the curve, it will meet the tangent on the axis of Y.

Also, from the two similar right-angled triangles, FBV and FBT, we have

$$TF : FB : : FB : FV.$$

Whence $\qquad BF^2 = TF \cdot FV.$

But FV *is constant, therefore* $(BF)^2$ *varies as* TF, *or as its equal* PF.

Scholium.—Conceive a line drawn parallel to the axis of the parabola to meet the curve at P; that line will make an angle with the tangent equal to the angle FTP. But the angle FTP is equal to the angle FPT; hence the $\angle LPA=$the

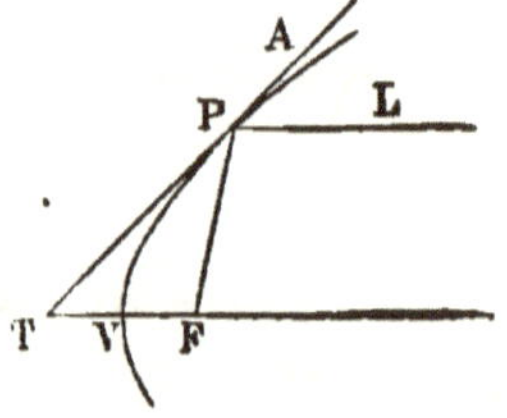

$\llcorner FPT$. Now, since light is incident upon and reflected from surfaces under equal angles, if we suppose LP to be a ray of light incident at P, the reflected ray will pass through the focus F, and this will be true for rays incident on every point in the curve; hence, if a reflecting mirror have a parabolic surface, all the rays of light that meet it parallel with the axis will be reflected to the focus; and for this reason many attempts have been made to form perfect parabolic mirrors for reflecting telescopes.

If a light be placed at the focus of such a mirror, it will reflect all its rays in one direction; hence, in certain situations, parabolic mirrors have been made for lighthouses for the purpose of throwing all the light seaward.

PROPOSITION V.

If two tangents be drawn to a parabola at the extremities of any chord passing through the focus, these tangents will be perpendicular to each other, and their point of intersection will be on the directrix.

Let PP' be any chord through the focus of the parabola, and PT, $P'T$ the tangents drawn through its extremities. Through T, their intersection, draw BB' perpendicular to the axis HF, and from the focus let fall the perpendiculars Ft, Ft' upon the tangents producing them to intersect BB' at B and B'. Draw, also, the lines PB, $P'B'$, and tt'.

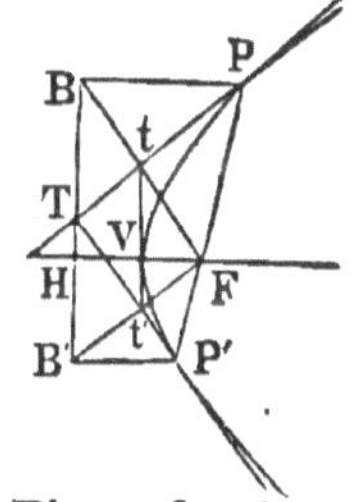

First.—The equation of the chord is

$$y = a\left(x - \frac{p}{2}\right) \tag{1}$$

and of the parabola

$$y^2 = 2px \tag{2}$$

Combining eqs. (1) and (2) and eliminating x, we find that the ordinates of the extremities of the chord are the roots of the equation

$$y^2 - \frac{2p}{a}y = p^2$$

Whence

$$y'=\frac{p+p\sqrt{a^2+1}}{a}\quad\text{and}\quad y''=\frac{p-p\sqrt{a^2+1}}{a}$$

Therefore the tangents of the angles that the tangent lines at the extremities of the chord make with the axis are

$$\frac{p}{y'}=\frac{a}{1+\sqrt{a^2+1}}\quad\text{and}\quad\frac{p}{y''}=\frac{a}{1-\sqrt{a^2+1}}$$

The product of these tangents is

$$\frac{a}{1+\sqrt{a^2+1}}\times\frac{a}{1-\sqrt{a^2+1}}=-1$$

Whence we conclude that the tangent lines are perpendicular to each other.

Second.—Because the $\triangle t F t'$ is right-angled and FV is a perpendicular let fall from the vertex of the right angle upon the hypothenuse, we have (Th. 25, B. II, Geom.)

$$\overline{Ft}^2 : \overline{Ft'}^2 : : Vt : Vt'$$

and because tt' and BB' are parallel, (Cor. 3, Prop. 4), we also have

$$\overline{Ft}^2 : \overline{Ft'}^2 : : \overline{FB}^2 : \overline{FB'}^2$$
$$: : HB : HB'$$

But (Cor. 3, Prop. 4,)

$$\overline{Ft}^2 : \overline{Ft'}^2 : : FP : FP'$$

Therefore

$$FP : FP' : : HB : HB'$$

Hence the lines PB, $P'B'$ are parallel to the axis of the parabola, and (Cor. 2, Prop. 4,) the angles BPt and tPF are equal. Therefore the right-angled triangles BPt and tPF are equal, and $PB=PF$. In the same way we prove that $P'B'=P'F$. The line BB' is therefore the directrix of the parabola.

Cor. Conversely : *If two tangents to the parabola are perpendicular to each other, the chord joining the points of contact passes through the focus.*

For, if not, draw a chord from one of the points of contact through the focus, and at the extremity of this chord draw a third tangent. Then the second and third tangents being both perpendicular to the first, must be parallel.

But a tangent line to a parabola, at a point whose ordinate is y', makes with the axis an angle having $\dfrac{p}{y'}$ for its tangent; and as no two ordinates of the parabola are algebraically equal, it is impossible that the curve should have parallel tangent lines.

PROPOSITION VI.

To find the equation of the parabola referred to a tangent line and the diameter passing through the point of contact as the co-ordinate axes.

Let V be the vertex and VX the axis of the parabola. Through any point of the curve, as P, draw the tangent PY and the diameter PR, and take these lines for a system of oblique co-ordinate axes. From a point M, assumed at pleasure, on the parabola, draw MR

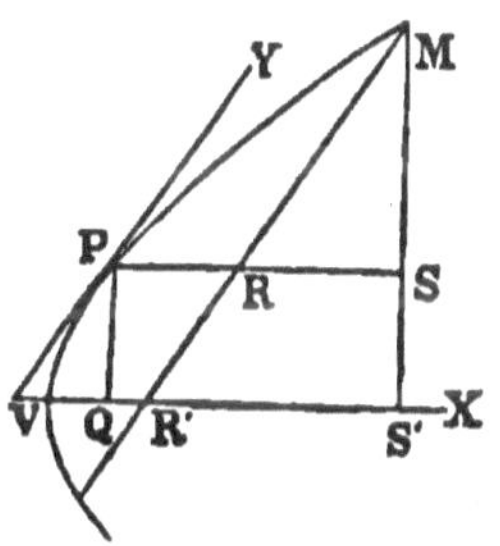

parallel to PY and MS perpendicular to VX; also, draw PQ perpendicular to VX.

Let our notation be $VQ=c$, $PQ=b$, $VS'=x$, $MS'=y$, $PR=x'$, $MR=y'$ and $\lfloor MRS=\lfloor MR'S'=m$; then the formulas for changing the reference of points from a system of rectangular to a system of oblique co-ordinate axes having a different origin, give, by making $\lfloor n=0$,

$$VS'=x=c+x'+y'\cos.m$$
$$MS'=y=b+y'\sin.m$$

These values of x and y substituted in the equation of the parabola referred to V as the origin which is

$$y^2 = 2px \qquad (1)$$

will give

$$b^2 + 2by'\sin.m + y'^2\sin.^2m = 2pc + 2px' + 2py'\cos.m \qquad (2)$$

Because P is on the curve, $b^2 = 2pc$, and because RM is parallel to the tangent PY, we also have (Prop. 3,)

$$\frac{\sin.m}{\cos.m} = \frac{p}{b}$$

Whence $\qquad 2by'\sin.m = 2py'\cos.m$

By means of these relations we can reduce eq. (2) to

$$y'^2\sin.^2m = 2px'$$

Or $\qquad\qquad y'^2 = \frac{2p}{\sin.^2m}x'$

If we denote $\dfrac{2p}{\sin.^2m}$ by $2p'$ the equation of the curve referred to the origin P and the oblique axes PX, PY, becomes

$$y'^2 = 2p'x'$$

an equation of the same form as that before found when the vertex V was the origin and the axes rectangular.

Cor. 1. Since the equation gives $y' = \pm\sqrt{2p'x'}$, that is for every value of x' two values of y', numerically equal, it follows that *every diameter of the parabola bisects all chords of the curve drawn parallel to a tangent through the vertex of the diameter.*

Cor. 2. *The squares of the ordinates to any diameter of the parabola are to each other as their corresponding abscissas.*

Let x, y and x', y' be the co-ordinates of any two points in the curve, then

$$y^2 = 2p'x$$
$$y'^2 = 2p'x'$$

Whence $\qquad\qquad \dfrac{y^2}{y'^2} = \dfrac{x}{x'}$

Or $$y^2 : y'^2 :: x : x'$$

Cor. 3. By a process in no respect differing from that followed in proposition 3 we shall find

$$yy'=p'(x+x')$$

for the equation of a tangent line to the parabola when referred to any diameter and the tangent drawn through its vertex as the co-ordinates axes.

If, in this equation, we make $y=0$ we get

$$x+x'=0 \text{ or } x=-x'.$$

That is, *the subtangent on any diameter of the parabola is bisected at the vertex of that diameter.*

SCHOLIUM.—*Projectiles, if not disturbed by the resistance of the atmosphere, would describe parabolas.*

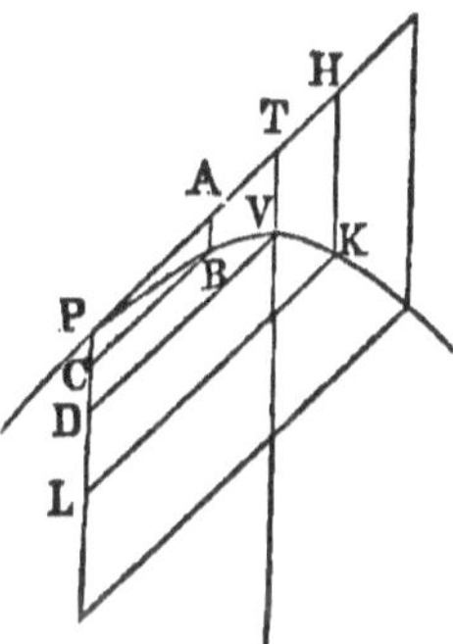

Let P be the point from which a projectile is thrown in any direction *PH*. Undisturbed by the atmosphere and by gravity, it would continue to move in that line, describing equal spaces in equal times. But gravity causes bodies to fall through spaces proportional to the squares of the times.

From P draw *PL* in the direction of a plumb line, the direction in which bodies fall when acted upon by gravity alone, and draw from A, T, H, etc., points taken at pleasure on *PH*, lines parallel to *PL*. Make AB equal to the distance through which a body starting from rest, would fall while the undisturbed projectile would move through the space PA, and lay off TV to correspond to the proportion

$$\overline{PA}^2 : \overline{PT}^2 :: AB : TV \qquad (1)$$

Also lay off HK to correspond to the proportion

$$\overline{PA}^2 : \overline{PH}^2 :: AB : HK \qquad (2)$$

In the same way we may construct other distances on lines drawn from points of *PH* parallel to *PL*.

Now through the points B, V, K, etc., draw parallels to *PH*, intersecting *PL* in C, D, L, etc., and through the points B, V,

K, etc., trace a curve. This curve will represent the path described by a projectile in vacuo, and will be a parabola.

Because AB is parallel to PC, and PA parallel to BC, the figure $PABC$ is a parallelogram, and so are each of the other figures, $PTVD$, $PHKL$, etc.

$$\text{Let } PA=y, PT=y', PH=y'' \text{ etc.}$$
$$\text{and } PC=x, PD=x', PL=x'' \text{ etc.}$$

Then proportions (1) and (2) become respectively

$$y^2 : y'^2 :: x : x'$$
$$y^2 : y''^2 :: x : x''$$

But by corollary 2 of this proposition, the curve that possesses the property expressed by these proportions is the parabola, and we therefore conclude that the path described by a projectile in vacuo is that curve.

PROPOSITION VII.

The parameter of any diameter of the parabola is four times the distance from the vertex of that diameter to the focus.

We are to prove that $2p'=4PF$.
Let the angle $YPR=m$ as before.
Then by (Prop. 3,)

$$\frac{\sin. m}{\cos. m}=\frac{p}{b}. \qquad (1)$$

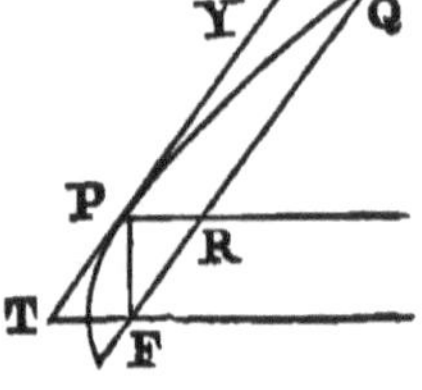

The co-ordinates of the point P being c, b, as in the last proposition, we have

$$b^2=2pc. \qquad (2)$$

From eq. (1)　　　$b^2\sin.^2 m=p^2\cos.^2 m.$
$$=p^2(1-\sin.^2 m)=p^2-p^2\sin.^2 m.$$

Or　　　$$\sin.^2 m=\frac{p^2}{b^2+p^2}=\frac{p^2}{2pc+p^2}=\frac{p}{2c+p}.$$

But in the last proposition $\dfrac{2p}{\sin.^2 m}=2p'$. Whence

$$\sin.^2 m=\frac{p}{p'}.$$

Therefore $p'=2c+p$.

Or $2p'=4\left(c+\dfrac{p}{2}\right)$

But $\left(c+\dfrac{p}{2}\right)=PF$. (Prop. 1.) Hence $2p'$, the parameter of the diameter PR, is *four times the distance of the vertex of the diameter from the focus.*

SCHOLIUM.—Through the focus F draw a line parallel to the tangent PY. Designate PR by x, and RQ by y. Then, by (Prop. 6),

$$y^2=2p'x.$$

But $PF=FT$, (Prop. 4, Cor. 2.) And $PR=TF$, because $TFRP$ is a parallelogram. Whence $PR=PF$; and, since $PR=x$, and $PF=c+\dfrac{p}{2}$,

$$x=\left(c+\dfrac{p}{2}\right)$$

Therefore $4x=4\left(c+\dfrac{p}{2}\right)=2p'$, or $x=\dfrac{p'}{2}$

This value of x put in the equation of the curve gives

$$y=p', \text{ or } 2y=2p'.$$

That is, the quantity $2p'$, which has been called the parameter of the diameter PR, is equal to the double ordinate passing through the focus.

PROPOSITION VIII.

If an ordinate be drawn to any diameter of the parabola, the area included between the curve, the ordinate and the corresponding abscissa, is two-thirds of the parallelogram constructed upon these co-ordinates.

Let $V'P'PQ$ be a portion of a parabola included between the arc $V'P'P$, and the co-ordinates $V'Q$, PQ of the extreme point P, referred to the diameter $V'Q$ and the tangent through its vertex.

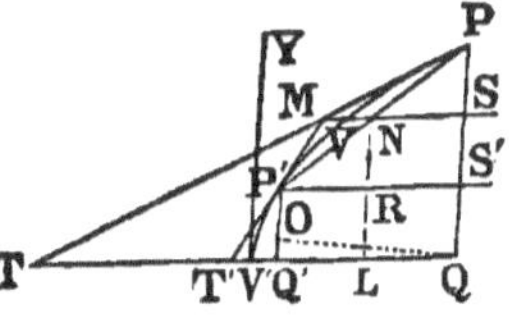

16

Take a point, P', on the curve between P and V'; draw the chord PP' and the ordinates PQ, $P'Q'$. Through N, the middle point of PP', draw the diameter NS, and at P and P' draw tangents to the parabola intersecting each other at M and the diameter $V'Q$ produced at T and T'. The tangents at the points P and P' have a common subtangent on the diameter VS, because these points, when referred to this diameter and the tangent at its vertex, have the same abscissa, VN, (Cor. 3, Prop. 6). The point M is therefore common to the two tangents and the diameter VS produced.

By this construction we have formed the trapezoid $PQQ'P'$ within, and the triangle TMT' without, the parabola, and we will now compare the areas of these figures. From N draw NL parallel to PQ, and from Q draw QO perpendicular to $P'Q'$, and let us denote the angle $YV'Q$ that the tangent at V' makes with the diameter $V'Q$ by m.

By the corollary just referred to we have

$$V'T = V'Q \text{ and } V'T' = V'Q'.$$

Whence $T'T = Q'Q$; and because N is the middle point of PP' we also have

$$NL = \frac{PQ + P'Q'}{2}$$

Therefore (Th. 34, B. I, Geom.,) the area of the trapezoid $PQQ'P$ is measured by

$$NL \times QO = NL \times Q'Q \sin.m = Q'Q \times NL \sin.m.$$

But $NL \sin.m$ is equal to the perpendicular let fall from N upon $Q'Q$ which is equal to that from M upon the same line. Hence the area of the triangle TMT' is measured by

$$\tfrac{1}{2}T'T \times NL \sin.m = \tfrac{1}{2}Q'Q \times NL \sin.m.$$

The area of the trapezoid is, therefore, twice that of the triangle.

If another point be taken between P' and V', and we make with reference to it and P' the construction that has just been made with reference to P' and P, we shall have another trapezoid within, and triangle without, the parabola, and the area of the trapezoid will be twice that of the triangle.

Let us suppose this process continued until we have inscribed a polygon in the parabola between the limits P and V'; then, if the distance of the consecutive points P, P', etc., be supposed indefinitely small, it is evident that the sum of the trapezoids will become the interior curvilinear area $PP'V'Q$, and the sum of the triangles the exterior curvilinear area $TPV'V$.

Since any one of these trapezoids is to the corresponding triangle as two is to one, the sum of the trapezoids will be to the sum of the triangles in the same proportion. But the interior and exterior area together make up the triangle PQT.

Therefore   interior area $=\frac{2}{3}\triangle PQT$,

and $\triangle PQT=\frac{1}{2}TQ\times PQ\sin.m= V'Q\times PQ\sin.m.$

But $V'Q\times PQ\sin. m$ measures the area of the parallelogram constructed upon the abscissa $V'Q$ and the ordinate PQ. We will denote $V'Q$ by x and PQ by y. Then the expression for the area in question becomes

$$\tfrac{2}{3}xy.\sin.m$$

Cor. When the diameter is the axis of the parabola, then $m=90°$, and sin. $m=1$, and the expression for the area becomes $\frac{2}{3}xy$. That is, *every segment of a parabola at right angles with the axis is two-thirds of its circumscribing rectangle.*

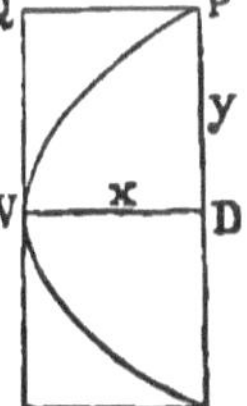

PROPOSITION IX.

To find the general polar equation of the parabola.

Let P be the polar point whose co-ordinates referred to the principal vertex, V, are c and b. Put $VD=x$, and DM

$=y$; then by the equation of the curve we have

$$y^2=2px. \qquad (1)$$

Put $PM=R$, the angle $MPX=m$, then we shall have

$$VD=x=c+R\cos. m.$$
$$DM=y=b+R\sin. m.$$

These values of x and y substituted in eq. (1) will give

$$(b+R\sin. m)^2=2p(c+R\cos. m). \qquad (2)$$

Expanding and reducing this equation, (R being the variable quantity), we find

$$R^2\sin.^2 m+2R(b\sin. m-p\cos. m)=2pc-b^2$$

for the general polar equation of the parabola required.

Cor. 1. When P is on the curve, $b^2=2pc$, and the general equation becomes

$$R^2\sin.^2 m+2R(b\sin. m-p\cos. m)=0.$$

Here one value of R is 0, as it should be, and the other value is

$$R=\frac{2(p\cos. m-b\sin. m)}{\sin.^2 m}$$

When $m=270°$, cos. $m=0$ and sin. $m=-1$. Then this last equation becomes

$$R=2b,$$

a result obviously true.

Cor. 2. When the pole is at the focus F, then $b=0$, and $c=\frac{p}{2}$, and these values reduce the general equation to

$$R^2\sin.^2 m-2Rp\cos. m=p^2.$$

But　　　　　　$\sin.^2 m=1-\cos.^2 m.$

Whence　　$R^2-R^2\cos.^2 m-2Rp\cos. m=p^2.$

Or　　　　　$R^2=p^2+2Rp\cos. m+R^2\cos.^2 m.$

Or　　　　　$R=p+R\cos. m.$

Whence　　　$R=\frac{p}{1-\cos. m},$

and this is the polar equation *when the focus is the pole.*

When $m=0$, cos.$m=1$, and then the equation becomes

$$R=\frac{p}{1-1}, \quad \text{or} \quad R=\frac{p}{0}= \text{infinite},$$

showing that there is no termination of the curve at the right of the focus on the axis.

When $m=90°$, cos.$m=0$, then $R=p$, as it ought to be, because p is the ordinate passing through the focus.

When $m=180°$, cos.$m=-1$, then $R=\frac{1}{2}p$; that is, the distance from the focus to the vertex is $\frac{1}{2}p$.

As m can be taken both above and below the axis and the cos.m is the same to the same arc above and below, it follows that the curve must be symmetrical in respect to the axis.

SCHOLIUM 1.—If we take p for the unit of measure, that is, assume $p=1$, then the general polar equation will become

$$R^2\text{sin.}^2m+2R(b\,\text{sin.}m-\text{cos.}m)=2c-b^2.$$

Now if we suppose $m=90°$, then sin.$m=1$, cos.$m=0$, and R would be represented by the line PM', and the equation would become

$$R^2+2bR=(2c-b^2),$$

and this equation is in the common form of a quadratic.

Hence, a parabola in which $p=1$ will solve any quadratic equation by making $c=VB$, $BP=b$, then PM' will give one value of the unknown quantity.

To apply this to the solution of equations, we construct a parabola as here represented.

Now, suppose we require the value of y, by construction, in the following equation,

$$y^2+2y=8.$$

Here $2b=2$, and $2c-b^2=8$.

Whence $b=1$, and $c=4.5$.

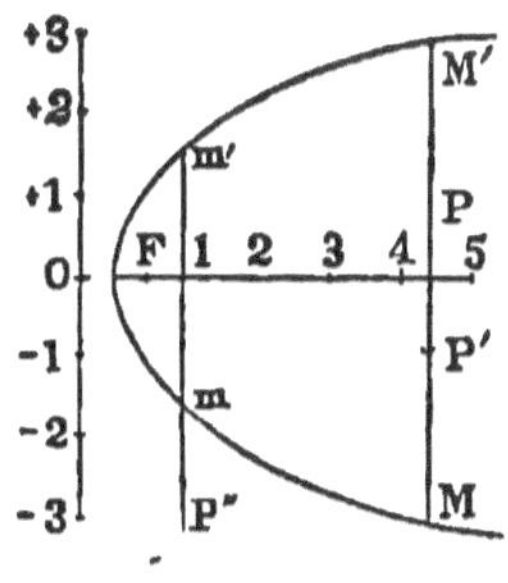

Lay off c on the axis, and from the extremity lay off b at right angles, above the axis if b is *plus*, and below if *minus*.

This being done, we find P is the polar point corresponding to

16*

this example, and $PM'=2$ is the *plus* value of y, and $PM=-4$ is the *minus* value.

Had the equation been

$$y^2-2y=8,$$

then P' would have been the polar point, and $P'M'=4$ the plus value, and $P'M=-2$ the minus value.

For another example let us construct the roots of the following equation :

$$y^2-6y=-7.$$

Here $b=-3$, and $2c-b^2=-7$.   Whence $c=1$.

From 1 on the axis take 3 downward, to find the polar point P''. Now the roots are $P''m$ and $P''m'$, both *plus*. $P''m=1.58$, and $P''m'=4.414$.

Equations having two *minus* roots will have their polar points above the curve.

When c comes out negative, the ordinates cannot meet the curve, showing that the roots would not be *real* but *imaginary*.

The roots of equations having large numerals cannot be constructed unless the numerals are first reduced.

To reduce the numerals in any equation, as

$$y^2+72y=146,$$

we proceed as follows :

Put $y=nz$, then

$$n^2z^2+72nz=146$$

$$z^2+\frac{72}{n}z=\frac{146}{n^2}.$$

Now we can assign any value to n that we please. Suppose $n=10$, then the equation becomes

$$z^2+7\cdot2z=1.46.$$

The roots of this equation can be *constructed*, and the values of y are *ten* times those of z.

Scholium 2.—The method of solving quadratic equations employed in Scholium 1 may be easily applied to the construction of the square roots of numbers.

Thus, if the square root of 20 were required, and we represent it by y, we shall have

$$y^2=20,$$

an incomplete quadratic equation; but it may be put under the form of a complete quadratic by introducing in the first number the term $\pm 0 \times y$, and we shall then have

$$y^2 \pm 0 \times y = 20.$$

Here $2b=0$, and $2c-b^2=20$; whence $c=10$; which shows that the ordinate corresponding to the abscissa 10 on the axis of the parabola will represent the square root of 20. In the same way the square roots of other numbers may be determined

EXAMPLES.

1. What is the square root of 50 ?

Let each unit of the scale represent 10, then 50 will be represented by 5. The half of 5 is $2\frac{1}{2}$. An ordinate drawn from $2\frac{1}{2}$ on the axis of X will be about 2.24, and the square root of 10 will be represented by an ordinate drawn from 5, which will be about 3,16. Hence, the square root of 50 cannot differ much from $(2.24)(3.16) = 7,0786$.

ANOTHER SOLUTION.

$50=25\times2$, $\sqrt{50}=5\sqrt{2}$. From 1 on the axis of X draw an ordinate; it will measure $1.4+$.

Hence, $\qquad \sqrt{50}=5(1.4+)=7,+$.

What is the square root of 175?

$$175=25\times7, \sqrt{175}=5\sqrt{7}.$$

An ordinate drawn from 3.5 the half of 7 will measure 2.65. Therefore $\sqrt{175}=5(2.65)=13.25$ nearly.

3. Given $x^2-\frac{2}{11}x=8$ to find x.   Ans. $x=2.9.+$
4. Given $\frac{3}{4}x^2+\frac{1}{3}x=\frac{7}{11}$ to find x.   Ans. $x=0.60+$.
5. Given $\frac{1}{4}y^2-\frac{1}{6}y=2$ to find y.   Ans. $y=3.17$, or $-2.5+$.

CHAPTER V.

THE HYPERBOLA.

To describe an hyperbola.

The definition of this curve suggests the following method of describing it mechanically:

Take a ruler $F'H$, and fasten one end at the point F', on which the ruler may turn as a hinge. At the other end of the ruler attach a thread, and let its length be less than that of the ruler by the given line $A'A$. Fasten the other end of the thread at F.

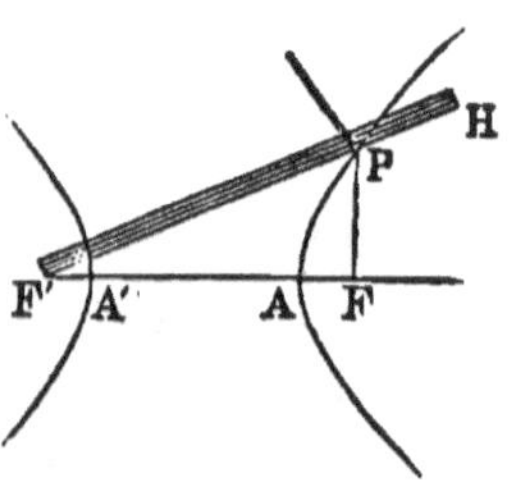

With a pencil, P, press the thread against the ruler and keep it at equal tension between the points H and F. Let the ruler turn on the point F', keeping the pencil close to the ruler and letting the thread slide round the pencil; the pencil will thus describe a curve on the paper.

If the ruler be changed and made to revolve about the other focus as a fixed point, the opposite branch of the curve can be described.

In all positions of P, except when at A or A', PF' and PF will be two sides of a triangle, and the difference of these two sides is constantly equal to the difference between the ruler and the thread; but that difference was made equal to the given line $A'A$; hence, by definition, the curve thus described must be an hyperbola.

PROPOSITION I.

To find the equation of the hyperbola referred to its center and axes.

Let C be the center, F' and F' the foci, and AA' the transverse axis of an hyperbola. Draw CC' at right angles to AA', and take these lines for the co-ordinate axes. From P,

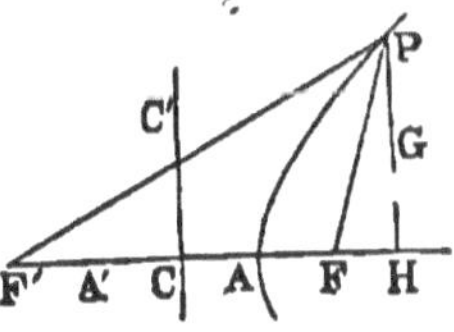

any point of the curve, draw PF, PF' to the foci, and PH perpendicular to AA'.

Make $CF=c$, $CA=A$, $CH=x$, and $PH=y$; then the equation which expresses the relation between the variables x and y, and the constances c and A, will be the equation of a hyperbola.

By the definition of the curve we have

$$r'-r=2A. \qquad (1)$$

The right-angled $\triangle PHF$ gives

$$r^2=(x-c)^2+y^2. \qquad (2)$$

The right-angled $\triangle PHF'$ gives

$$r'^2=(x+c)^2+y^2. \qquad (3)$$

Subtracting eq. (2) from eq. (3) we get

$$r'^2-r^2=4cx. \qquad (4)$$

Dividing eq. (4) by eq. (1) we have

$$r'+r=\frac{2cx}{A}. \qquad (5)$$

Combining eqs. (1) and (5) we find

$$r'=A+\frac{cx}{A}, \quad \text{and} \quad r=-A+\frac{cx}{A}.$$

This value of r substituted in eq. (2) gives

$$A^2-2cx+\frac{c^2x^2}{A^2}=x^2-2cx+c^2+y^2.$$

Reducing, we find

$$A^2y^2+(A^2-c^2)x^2=A^2(A^2-c^2),$$

for the equation sought.

Scholium.—As c is greater than A, it follows that (A^2-c^2) must be negative; therefore we may assume this value equal to $-B^2$. Then the equation becomes

$$A^2y^2-B^2x^2=-A^2B^2.$$

This form is preferred to the former one on account of its similarity to the equation of the ellipse, the difference being only in the negative value of B^2.

Because $A^2-c^2=-B^2$, $A^2+B^2=c^2$

Now to show the geometrical magnitude of B, take C as a center, and CF as a radius, and describe the circle FHF'. From A draw AH at right angles to CF. Now $CH=c$, $CA=A$, 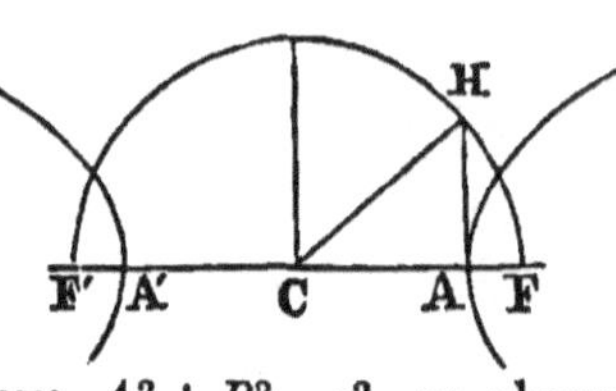 and if we put $AH=B$, we shall have $A^2+B^2=c^2$, as above. Whence AH must equal B.

PROPOSITION II.

To determine the figure of the hyperbola from its equation.

Resuming the equation

$$A^2y^2-B^2x^2=-A^2B^2,$$

and solving it in respect to y, we find

$$y=\pm\frac{B}{A}\sqrt{x^2-A^2}.$$

If we make $x=0$, or assign to it any value less than A, the corresponding value of y will be imaginary, showing that the curve does not exist above or below the line $A'A$.

If we make $x=A$, then $y=\pm0$, showing two points in the curve, both at A.

If we give to x any value greater 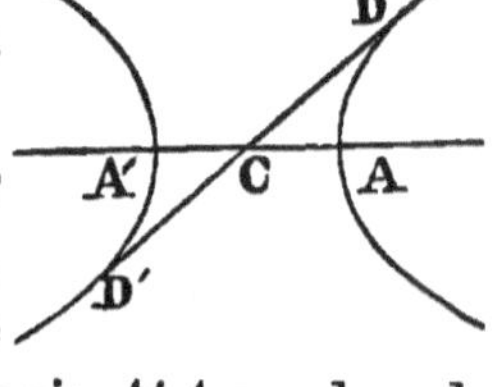 than A, we shall have two values of y, numerically equal, showing that the curve is symmetrically divided by the axis $A'A$ produced.

If we now assign the same value to x taken *negatively*, that is, make $x\ (-x)$, we shall have two other values of y, the same as before, corresponding to the left branch of the curve. Therefore, *the two branches of the curve are equal in magnitude, and are in all respects symmetrical but opposite in position.*

Hence every diameter, as DD′, *is bisected in the center, for any other hypothesis would be absurd.*

SCHOLIUM 1.—If through the center, C, we draw CD, CD', at right angles to $A'A$, and each equal to B, we can have two opposite branches of an hyperbola passing through D and D' above and below C. as the two others which pass through the points A' and A, at the right and left of C.

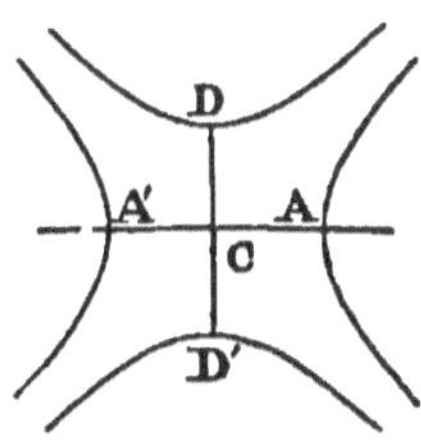

The hyperbola which passes through D and D' is said to be *conjugate* to that which passes through A and A', or the two *are conjugate to each other*.

DD' is the conjugate diameter to $A'A$, and DD' may be *less* than, *equal* to, or *greater* than $A'A$, according to the relative values of c and A in Prop. 1.

When B is numerically equal to A, the equation of the curve becomes

$$y^2 - x^2 = -A^2,$$

and $DD' = AA'$. In this case the hyperbola is said to be *equilateral*.

SCHOLIUM 2.—To find the value of the double ordinate which passes through the focus, we must take the equation of the curve

$$A^2 y^2 - B^2 x^2 = -A^2 B^2,$$

and make $x = c$, then

$$A^2 y^2 = B^2(c^2 - A^2).$$

But we have shown that $A^2 + B^2 = c^2$, or $B^2 = c^2 - A^2$.

Whence $$A^2 y^2 = B^4.$$

Or $$Ay = B^2, \text{ or } 2y = \frac{2B^2}{A}.$$

That is, $$2A : 2B :: 2B : 2y,$$

showing that *the parameter of the hyperbola is equal to the double ordinate, to the major axis, that passes through the focus.*

SCHOLIUM 3.—To find the equation for the conjugate hyperbola which passes through the points D, D', we take the general equation

$$A^2 y^2 - B^2 x^2 = -A^2 B^2,$$

and change A into B and x into y, the equation then becomes
$$B^2x^2-A^2y^2=-A^2B^2,$$
which is the equation for conjugate hyperbola.

PROPOSITION III.

To find the equation of the hyperbola when the origin is at the vertex of the transverse axis.

When the origin is at the center, the equation is
$$A^2y^2-B^2x^2=-A^2B^2.$$

And now, if we move the origin to the vertex at the right, we must put

$$x=A+x'.$$

Substituting this value of x in the equation of the hyperbola referred to its center and axes, we have
$$A^2y^2-B^2x'^2-2B^2Ax'=0.$$

We may now omit the accents, and put the equation under the following form :

$$y^2=\frac{B^2}{A^2}(x^2+2Ax),$$

which is the equation of the hyperbola when the origin is the vertex and the co-ordinates rectangular.

PROPOSITION IV.

To find the equation of a tangent line to the hyperbola, the origin being the center.

In the first place, conceive a line cutting the curve in two points, P and Q. Let x and y be co-ordinates of any point on the line, as S, x' and y' co-ordinates of the point P on the curve, and x'' and y'' the co-ordinates of the point Q on the curve.

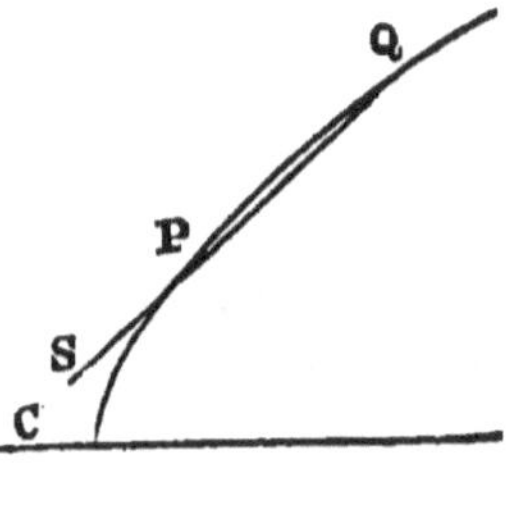

The student can now work through the proposition in precisely the same manner as Prop. 6, of the ellipse was worked, using the equation for the hyperbola in place of that of the ellipse, and in conclusion he will find

$$A^2 yy' - B^2 xx' = -A^2 B^2,$$

for the equation sought.

Cor. To find the point in which a tangent line cuts the axis of X, we must make $y=0$, in the equation for the tangent; then

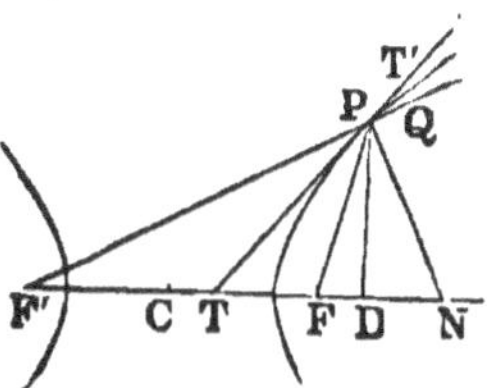

$$x = \frac{A^2}{x'} = CT.$$

If we subtract this from $CD\,(x')$ we shall have the sub-tangent

$$TD = x' - \frac{A^2}{x'} = \frac{x'^2 - A^2}{x'}$$

PROPOSITION V.

To find the equation of a normal to the hyperbola.

Let a be the tangent of the angle that the line TP makes with the transverse axis, (see last figure), and a' the same with reference to the line PN. Then if PN is a normal, it must be at right angles to PT, and hence we must have

$$aa' + 1 = 0. \qquad (1)$$

Let x' and y' be the cor-ordinates of the point P on the curve, and x, y, the co-ordinates of any point on the line PN, then we must have

$$y - y' = a'(x - x'). \qquad (2)$$

In working the last proposition, for the tangent line PT we should have found

$$a = \frac{B^2 x'}{A^2 y'}.$$

This value of a put in eq. (1) will show us that

$$a' = -\frac{A^2 y'}{B^2 x'}.$$

And this value of a' put in eq. (2) will give us

$$y-y'=-\frac{A^2y'}{B^2x'}(x-x'),$$

for the equation of the normal required.

Cor. To find the point in which the normal cuts the axis of X, we must make $y=0$.

This reduces the equation to

$$1=\frac{A^2}{B^2x'}(x-x').$$

Whence $x=\left(\frac{A^2+B^2}{A^2}\right)x'=CN.$

If we subtract CD, (x'), from CN, we shall have DN, the *sub-normal*.

That is, $\left(\frac{A^2+B^2}{A^2}\right)x'-x'=\frac{B^2x'}{A^2}$, the *sub-normal*.

PROPOSITION VI.

A tangent to the hyperbola bisects the angle contained by lines drawn from the point of contact to the foci.

If we can prove that

$$F'P:PF::F'T:TF, \quad (1)$$

it will then follow (Th. 24, B. II, Geom.,) that the angle $F'PT=$the angle TPF.

In Prop. 1, of the hyperbola, we find that

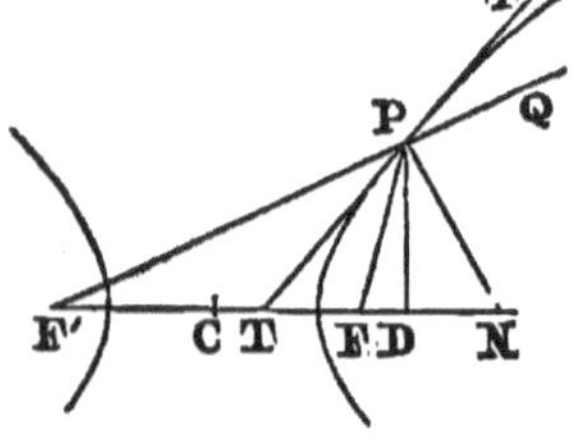

$$F'P=r'=A+\frac{cx}{A}, \text{ and } PF=r=-A+\frac{cx}{A};$$

and by corollary to Prop. 4

$$F'T=F'C+CT=c+\frac{A^2}{x}, \text{ and } TF=c-\frac{A^2}{x},$$

We will now assume the proportion

$$\left(A+\frac{cx}{A}\right):\left(-A+\frac{cx}{A}\right)::\left(c+\frac{A^2}{x}\right):z. \quad (2)$$

Multiply the terms of the first couplet by A, and those of the last couplet by x, then we shall have

$$(A^2+cx) : (-A^2+cx) :: (cx+A^2) : xz.$$

Observing that the first and third terms of this proportion are equal, therefore

$$xz = cx - A^2.$$

Or $\qquad\qquad z = c - \dfrac{A^2}{x} = TF.$

Now the first three terms of proportion (2) were taken equal to the first three terms of proportion (1), and we have proved that the fourth term of proportion (2) must be equal to the fourth term of proportion (1), therefore proportion (1) is true, and consequently

$$F'PT = TPF.$$

Cor. 1. As TT' is a tangent, and PN its normal. it follows that the angle $TPN=$ the angle $T'PN$, for each is a right angle. From these equals take away the equals TPF, $T'PQ$, and the remainder FPN must equal the remainder QPN. That is, *the normal line at any point of the hyperbola bisects the exterior angle formed by two lines drawn from the foci to that point.*

Cor. 2. The value of CT we have found to be $\dfrac{A^2}{x}$, and the value of CD is x, and it is obvious that

$$\frac{A^2}{x} : A :: A : x,$$

is a true proportion. *Therefore* (A) *is a mean proportional between* CT *and* CD.

A tangent line can never meet the axis in the center, because the above proportion must always exist, and to make the first term *zero in value*, we must suppose x to be *infinite. Therefore a tangent line passing through the center cannot meet the hyperbola short of an infinite distance therefrom.*

Such a line is called an *asymptote.*

OF THE CONJUGATE DIAMETERS OF THE HYPERBOLA.

DEFINITION.— *Two diameters of an hyperbola are said to be conjugate when each is parallel to a tangent line drawn through the vertex of the other.*

According to this definition, GG' and HH' in the adjoining figure are conjugate diameters.

EXPLANATION. 1.—The tangent line which passes through the point H is parallel to CG. Hence CG makes the same angle with the axis as that tangent line does.

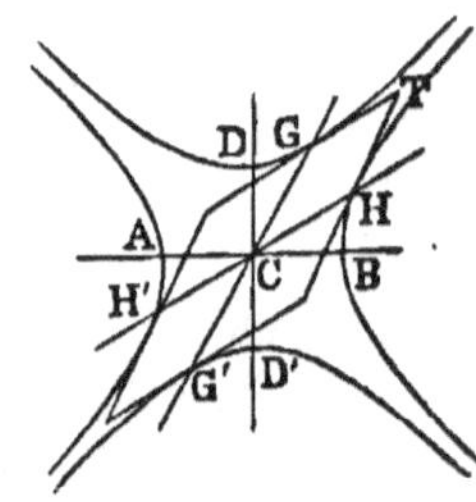

If we designate the co-ordinates of the point H, in reference to the center and axes by x' and y', and by a the tangent of the angle made by the inclination of CG with the axis, then in the investigation (Prop. 6,) we find

$$a = \frac{B^2 x'}{A^2 y'}. \tag{1}$$

Now if we designate the tangent of the angle which CH makes with the axis by a', the equation of CH must be of the form

$$y' = a'x',$$

because the line passes through the center.

Whence
$$a' = \frac{y'}{x'}. \tag{2}$$

Multiplying eqs. (1) and (2) together member by member, and we find

$$aa' = \frac{B^2}{A^2},$$

to which equation *all conjugate diameters must correspond.*

EXPLANATION 2.—If we designate the angle GCB by n, and HCB by m, we shall have

$$\frac{\sin. m}{\cos. m} = a', \quad \frac{\sin. n}{\cos. n} = a.$$

And
$$\tan. m \tan. n = \frac{B^2}{A^2}.$$

PROPOSITION VII.

To find the equation of the hyperbola referred to its center and conjugate diameters.

The equation of the curve referred to the center and axes is

$$A^2y^2 - B^2x^2 = -A^2B^2.$$

Now, to change rectangular co-ordinates into oblique, the origin being the same, we must put

$$\left. \begin{array}{l} x = x' \cos. m + y' \cos. n \\ y = x' \sin. m + y' \sin. n \end{array} \right\} \text{ Chap. 1, Prop. 9.}$$

And

These values of x and y, substituted in the above general equation, will produce

$$\left\{ \begin{array}{l} (A^2 \sin.^2 n - B^2 \cos.^2 n)y'^2 + (A^2 \sin.^2 m - B^2 \cos.^2 m)x'^2 \\ \quad + 2(\sin. m \sin. nA^2 - \cos. m \cos. nB^2)x'y' \end{array} \right\}$$
$$= -A^2B^2. \tag{1}$$

Because the diameters are conjugate, we must have

$$\frac{\sin. m}{\cos. m} \cdot \frac{\sin. n}{\cos. n} = \frac{B^2}{A^2}.$$

Whence $(\sin. m \sin. n\, A^2 - \cos. m \cos. n\, B^2) = 0$ (k)

This last equation reduces eq. (1) to

$$(A^2\sin.^2 n - B^2\cos.^2 n)y'^2 + (A^2\sin.^2 m - B^2\cos.^2 m)x'^2 = -A^2B^2 \tag{2}$$

which is the equation of the hyperbola referred to the center and conjugate diameters.

If we make $y' = 0$, we shall have

$$x'^2 = \frac{-A^2B^2}{(A^2\sin.^2 m - B^2\cos.^2 m)} = \overline{CH}^2 \tag{3}$$

If we make $x' = 0$, we shall have

$$y'^2 = \frac{A^2B^2}{(A^2\sin.^2 n - B^2\cos.^2 n)} = \overline{CG}^2 \tag{4}$$

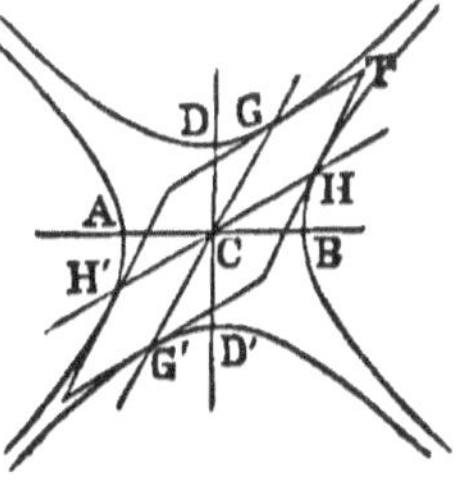

If we put A'^2 to represent $\overline{CH}^2$, and regard it as *positive*, the denominator in eq. (3) must be negative, the nu-

17*

merator being negative.   That is, $A^2\sin^2 m$ must be less than $B^2\cos^2 m$.

That is,
$$A^2\sin^2 m < B^2\cos^2 m.$$
$$\tan. m < \frac{B}{A}.$$

But
$$\tan. m \tan. n = \frac{B^2}{A^2}.$$

Whence
$$\tan. n > \frac{B}{A}, \text{ or, } A^2\sin^2 n > B^2\cos^2 n.$$

Therefore the denominator in eq. (4) is positive, but the numerator being negative, therefore $\overline{CG}^2$ must be negative.   Put it equal to $-B'^2$.

Now the equations (3) and (4) become

$$A'^2 = \frac{-A^2 B^2}{(A^2\sin^2 m - B^2\cos^2 m)}, \quad -B'^2 = \frac{A^2 B^2}{(A^2\sin^2 n - B^2\cos^2 n)},$$

Or
$$(A^2\sin^2 m - B^2\cos^2 m) = \frac{-A^2 B^2}{A'^2},$$

$$(A^2\sin^2 n - B^2\cos^2 n) = \frac{A^2 B^2}{B'^2}.$$

Comparing these equations with eq. (2) we perceive that eq. (2) may be written thus :

$$\frac{A^2 B^2}{B'^2} y'^2 - \frac{A^2 B^2}{A'^2} x'^2 = -A^2 B^2.$$

Whence
$$A'^2 y'^2 - B'^2 x'^2 = -A'^2 B'^2.$$

Omitting the accents of x' and y', since they are general variables, we have

$$A'^2 y^2 - B'^2 x^2 = -A'^2 B'^2,$$

for the equation of the hyperbola referred to its center and *conjugate diameters*.

Scholium 1.—As this equation is precisely similar to that referred to the center and axes, it follows that all results hitherto determined in respect to the latter will apply to conjugate diameters by changing A to A' and B to B',

For instance, the equation for a tangent line in respect to the center and axes has been found to be

$$A^2 yy' - B^2 xx' = -A^2 B^2.$$

Therefore, in respect to conjugate diameters it must be

$$A'^2 yy' - B'^2 xx' = -A'^2 B'^2,$$

and so on for normals, sub-normals, tangents and sub-tangents.

SCHOLIUM 2.—If we take the equation

$$A'^2 y^2 - B'^2 x^2 = -A'^2 B'^2,$$

and resolve it in relation to y, we shall find that for every value of x greater than A' we shall find two values of y numerically equal, which shows that ON bisects MM and every line drawn parallel to MM, or parallel to a tangent drawn through L, the vertex of the diameter LL'.

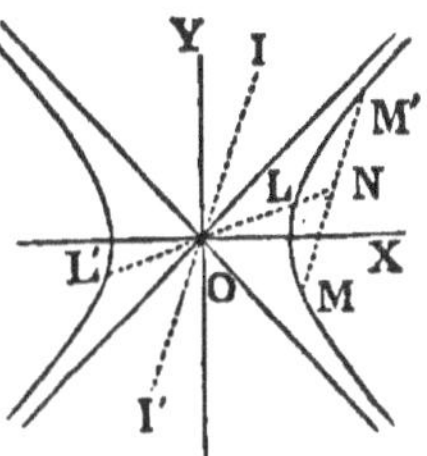

Let the student observe that these several geometrical truths were discovered by changing rectangular to oblique co-ordinates. We will now take the reverse operation, in the hope of discovering other geometrical truths.

Hence the following :

PROPOSITION VIII.

To change the equation of the hyperbola in reference to any system of conjugate diameters, to its equation in reference to the axes.

The equation of the hyperbola referred to conjugate diameters is

$$A'^2 y'^2 - B'^2 x'^2 = -A'^2 B'^2.$$

To change oblique to rectangular co-ordinates, the formulas are (Chap. 1, Prop. 10,)

$$x' = \frac{x \sin. n - y \cos. n}{\sin. (n-m)}, \qquad y' = \frac{y \cos. m - x \sin. m}{\sin. (n-m)}.$$

Substituting these values of x' and y' in the equation, we shall have

$$\frac{A'^2 (y \cos. m - x \sin. m)^2}{\sin.^2 (n-m)} - \frac{B'^2 (x \sin. n - y \cos. n)^2}{\sin.^2 (n-m)} = -A'^2 B'^2.$$

By expanding and reducing, we shall have

$$\left\{ \begin{array}{c} (A'^2\cos.^2m - B'^2\cos.^2n)y^2 + (A'^2\sin.^2m - B'^2\sin.^2n)x^2 \\ 2(-A'^2\sin.m\cos.m + B'^2\sin.n\cos.n)xy \end{array} \right\}$$
$$= -A'^2B'^2\sin.^2(n-m).$$

which, to be the equation of the hyperbola when referred to the center and axes, must take the well known form,

$$A^2y^2 - B^2x^2 = -A^2B^2.$$

Or in other words, these two equations must be, in fact, identical, and we shall therefore have

$$A'^2\cos.^2m - B'^2\cos.^2n = A^2. \qquad (1)$$
$$A'^2\sin.^2m - B'^2\sin.^2n = -B^2. \qquad (2)$$
$$-A'^2\sin.m\cos.m + B'^2\sin.n\cos.n = 0. \qquad (3)$$
$$-A'^2B'^2\sin.^2(n-m) = -A^2B^2. \qquad (4)$$

By adding eqs. (1) and (2), observing that $(\cos.^2m + \sin.^2m) = 1$, we shall have

$$A'^2 - B'^2 = A^2 - B^2.$$

Or $\qquad\qquad 4A'^2 - 4B'^2 = 4A^2 - 4B^2,$

which equation shows this general *geometrical truth:*

That the difference of the squares of any two conjugate diameters is equal to the difference of the squares of the axes.

Hence, there can be no equal conjugate diameters unless $A = B$, and then every diameter *will be equal to its conjugate:* that is, $A' = B'$.

Equation (3) corresponds to $\tan.m\tan.n = \dfrac{B^2}{A^2},$ the equation of condition for conjugate diameters.

Equation (4) reduces to

$$A'B'\sin.(n-m) = AB.$$

The first member is the measure of the parallelogram $GCHT$, and it being equal to $A \times B$, shows this geometrical truth :

That the parallelogram formed by drawing tangent lines through the vertices of any system of conjugate diameters of the hyperbola, is equivalent to the rectangle formed by drawing tangent lines through the vertices of the axes.

REMARK.—The reader should observe that this proposition is similar to (Prop. 13,) of the ellipse, and the general equation here found, and the incidental equations (1), (2), (3), and (4), might have been directly deduced from the ellipse by changing B into $B\sqrt{-1}$, and B' into $B'\sqrt{-1}$.

OF THE ASYMPTOTES OF THE HYPERBOLA.

DEFINITION.—If tangent lines be drawn through the vertices of the axes of a system of conjugate hyperbolas, the diagonals of the rectangle so formed, produced indefinitely, are called *asymptotes* of the hyperbolas.

Let AA', BB', be the axes of conjugate hyperbolas, and through the vertices A, A', B, B', let tangents to the curves be drawn forming the rectangle, as seen in the figure. The diagonals of this rectangle produced, that is, DD' and EE', are the *asymptotes* to the curves corresponding to the definition.

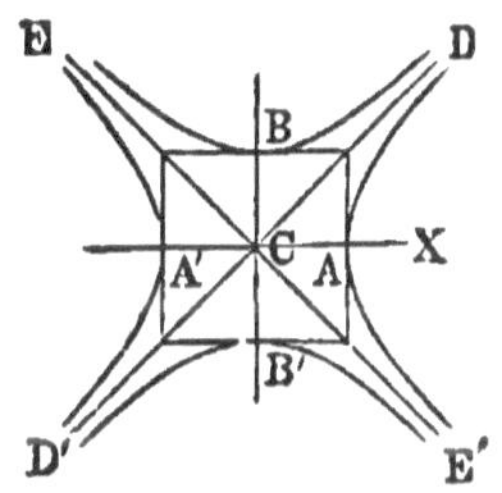

If we represent the angle DCX by m, $E'CX$ will be m also, for these two angles are equal because $CB = CB'$.

It is obvious that

$$\tan. m = \frac{B}{A}.$$

Whence

$$\frac{\sin.^2 m}{\cos^2. m} = \frac{B^2}{A^2}$$

But $\cos.^2 m = 1 - \sin.^2 m$. Therefore

$$\frac{\sin.^2 m}{1 - \sin.^2 m} = \frac{B^2}{A^2}.$$

Consequently $\sin.^2 m=\dfrac{B^2}{A^2+B^2}$, and $\cos.^2 m=\dfrac{A^2}{A^2+B^2}$, which equations furnish the value of the angle which the asymptotes form with the transverse axis.

PROPOSITION IX.

To find the equation of the hyperbola, referred to its center and asymptotes.

Let $CM=x$, and $PM=y$. Then the equation of the curve referred to its center and axes is

$$A^2y^2-B^2x^2=-A^2B^2. \qquad (1)$$

From P draw PH parallel to CE, and PQ parallel to CM. Let $CH=x'$, and $HP=y'$.

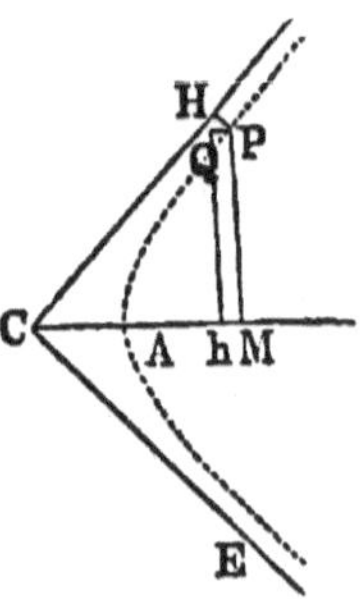

Now the object of this proposition is to find the values of x and y in terms of x' and y', to substitute them in eq. (1). The resulting equation reduced to its most simple form will be the equation sought.

The angle HCM is designated by m, and because HP is parallel to CE, and PQ parallel to CM, the angle HPQ is also equal to m.

Now in the right angled triangle CHh we have $Hh=x'$ sin. m, and $Ch=x'$ cos. m.

In the right angled triangle PQH we have $HQ=y'$ sin. m, and $PQ=y'$ cos. m.

Whence $Hh-HQ=Qh=PM=y=x'$ sin. $m-y'$ sin. m.

Or $\qquad\qquad y=(x'-y')$ sin. m. $\qquad\qquad$ (2)

$\qquad Ch+QP=CM=x=x'$ cos. $m+y'$ cos. m.

Or $\qquad\qquad x=(x'+y')$ cos. m. $\qquad\qquad$ (3)

These values of y and x found in eqs. (2) and (3) substituted in eq. (1) will give

$$A^2(x'-y')^2 \sin.^2 m - B^2 (x'+y')^2 \cos.^2 m = -A^2 B^2.$$

Placing in this equation the values of $\sin.^2 m$ and $\cos.^2 m$, previously determined, we have

$$\frac{A^2 B^2}{A^2+B^2}(x'-y')^2 - \frac{A^2 B^2}{A^2+B^2}(x'+y')^2 = -A^2 B^2.$$

Dividing through by $A^2 B^2$, and at the same time multiplying by (A^2+B^2), we get

$$(x'-y')^2-(x'+y')^2=-(A^2+B^2).$$

Or
$$-4x'y'=-(A^2+B^2).$$

Or
$$x'y'=\frac{A^2+B^2}{4},$$

which is the equation of the hyperbola referred to its center and asymptotes.

Cor. As x' and y' are general variables, we may omit the accents, and as the second member is a constant quantity, we may represent it by M^2. Then

$$xy=M^2, \text{ or } x=\frac{M^2}{y}.$$

This last equation shows that x increases as y decreases; that is, *the curve approaches nearer and nearer the asymptote as the distance from the center becomes greater and greater.*

But x can never become infinite until y becomes 0; that is, *the asymptote meets the curve at an infinite distance,* corresponding to Cor. 2, Prop. 6.

PROPOSITION X.

All parallelograms constructed upon the abscissas, and ordinates of the hyperbola referred to its asymptotes are equivalent, each to each, and each equivalent to $\frac{1}{2}$AB.

Let x and y be the co-ordinates corresponding to any point in the curve, as P. Then by the equation of the curve in relation to the center and asymptotes, we have

$$xy=M^2. \tag{1}$$

Also let x', y', represent the co-ordinates of the point Q. Then

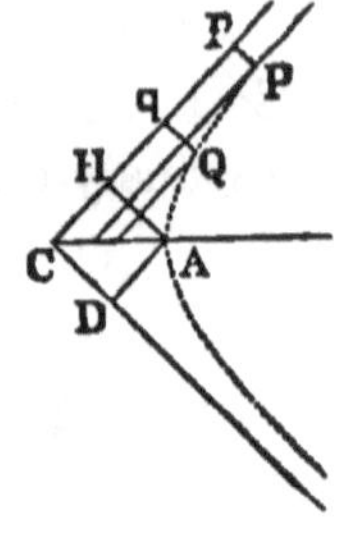

$$x'y' = M^2. \qquad (2)$$

The angle pCD between the asymptotes we will represent by $2m$. Now multiply both members of equations (1) and (2) by sin. $2m$.

Then we shall have

$$xy \text{ sin. } 2m = M^2 \text{ sin. } 2m. \qquad (3)$$
$$x'y' \text{ sin. } 2m = M^2 \text{ sin. } 2m. \qquad (4)$$

The first member of eq. (3) represents the parallelogram CP, and the first member of eq. (4) represents the parallelogram CQ; and as each of these parallelograms is equivalent to the same constant quantity, *they are equivalent to each other.*

Now A is another point in the curve, and therefore the parallelogram $AHCD$ is equal to (M^2 sin. $2m$), and therefore equal to CQ, or CP. Hence all parallelograms bounded by the asymptotes and terminating in a point in the curve, are equivalent to one another, and each equivalent to the parallelogram $AHCD$, which has for one of its diagonals half of the transverse axis of A.

We have now to find the analytical expression for this parallelogram.

The angle $HCA = m$, $ACD = m$, and because AH is parallel to CD, $CAH = m$. Hence, the triangle CAH is isosceles, and $CH = HA$. The angle $AHq = 2m$. Now by trigonometry

$$\text{sin. } 2m : A :: \text{sin. } m : CH.$$

But sin. $2m = 2$ sin. m cos. m. Whence

$$2 \text{ sin. } m \text{ cos. } m : A :: \text{sin. } m : CH.$$

$$CH = \frac{A}{2 \cos. m}.$$

Multiply each member of this equation by $CA = A$, and sin. m, then

$$A.(CH)\sin. m = \frac{A^2}{2}\frac{\sin. m}{\cos. m} = \frac{A^2}{2}\tan. m.$$

The first member of this equation represents the area of the parallelogram $CHAD$, and the $\tan. m = \dfrac{B}{A}.$ Hence, the parallelogram is equal $\dfrac{A^2}{2}\cdot\dfrac{B}{A} = \tfrac{1}{2}AB,$ which is the value also of all the other parallelograms, as CQ, CP, etc.

PROPOSITION XI.

To find the equation of a tangent line to the hyperbola referred to its center and asymptotes.

Let P and Q be any two points on the curve, and denote the co-ordinates of the first by x', y', and of the second by x'', y''.

The equation of a straight line passing through these points will be of the form

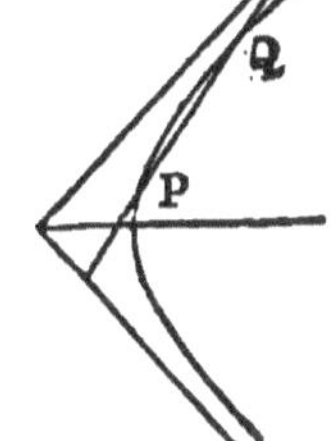

$$y - y' = a(x - x') \qquad (1)$$

in which $a = \dfrac{y' - y''}{x' - x''}.$

We are now to find the value of a when the line becomes a tangent at the point P.

Because P and Q are points in the curve, we have

$$x'y' = x''y''.$$

From each member of this last equation subtract $x'y''$, then

$$x'y' - x'y'' = x''y'' - x'y''.$$

Or $\qquad\qquad x'(y' - y'') = -y''(x' - x'').$

Whence $\qquad a = \dfrac{y' - y''}{x' - x''} = -\dfrac{y''}{x'}.$

This value of a put in eq. (1) gives

$$y - y' = -\frac{y''}{r'}(x - x'). \qquad (2)$$

Now if we suppose the line to revolve on the point P as a center until Q coincides with P, then the line will be a tangent, and $x'=x''$, and $y'=y''$, and eq. (2) will become

$$y-y'=-\frac{y'}{x'}(x-x'),$$

which is the equation sought.

Cor. To find the point in which the tangent line meets the axis of X, we must make $y=0$; then

$$x=2x'.$$

That is, Ct is twice CR, and as RP and CT are parallel, $tP=PT$.

A tangent line included between the asymptotes is bisected by the point of tangency.

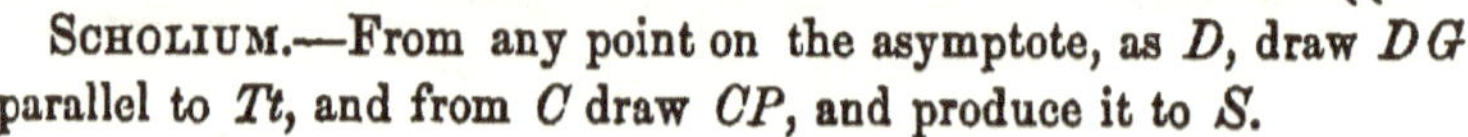

SCHOLIUM.—From any point on the asymptote, as D, draw DG parallel to Tt, and from C draw CP, and produce it to S.

By scholium 2 to Prop. 7 we learn that CP produced will bisect all lines parallel to tT and within the curve; hence gd is bisected in S.

But as CP bisects tT, it bisects all lines parallel to tT within the asymptotes, and DG is also bisected in S; hence $dD=Gg$.

In the same manner we might prove $dh=kv$, because hk is parallel to some tangent which might be drawn to the curve, the same as DG is parallel to the *particular* tangent tT.

Hence, *If any line be drawn cutting the hyperbola, the parts between the asymptotes and the curve are equal.*

This property enables us to describe the hyperbola by points, when the asymptotes and one point in the curve are given.

Through the given point d, draw any line, as DG, and from G set off $Gg=dD$, and then g will be a point in the curve. Draw any other line, as hk, and set off $kv=dh$; then v is another point in the curve. And thus we might find other points between v and g, or on either side of v and g.

PROPOSITION XII.

To find the polar equation of the hyperbola, the pole being at either focus.

Take any point P in the hyperbola, and let its distance from the nearest focus be represented by r, and its distance from the other focus be represented by r'.

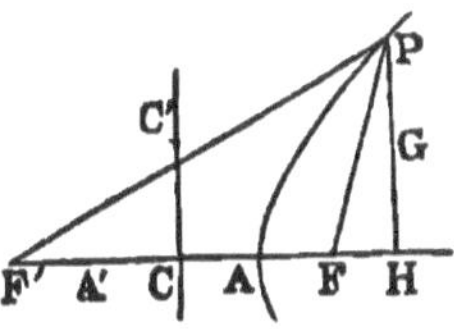

Put $CH=x$, $CF=c$, and $CA=A$. Then, by Prop. 1, we have

$$r=-A+\frac{cx}{A}, \qquad (1)$$

$$r'=A+\frac{cx}{A}. \qquad (2)$$

Now the problem requires us to replace the symbol x, in these formulas, by its value, expressed in terms of r and r', and some function of the angle that these lines make with the transverse axis.

First.—In the right-angled triangle PFH, if we designate the angle PFH by v, we shall have

$$1 : r :: \cos. v : FH = r \cos. v.$$

$CH = CF + FH$. That is, $x = c + r \cos. v.$

The value of x put in eq. (1) gives

$$r = -A + \frac{c^2 + cr \cos. v}{A}.$$

Whence $\qquad r = \dfrac{c^2 - A^2}{A - c \cos. v}. \qquad (3)$

Second.—In the right-angled triangle $F'PH$, if we designate the angle $PF'H$ by v', we shall have

$$1 : r' :: \cos. v' : F'H = r' \cos. v'.$$

But $F'H = F'C + CH$. That is, $r' \cos. v' = c + x.$

Or $\qquad x = r' \cos. v' - c,$

and this value of x put in eq. (2) gives

$$r'=A+\frac{cr'\cos.v'-c^2}{A}.$$

Whence
$$r'=\frac{A^2-c^2}{A-c\cos.v'}.\qquad(4)$$

Equations (3) and (4) are the polar equations required.

Let us examine eq. (3). Suppose $v=0$, then $\cos.v=1$, and

$$r=\frac{c^2-A^2}{A-c}=-A-c.$$

But a radius vector can never be a *minus* quantity, therefore there is no portion of the curve on the axis to the right of F.

To find the length of r when it first strikes the curve, we find the value of the denominator when its value first becomes positive, which must be when A becomes equal to $c\cos.v$; that is, when the denominator is 0. the value of r will be real and infinite.

If
$$A-c\cos.v=0,$$

then
$$\cos.v=\frac{A}{c}.$$

This equation shows that when r first meets the curve it is parallel to the asymptote, and infinite.

When $v=90°$, $\cos.v=0$, and then r is perpendicular at the point F, and equal to $\frac{c^2-A^2}{A}$, or $\frac{B^2}{A}$, half the parameter of the curve, as it ought to be.

When $v=180°$, then $\cos.v=-1$, and $-c\cos.v=c$; then

$$r=\frac{c^2-A^2}{c+A}=c-A=FA,$$

a result obviously true.

As v increases, the value of r will remain positive, and, consequently, give points of the hyperbola until $\cos.v$ again becomes equal to $\frac{A}{c}$, which will be when the radius vector makes with the transverse axis an angle equal to $360°$ minus that whose cosine is $\dfrac{A}{c}$. Equation (3) will therefore determine all points in the right hand branch of the hyperbola.

Now let us examine equation (4). If we make $v'=0$, then

$$r'=\frac{A^2-c^2}{A-c}=A+c=F'A,$$

as it ought to be.

To find when r' will have the greatest possible value, we must put

$$A-c\cos. v'=0.$$

Whence
$$\cos. v'=\frac{A}{c}.$$

This shows that v' is then of such a value as to make r' parallel to the *asymptote* and infinite in length. If we increase the value of v' from this point, the denominator will become positive, while the numerator is negative, which shows that then r' will become negative, indicating that it will not meet the curve.

The value of r will continue negative until the radius vector falls below the transverse axis, and makes with it an angle having $+\dfrac{A}{c}$ for its cosine. Values of v between this and $360°$ will render r positive and give points of the hyperbola. Equation (4) will, therefore, also determine all the points in the right hand branch of the hyperbola.

By changing the sign of c, we change the pole to the focus F', and eqs. (3) and (4), which then determine the left hand branch of the hyperbola, become

$$r=\frac{c^2-A^2}{A+c\cos. v}, \qquad (3')$$

and
$$r'=\frac{A^2-c^2}{A+c\cos. v'}. \qquad (4')$$

10*

GENERAL REMARKS.—When the origin of co-ordinates is at the circumference of a circle, its equation is

$$y^2 = 2Rx - x^2.$$

When the origin of a parabola is at its vertex, its equation is

$$y^2 = 2px.$$

When the origin of co-ordinates of the ellipse is at the vertex of the major axis, the equation of the curve is

$$y^2 = \frac{B^2}{A^2}(2Ax - x^2).$$

When the origin of co-ordinates is at the vertex of the hyperbola, the equation for that curve is

$$y^2 = \frac{B^2}{A^2}(2Ax + x^2).$$

But all of these are comprised in the general equation

$$y^2 = 2px + qx^2.$$

In the circle and the ellipse, q is negative; in the hyperbola it is positive, and in the parabola it is 0.

CHAPTER VI.

ON THE GEOMETRICAL REPRESENTATION OF EQUATIONS OF THE SECOND DEGREE BETWEEN TWO VARIABLES.

1.—It has been shown in Chap. 1, that every equation of the first degree between two variables may be represented by a straight line.

It has also been shown that the equations of the circle, the ellipse, the parabola and the hyperbola were all some of the different forms of an equation of the second degree between two variables. It is now proposed to prove that, when an equation of the second degree between two variables represents any geometrical magnitude, it is some one of these curves.

The limits assigned to this work compel us to be as brief in this investigation as is consistent with clearness. We shall, therefore, restrict ourselves to a demonstration of this proposition; the determination of the criteria by which it may be decided in every case presented, to which of the conic sections the curve represented by the equation belongs, and the indication of the processes by which the curve may be constructed.

2.—The equation of the second degree between two variables, in its most general form, is

$$Ay^2 + Bxy + Cx^2 + Dy + Ex + F = 0,$$

for, by giving suitable values to the arbitrary constants, A, B, C, etc., every particular case of such equation may be deduced from it.

The formulas for the transformation of co-ordinates being of the first degree in respect to the variables, the degree of an equation will not be changed by changing the reference of the equation from one system of co-ordinate axes to another. We may therefore assume that our co-ordinate axes are rectangular without impairing the generality of our investigation.

The resolution, in respect to y, of the general equation gives

$$y = -\frac{B}{2A}x - \frac{D}{2A} \pm \frac{1}{2A}\sqrt{\begin{array}{c}B^2 \\ -4AC\end{array}\Big|x^2 + \begin{array}{c}2BD \\ -4AE\end{array}\Big|x + \begin{array}{c}D^2 \\ -4AF\end{array}}$$

Now let AX, AY be the co-ordinate axes, and draw the straight line MQ, whose equation is

$$y = -\frac{B}{2A}x - \frac{D}{2A}.$$

For any value, AD, of x, the ordinate, DC, of this line, is expressed by

$$-\frac{B}{2A}x - \frac{D}{2A},$$

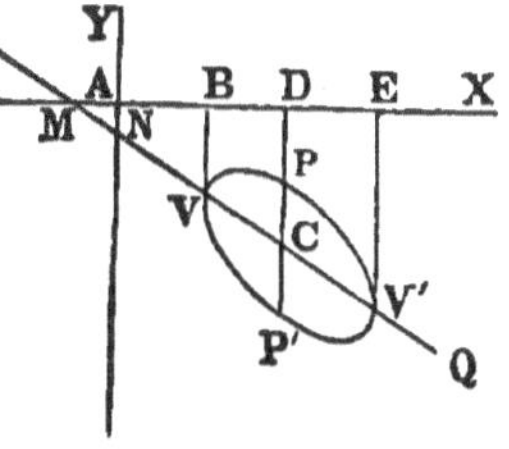

and this ordinate, increased and diminished successively by what the radical part, when real, of the general value of y becomes for the same substitution for x, will give two ordinates, DP, DP', corresponding to the abscissa AD.

Since P and P' are two points whose co-ordinates, when substituted for x and y, will satisfy the equation, $Ay^2 + Bxy + Cx^2 +$, etc., $= 0$, they are points in the line that this equation represents. By thus constructing the values of y answering to assumed values of x, we may determine any number of points in the curve.

In getting the points P and P', we laid off, on a parallel to the axis of y, equal distances above and below the point C; PP' is, therefore, a chord of the curve parallel to that axis, and is bisected at the point C.

The solution of the general equation in respect to x, gives

$$x = -\frac{B}{2C}y - \frac{E}{2C} \pm \frac{1}{2C}\sqrt{\left|\frac{B^2}{-4AC}\right|y^2 + \left|\frac{2BE}{-4CD}\right|y + \left|\frac{E^2}{-4CF}\right|}$$

The equation

$$x = -\frac{B}{2C}y - \frac{E}{2C},$$

is that of a straight line, making, with the axis of y, an angle whose tangent is $-\dfrac{B}{2C}$, and intersecting the axis of X at a distance from the origin equal to $-\dfrac{E}{2C}$.

As above, it may be shown that any value of y that makes the radical part of the general value of x real, responds to two points of the curve, and that the straight line whose equation is

$$x = -\frac{B}{2C}y - \frac{E}{2C},$$

bisects the chord, parallel to the axis of X, that joins these points.

By placing the quantity under the radical sign in the value of y equal to 0, we have an equation of the second degree in respect to x, which will give two values for x.

If these values are real the corresponding points of the curve are on the line MQ; that is, they are the intersections of this line with the curve, since, for each of these values, there will be but one value of y, which, in connection with that of x, will satisfy the general equation, and these values also satisfy the equation,

$$y=-\frac{B}{2A}x-\frac{D}{2A}.$$

In like manner, placing the quantity under the radical sign in the value of x equal to 0, we shall find two values of y, to each of which there will respond a single value of x, and the points of the curve answering to these values of y will be the intersections of the curve with the line whose equation is

$$x=-\frac{B}{2C}y-\frac{E}{2C}$$

A diameter of a curve is defined to be any straight line that bisects a system of parallel chords of the curve. From the preceding discussion we therefore conclude,

1. *That if an equation of the second degree between two variables be resolved in respect to either variable, the equation that results from placing this variable equal to that part of its value which is independent of the radical sign will be the equation of that diameter of the curve which bisects the system of chords parallel to the axis of the variable.*

2. *The values of the other variable found from the equation which results from placing the quantity under the radical sign equal to zero, in connection with the corresponding values of the first variable, will be the co-ordinates of the vertices of the diameter.*

3. The formulas for changing the reference of points from a system of rectangular co-ordinate axes to any other system having a different origin are

$$x=a+x'\cos. m+y'\cos. n.$$
$$y=b+x'\sin. m+y'\sin. n.$$

Substituting these values of x and y in the equation
$$Ay^2 + Bxy + Cx^2 + Dy + Ex + F = 0$$
developing, and arranging the terms of the resulting equation with reference to the powers of y' and x' and their product, we find

$$\left. \begin{array}{l} (A \sin^2 n + B \sin. n \cos. n + C \cos^2 n)\, y'^2 \\ +(A \sin^2 m + B \sin. m \cos. m + C \cos^2 m)\, x'^2 \\ +[2A \sin. m \sin. n + B (\sin. m \cos. n \\ \qquad +\sin n \cos. m) + 2C \cos. m \cos. n]\, x'\, y' \\ +[(2Ab + Ba + D) \sin. n + (2Ca + Bb + E) \\ \qquad \cos. n]y' \\ +[2Ab + Ba + D) \sin. m + (2Ca + Bb + E) \\ \qquad \cos. m]x' \\ +Ab^2 + Bab + Ca^2 + Db + Ea + F. \end{array} \right\} = 0 \quad (1)$$

Since we have four arbitrary quantities, a, b, m, and n entering this equation we may cause them to satisfy any four reasonable conditions. Let us see if, by means of them, it be possible to reduce the coefficients of the first powers, and of the product of the variables, separately to zero.

We should then have

$$\left. \begin{array}{l} 2A \sin. m \sin. n + B (\sin. m \cos. n + \sin. n \\ \qquad \cos. m) + 2C \cos. m \cos. n. \end{array} \right\} = 0 \quad (2)$$

$$(2Ab + Ba + D) \sin. n + (2Ca + Bb + E) \cos. n = 0 \quad (3)$$
$$(2Ab + Ba + D) \sin. m + (2Ca + Bb + E) \cos. m = 0 \quad (4)$$

These equations may be put under the form
$$2A \tan. m \tan. n + B (\tan. m + \tan. n) + 2C = 0 \quad (2')$$
$$(2Ab + Ba + D) \tan. n + 2Ca + Bb + E = 0 \quad (3')$$
$$(2Ab + Ba + D) \tan. m + 2Ca + Bb + E = 0 \quad (4')$$

Now, since it is necessary that m and n should differ in value, it is evident that, in order to satisfy eqs. (3') and (4'), we must have

$$2Ab + Ba + D = 0 \quad (5)$$

And $\qquad 2Ca + Bb + E = 0 \quad (6)$

Whence

$$a=\frac{2AE-BD}{B^2-4AC}$$

And

$$b=\frac{2CD-BE}{B^2-4AC}$$

These values of a and b are infinite when $B^2-4AC=0$, and it will then be impossible to satisfy both eqs. (3′) and (4′), and consequently to destroy the co-efficients of the first powers of the two variables in eq. (1); we shall, for the present, assume that B^2-4AC is either greater or less than zero.

By transposition and division eqs. (5) and (6) become

$$b=-\frac{B}{2A}a-\frac{D}{2A}$$

And

$$a=-\frac{B}{2C}b-\frac{E}{2C}$$

the first of which, if a and b be regarded as variables, is the equation of the diameter that bisects the chords of the curve which are parallel to the axis of y, and the second, that of the diameter which bisects the chords which are parallel to the axis of X. The values of a and b, given above, are, therefore, the co-ordinates of the intersection of these diameters.

Since eq. (2′) contains both of the undetermined quantities, m and n, we are at liberty to assume the value of either, and the equation will then give the value of the other. Let us take for the new axis of X the diameter whose equation is

$$y=-\frac{B}{2A}x-\frac{D}{2A}$$

then tan. $m=-\dfrac{B}{2A}$. This value of tan. m substituted in eq. (2′) gives

$$2A(B-B)\ \text{tan.}\ n=B^2-4AC,$$

Or

$$\text{tan.}\ n=\frac{B^2-4AC}{0}=\infty$$

That is, the new axis of y is at right angles to the primitive axis of X.

The values of a, b, and tan. n which we have thus found, in connection with the assumed value of tan. m, will reduce the co-efficients of the first powers and of the product of the variables in eq. (1) to zero.

To find what the co-efficients of y'^2 and x'^2 become, we must first get the values of the sines and cosines of the angles m and n from the values of tan. m and tan. n.

Since tan. $m = -\dfrac{B}{2A}$, and $n = 90°$ we have

$$\text{sin. } m = \pm \frac{B}{\sqrt{4A^2 + B^2}} \quad \text{cos. } m = \mp \frac{2A}{\sqrt{4A^2 + B^2}}$$

$$\text{sin. } n = 1 \quad \text{cos. } n = 0.$$

The sign $\pm$ is written before the value of sin. m, and the sign $\mp$ before that of cos. m, because if the essential sign of tan. m is minus, which will be the case when A and B have the same sign, sin. m and cos. m must have opposite signs; but if the essential sign of tan. m is plus, then A and B have opposite signs, and sin. m and cos. m must have like signs.

Making these substitutions in eq. (1) it will become, whether the signs of A and B are like or unlike,

$$Ay'^2 - A\left(\frac{B^2 - 4AC}{4A^2 + B^2}\right) x'^2 = - (Ab^2 + Bab + Ca^2 + Db + Ea + F. \tag{1'}$$

Now, since the first term of the general equation may always be supposed positive, the two terms in the first member of equation (1') will have like signs when $B^2 - 4AC < 0$, and unlike signs when $B^2 - 4AC > 0$. In the first case the form of the equation is that of the equation of the ellipse, and in the second, the form is that of the equation of the hyperbola, referred in either case, to the center and conjugate diameters.

Hence, when the transformation by which eq. (1′) was derived from the general equation

$$Ay^2 + Bxy + Cx^2 + Dy + Ex + F = 0$$

is possible, we conclude that the latter equation will represent either the ellipse, or hyperbola, according as

$$B^2 - 4AC < 0, \text{ or } B^2 - 4AC > 0.$$

4.—Let us now examine the case in which

$$B^2 - 4AC = 0.$$

Since, under this hypothesis, the co-efficients of the first powers of both variables in eq. (1) cannot be destroyed, we will see if it be possible to destroy the absolute term of the equation, and the co-efficients of the product of the variables, the second power of one variable and the first power of the other variable.

Then the equations to be satisfied are

$$Ab^2 + Bab + Ca^2 + Db + Ea + F = 0. \qquad (7)$$

$$\left\{ \begin{array}{l} 2A\sin.m\sin.n + B(\sin.m\cos.n + \sin.n\cos.m) \\ + 2C\cos.m\cos.n \end{array} \right\} = 0. \qquad (2)$$

$$A\sin.^2 m + B\sin.m\cos.m + C\cos.^2 m = 0. \qquad (8)$$

$$(2Ab + Ba + D)\sin.n + (2Ca + Bb + E)\cos.n = 0. \qquad (3)$$

when it is required that the co-efficients of x'^2 and y' should reduce to zero in connection with the absolute term and the co-officient of $x'y'$, in eq. (1). To reduce the co-efficients of y'^2 and x' to zero, instead of those of x'^2 and y', it would be necessary to replace eqs. (8) and (3) by

$$A\sin.^2 n + B\sin.n\cos.n + C\cos.^2 n = 0. \qquad (9)$$

$$(2Ab + Ba + D)\sin.m + (2Ca + Bb + E)\cos.m = 0. \qquad (4)$$

Equations (2) and (8) may be written

$$2A\tan.m\tan.n + B(\tan.m + \tan.n) + 2C = 0. \qquad (2')$$

$$A\tan.^2 m + B\tan.m + C = 0. \qquad (8')$$

From eq. (8′) we find

$$\tan.m = -\frac{B}{2A} \pm \frac{1}{2A}\sqrt{B^2 - 4AC} = -\frac{B}{2A},$$

18

and this value of tan. m substituted in eq. (2') gives
$$2A(B—B)\tan.n = B^2—4AC,$$

or
$$\tan. n = \frac{0}{0}.$$

That is, when tan. m is equal to $—\dfrac{B}{2A}$, eq. (2') and, therefore, eq. (2), will be satisfied independently of the angle n.

Equation (7), being what the general equation becomes when a and b take the place of x and y respectively, shows that the new origin of co-ordinates must be on the curve. Solving this equation with reference to b, and introducing the condition $B^2—4AC=0$, we find

$$b = —\frac{B}{2A}a —\frac{D}{2A} \pm \frac{1}{2A}\sqrt{2(BD—2AE)a + D^2—4AF}$$

Now, because the imposed conditions require that the transformed equation shall be of the form
$$My'^2 = Nx',$$
it follows that every value of x' must give two numerically equal values of y'; hence, the new axis of Y must be parallel to the system of chords bisected by the new axis of X. That is, n must be equal to 90°, and, consequently, sin. $n=1$, cos. $n=0$.

Equation (3) will therefore become
$$2Ab + Ba + D = 0.$$

Whence $b = —\dfrac{B}{2A}a —\dfrac{D}{2A}$, and the radical part of the value of b will disappear, or we shall have
$$2(BD—2AE)a + D^2—4AF = 0.$$

From which we get
$$a = —\frac{D^2—4AF}{2(BD—2AE)}.$$

These values of a and b place the new origin at the vertex of the diameter whose equation is
$$y = —\frac{B}{2A}x —\frac{D}{2A},$$

and make the new axis of Y a tangent line to the curve at the vertex of this diameter.

The values of a, b, m and n which we have now found, substituted in eq. (1), will reduce it to

$$Ay'^2 + (2Ca + Bb + E)\cos. m\, x' = 0.$$

Or $$y'^2 + \frac{1}{A}(2Ca + Bb + E)\cos. m\ x' = 0.$$

Denoting the co-efficient of x' by $-2p'$, this last equation becomes

$$y'^2 = 2p'x', \qquad\qquad (10)$$

which is of the form of the equation of the parabola referred to a tangent line and the diameter passing through the point of contact.

The transformation by which eq. (10) was derived from the general equation is always possible when $B^2 - 4AC = 0$, unless we also have $BD - 2AE = 0$. If we suppose that both of these conditions are satisfied, the general value of y, which is

$$y = -\frac{B}{2A}x - \frac{D}{2A} \pm \frac{1}{2A}\sqrt{(B^2 - 4AC)x^2 + 2(BD - 2AE)x + D^2 - 4AF}$$

reduces to

$$y = -\frac{B}{2A}x - \frac{D}{2A} \pm \frac{1}{2A}\sqrt{D^2 - 4AF}$$

whence

$$y = -\frac{B}{2A}x - \frac{D}{2A} + \frac{1}{2A}\sqrt{D^2 - 4AF}$$

and $$y = -\frac{B}{2A}x - \frac{D}{2A} - \frac{1}{2A}\sqrt{D^2 - 4AF},$$

which are the equations of two parallel straight lines.

Under the suppositions just made, the general equation may be written under the form

$$(2Ay + Bx + D + \sqrt{D^2 - 4AF})(2Ay + Bx + D - \sqrt{D^2 - 4AF}) = 0,$$

which may be satisfied by making, first one, then the other factor of the first member, equal to zero. Each of the equations thus obtained, being of the first degree in respect to x and y, will represent a right line.

If the further condition, $D^2-4AF<0$, be imposed, the right lines will have no existence, and the general equation can be satisfied by no real values of x and y.

The value of $2p'$, the parameter of the diameter which becomes the new axis of X, will be found by substituting in the expression

$$-\frac{1}{A}(2Ca+Bb+E)\cos. m,$$

the values of a, b and $\cos. m$.   These values are

$$a=-\frac{D^2-4AF}{2(BD-2AE)}, \quad b=\frac{4ADE-4ABF-BD^2}{4A(BD-2AE)},$$

$$\cos. m=\pm\frac{2A}{\sqrt{4A^2+B^2}}.$$

To reduce eq. (1) to the form

$$x'^2=2p''y' \tag{11}$$

we must satisfy equations (7), (2), (9) and (4).

From eq. (9) we find $\tan. n=-\frac{B}{2A}$, and this value of $\tan. n$ substituted in eq. (2') gives $\tan. m=\frac{0}{0}$; results which might have been anticipated, since eqs. (3) and (4) are the same, except that m in the former takes the place of n in the latter.

Because eq. (11) will give two numerically equal values of x' for every value of y', the new axis of X must be parallel to the system of chords bisected by the new axis of Y; hence $m=0°$, $\sin. m=0$, $\cos. m=1$, and equation (4) therefore reduces to

$$2Ca+Bb+E=0$$

Whence
$$a=-\frac{B}{2C}b-\frac{E}{2C},$$

Solving eq. (7) with reference to a we have

$$a=-\frac{B}{2C}b-\frac{E}{2C}\pm\frac{1}{2C}\sqrt{2(BE-2CD)b+E^2-4CF}$$

By comparing this value of a with that which precedes we find

$$2(BE-2CD)b+E^2-4CF=0,$$

Whence $$b=-\frac{E^2-4CF}{2(BE-2CD)}$$

These values of a and b place the new origin at the vertex of the diameter whose equation is

$$x=-\frac{B}{2C}y-\frac{E}{2C}$$

Or $$y=-\frac{2C}{B}x-\frac{E}{B}$$

The transformation by which eq. (4) is derived from eq. (1) will be impossible when b is infinite; that is when $BE-2CD=0$.

It may be easily proved that when $B^2-4AC=0$, the condition $BD-2AE=0$ necessarily includes the condition $BE-2CD=0$; that is, when we cannot transform eq. (1) into eq. (10), it will also be impossible to transform it into eq. (11).

For $$BD-2AE=0 \text{ gives} \frac{B}{2A}-\frac{E}{D}=0.$$

And $$B^2-4AC=0 \text{ gives} \frac{B}{2A}=\frac{2C}{B}$$

Whence $$\frac{2C}{B}-\frac{E}{D}=0, \text{ or } BE-2CD=0.$$

5.—We have now established the following criteria for the interpretation of any equation of the second degree between two variables, viz:

For the ellipse, $B^2-4AC<0$.
For the hyperbola, $B^2-4AC>0$.
For the parabola, $B^2-4AC=0$.

It remains for us to indicate the construction of any of these curves from its equation, and in doing this, we

19*

shall follow the order in which the conditions are given above.

First, $B^2-4AC<0$, the ellipse.

6.—Let us resume the formulas.

$$a=\frac{2AE-BD}{B^2-4AC}$$

$$b=\frac{2CD-BE}{B^2-4AC}, \quad \text{tan. } m=-\frac{B}{2A}.$$

$$Ay'^2-A\left(\frac{B^2-4AC}{4A^2+B^2}\right)x'^2=-(A\,b^2+Bab+Ca+Db+Ea$$

$$+F, \tag{1'}$$

and suppose, for a particular case, $B=0$, and $A=C$.

We shall then have $a=-\dfrac{E}{2A}$, $b=-\dfrac{D}{2A}$

And $\qquad y'^2+x'^2=\dfrac{D^2+E^2-4AF}{4A^2}$

That is, the general equation, under the suppositions made, represents a circle having $a=-\dfrac{E}{2A}$, $b=-\dfrac{D}{2A}$ for the co-ordinates of its center, and $\sqrt{\dfrac{D^2+E^2-4AF}{4A^2}}$ for its radius.

Draw AX, AY for the primitive co-ordinate axes, lay off $AB=-\dfrac{E}{2A}$, $AD=-\dfrac{D}{2A}$, and through the points B and D draw the parallels BC and DC to the axes. Their intersection, C, is the center of the circle, and the circumference described with $CE=\sqrt{\dfrac{D^2+E^2-4AF}{4A^2}}$ as a radius, will be that represented by the given equation.

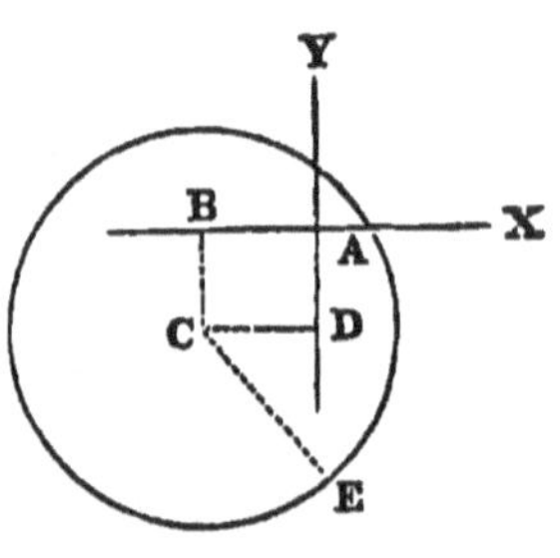

The general equation gives

$$y=-\frac{B}{2A}x-\frac{D}{2A}\pm\frac{1}{2A}\sqrt{(B^2-4AC)x^2+2(BD-2AE)x+D^2-4AF}.$$

Placing the quantity under the radical sign, in this value of y, equal to zero, we have

$$x^2+2\frac{(BD-2AE)}{B^2-4AC}x+\frac{D^2-4AF}{B^2-4AC}=0, \qquad \text{(p)}$$

and denoting the roots of this equation by x' and x'', the value of y may be written

$$y=-\frac{B}{2A}x-\frac{D}{2A}\pm\frac{1}{2A}\sqrt{(B^2-4AC)(x-x')(x-x'')}. \qquad \text{(q)}$$

Now x' and x'' are the abscissas of the vertices of the diameter whose equation is

$$y=-\frac{B}{2A}x-\frac{D}{2A}.$$

The corresponding values of y are

$$y'=-\frac{Bx'+D}{2A},$$

$$y''=-\frac{Bx''+D}{2A}.$$

Substituting these values of x', x'' and y', y'' in the formula

$$\sqrt{(x'-x'')^2+(y'-y'')^2},$$

we have $\dfrac{x''-x'}{2A}\sqrt{B^2+4A^2}$ for the length of the diameter.

The diameter which is conjugate to this is that which is parallel to the axis of y. We find the ordinates of its vertices by substituting $a=\dfrac{x'+x''}{2}$ for x in eq. (q), which then becomes

$$y=-\frac{B(x'+x'')}{4A}-\frac{D}{2A}\pm\frac{x'-x''}{4A}\sqrt{4AC-B^2}.$$

Denoting these two values of y by y_1, y_2, their difference, which is the length of the conjugate diameter, is

$$y_1-y_2=\frac{x'-x''}{2A}\sqrt{4AC-B^2}$$

To find the angle that the conjugate diameters make with each other, let VV' be the first diameter and QQ' the second. The angle that VV' makes with the axis of X is equal to $V'VR$, and its cosine is

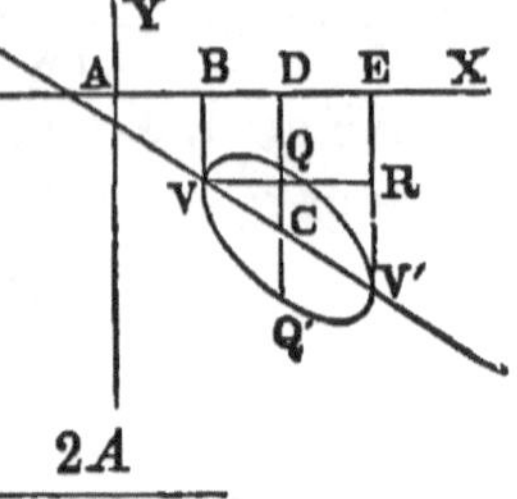

$$\frac{VR}{VV'}=\frac{x''-x'}{\dfrac{x''-x'}{2A}\sqrt{B^2+4A^2}}=\frac{2A}{\sqrt{B^2+4A^2}},$$

and the $\lfloor QCV'=$the $\lfloor BVV'=90°+$the $\lfloor V'VR$.

When the roots of eq. (p) are equal, the vertices of the first diameter, and also those of its conjugate, coincide, and the ellipse reduces to a point. Equation (q) may then be put under the form

$$y=-\frac{Bx+D}{2A}\pm\frac{x-x'}{2A}\sqrt{B^2-4AC}.$$

Because B^2-4AC is negative, this value of y will be imaginary for every value of x except the particular one, $x=x'$, which causes the radical to disappear.

When the roots of eq. (p) are real and unequal, that one of the factors $(x-x')$, $(x-x'')$ under the radical in eq. (q), which corresponds to the root which is algebraically the greater, will be negative, while the other will be positive, for all values of x included between the limits of the smaller and greater roots. The quantity under the radical, being then composed of the product of three factors, two of which are negative and one positive, will itself be positive and the corresponding values of y will therefore be real.

All values of x which exceed the greater, and, also, all values of x which are less than the smaller, of these roots, will render the quantity under the radical negative and the corresponding values of y imaginary. The roots x' and x'' are therefore the limits within which we would select values of x to substitute in the equation to get the co-ordinates of points of the curve.

When the roots of eq. (p) are imaginary, the product of the factors $(x-x')$, $(x-x'')$ under the radical in eq. (q) will remain positive for all real values of x; and because the other factor is $B^2-4AC<0$, the radical will always be imaginary: that is, no real value of x which will give a real value for y. There is, then, in this case, no point in the plane of the co-ordinate axes whose co-ordinates will satisfy eq. (q), and, consequently, the equation from which it was derived, and the curve, has no existence, or it is imaginary.

By the solution of eq. (p) it will be found that when the expression

$$(BD-2AE)^2-(B^2-4AC)(D^2-4AF)$$

is positive, the roots of the equation are real and unequal; when the expression is zero the roots are real and equal, and when negative the roots are imaginary.

If we solve the general equation with reference to x instead of y, and place the quantity under the radical sign equal to zero, we shall find that when the expression

$$(BE-2CD)^2-(B^2-4AC)(E^2-4CF)$$

is positive, the roots of the resulting equation are real and unequal; when zero, these roots are real and equal, and when negative they are imaginary.

It might be inferred that if these roots are real and unequal, equal, or imaginary when the general equation is resolved with reference to one variable, they would be like characterized when it is resolved with reference to the other. To prove this, we develope the first of the above expressions and find that it becomes

$$4A\left(A(E)^2+C(D)^2+F(B)^2-BDE-4ACF.\right)$$

The development of the second is

$$4C\left(A(E)^2+C(D)^2+F(B)^2-BDE-4ACF.\right)$$

The only difference in these developments is that the coefficient of the parenthesis in the first is $4A$, and in the second it is $4C$; but when $B^2-4AC<0$, A and C must have the same sign, hence these expressions must be positive, negative, or zero at the same time.

Second, $B^2-4AC>0$, the hyperbola.

7.—We will begin by supposing $B=0$, and $A=-C$. The formulas for a, b and tan. m will then give

$$a=\frac{E}{2A}, \quad b=-\frac{D}{2A}, \quad \text{tan.}m=0,$$

and eq. (1′) will become

$$y'^2-x'^2=\frac{D^2-E^2-4AF}{4A^2}$$

This is the equation of an equilateral hyperbola whose semi-axis is the square root of the numerical value of the expression $\dfrac{D^2-E^2-4AF}{4A^2}$. Since tan. $m=0$, $m=0$, and one of the axes of the hyperbola is parallel and the other perpendicular to the primitive axis of X. If the sign of $\dfrac{D^2-E^2-4AF}{4A^2}$ is negative, the transverse is the parallel axis; if negative, it is the perpendicular axis.

To construct the curve, let AX and AY be the primitive co-ordinate axes. Lay off the positive abscissa $AD=\dfrac{E}{2A}$, and the negative ordinate $AE=-\dfrac{D}{2A}$; the parallels to the axes

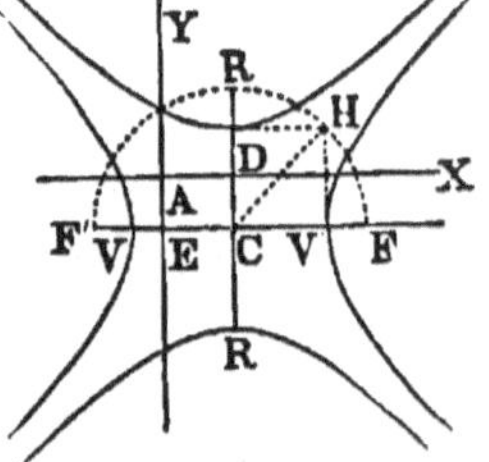

drawn through D and E will be the axes of the hyperbola, and C will be its center. On these axes, lay off from the center, the distances CV, CV', CR, CR', each equal to $\sqrt{\dfrac{D^2-E^2-4AF}{4A^2}}$, and we have the axes of conjugate equilateral hyperbolas. The foci may be found by describing a circumference with C as a center and CH, the hypothenuse of the isosceles right-angled triangle CVH, as a radius; the circumference will intersect the axes at the foci.

For another case, let us suppose $A=0$ and $C=0$; then the value $-\dfrac{B}{2A}$ which was assumed for tan. m becomes infinite, or the new axis of X is perpendicular to the primitive axis of X, and since tan. n is also infinite, the new co-ordinates axes would coincide; in other words, with this value of tan. m, it would be impossible, under the hypothesis, to transform the original equation into eq. (1'). But if $A=0$, and $C=0$, the co-efficient of $x'y'$ in eq. (1) becomes

$$B(\text{sin. } m \text{ cos. } n + \text{sin. } n \text{ cos. } m).$$

Placing this equal to zero, and dividing through by B cos. m cos. n, we have

$$\text{tan. } m + \text{tan. } n = 0,$$

Or $$\text{tan. } m = -\text{tan. } n.$$

Since we are at liberty to select a value for either m or n, let us make $n=45°$; then $m=-45°$. The values of a and b, which will destroy the co-efficients of x' and y' are,

$$a=-\dfrac{D}{B},\ b=-\dfrac{E}{B}.$$

Substituting these values in eq. (1), reducing and transposing, we have

$$y'^2-x'^2=\dfrac{2(DE-BF)}{B^2}$$

which is also the equation of the equilateral hyperbola, the co-ordinates of whose center are $a=-\dfrac{D}{B},\ b=-\dfrac{E}{B},$

and whose semi-axis is the square root of the numerical value of $\dfrac{2(DE-BF)}{B^2}$. The asymptotes of this hyperbola are parallel to the primitive axes, and if $\dfrac{2(DE-BF)}{B^2}$ is negative, the transverse axis makes a negative angle with the primitive axis of X, if positive, it makes a positive angle with that axis.

There is another case in which the transformation by which eq. (1′) was obtained, cannot be made with the value $-\dfrac{B}{2A}$ for tan m. It is that in which A becomes zero, and C does not. We then assume for tan. m the tangent of the angle that the diameter whose equation is

$$x=-\frac{B}{2C}y-\frac{E}{2C}$$

makes with the axis of X. That is, we make

$$\tan. \; m=-\frac{2C}{B}$$

Proceeding with this as with the value $-\dfrac{B}{2A}$, we shall find for the transformed equation

$$Cy'^2-C\left(\frac{B^2-4AC}{\sqrt{4C^2+B^2}}\right)x'^2=-(Ab^2+Bab+Ca^2+Db+Ea+F$$

By making $A=0$, this equation becomes

$$Cy'^2-\frac{CB^2}{\sqrt{4C^2+B^2}}x'^2=-(Bab+Ca^2+Db+Ea++F)$$

which is that of an hyperbola referred to a system of conjugate diameters, one of which bisects the chords which are parallel to the primitive axis of X.

In the general case the course to be pursued for the hyperbola differs so little from that already indicated for the ellipse, that it is unnecessary to dwell upon it at length.

The quantity under the radical in the general value of y placed equal to zero gives the equation

$$x^2 + \frac{2(BD - 2AE)}{B^2 - 4AC}x + \frac{D^2 - 4AF}{B^2 - 4AC} = 0,$$

The roots of this equation are the abscissas of the vertices of the diameter, whose equation is

$$y = -\frac{B}{2A}x - \frac{D}{2A}.$$

When these roots are real and unequal, the diameter terminates in the hyperbola; when imaginary, it terminates in the conjugate hyperbola.

Denoting these abscissas, when real, by x' and x'', and the corresponding ordinates by y' and y'', we have

$$y' = -\frac{Bx' + D}{2A}$$
$$y'' = -\frac{Bx'' + D}{2A}$$

By placing these values of x', x'' and y', y'' in the formula

$$\sqrt{(x' - x'')^2 + (y' - y'')}$$

we shall have the length of the diameter, and the angle included between it and its conjugate will be found precisely as in the ellipse.

If x' be the smaller and x'' the greater abscissa, then all values of x between x' and x'' will give imaginary values for y, and will answer to no points of the curve; but all values of x less than x', and also all values of x greater than x'' will give real values for y', and such values of x with the corresponding values of y will be the co-ordinates of points of the hyperbola.

When the roots x', x'' are imaginary, the diameter whose equation is

$$y = -\frac{B}{2A}x - \frac{D}{2A}$$

20

terminates in the hyperbola which is conjugated to that represented by the given equation, and the diameter which is conjugate to this diameter will terminate in the given hyperbola.

The conjugate diameter may be found in the case of both the ellipse and hyperbola by making first $y'=0$ in eq. (1'), and taking the square root of the corresponding numerical value of x'^2, and then $x'=0$, and taking the square root of the corresponding numerical value of y'^2.

8.—In the transformation of co-ordinates by which the original equation was changed into eq. (1) had the condition, that the new co-ordinate axes should be rectangular, been imposed, as it might, we would have had $n-m=90°$, $n=90°+m$. Sin. $n=\cos. m$, cos. $n=-\sin. m$.

These values being substituted in eq. (2) will give

$$2A \sin. m \cos. m - B \sin.^2 m + B \cos.^2 m - 2C \sin. m \cos. m = 0,$$

which, by dividing through by $\cos.^2 m$, and denoting $\dfrac{\sin. m}{\cos. m}$ by t, becomes

$$2At - Bt^2 + B - 2Ct = 0.$$

Whence $\quad t = \dfrac{A-C}{B} \pm \dfrac{1}{B}\sqrt{B^2+(A-C)^2}.$

Since the product of these two values of t is equal to —1, they are the tangents of the angles that two straight lines at right angles to each other make with the axis of X. Now, if eqs. (5) and (6) are satisfied at the same time; that is, if the new origin be placed at the point of which the co-ordinates are

$$a = \frac{2AE-BD}{B^2-4AC}, \quad b = \frac{2CD-BE}{B^2-4AC},$$

the values of t just found will be the tangents of the angles that the axes of the ellipse, or hyperbola, as the case may be, make with the primitive axis of X. Denoting these tangents by t' and t'', we shall have

$$y-b=t'(x-a),$$
$$y-b=t''(x-a),$$

for the equations of the axes, and by combining the equations of the axes with the original equation, we may find the co-ordinates of their vertices, and, consequently, their length.

9.—When the roots x' and x'' become equal, the value of y may be written

$$y=-\frac{Bx+D}{2A}\pm\frac{x-x'}{2A}\sqrt{B^2-4AC}.$$

For the hyperbola, $B^2-4AC>0$, and these values of y are real. We therefore have

$$y=-\frac{B}{2A}x-\frac{D}{2A}+\frac{x-x'}{2A}\sqrt{B^2-4AC}, \qquad \text{(r)}$$

and
$$y=-\frac{B}{2A}x-\frac{D}{2A}-\frac{x-x'}{2A}\sqrt{B^2-4AC}. \qquad \text{(s)}$$

These equations represent two right lines, and, since the co-efficients of x, when the second members are arranged with reference to it, are different, these lines will intersect. We see that by making $x=x'$, the two equations will give the same value for y. Hence, $x=x'$, and $y=-\dfrac{Bx'+D}{2A}$ are the co-ordinates of the intersection of the lines.

The line BE, whose equation is

$$y=-\frac{B}{2A}x-\frac{D}{2A},$$

still has the property of bisecting all lines drawn parallel to the axis of Y, which are limited by the lines BC and BD, whose equations are eqs. (r) and (s).

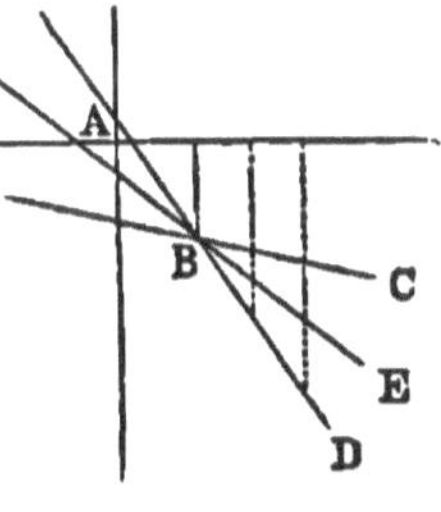

Third, $B^2-4AC=0$, the parabola.

10.—The equation of the diameter that bisects the chords of the curve which are parallel to the axis of Y is

$$y=-\frac{B}{2A}x-\frac{D}{2A},$$

and that of the diameter which bisects the chords parallel to the axis of X is

$$x=-\frac{B}{2C}y-\frac{E}{2C};$$

or

$$y=-\frac{2C}{B}x-\frac{E}{B}.$$

Since a tangent line drawn through the vertex of a diameter is parallel to the chords that the diameter bisects, it follows that the diameters represented by the above equations are perpendicular to each other, and, therefore, (Prop. 5, Chap. 4), their intersection, in the case of the parabola, is on the directrix.

The abscissa of the vertex of the first diameter is the value of x given by the equation .

$$2(BD-2AE)x+D^2-4AF=0,$$

the first member of which is the quantity under the radical in the general value of y, after we have made $B^2-4AC=0$.

Denoting this abscissa by x' we have

$$x'=-\frac{D^2-4AF}{2(BD-2AE)},$$

and

$$y'=-\frac{Bx'+D}{2A}.$$

If we denote the co-ordinates of the vertex of the second diameter by x'' and y'', we have

$$y''=-\frac{E^2-4CF}{2(BE-2CD)},$$

$$x''=-\frac{By''+E}{2C}.$$

Let P and P' be the two vertices thus found. Through the first draw PT parallel to the axis of Y, and through the second, $P'T$ parallel to the axis of X. These lines will be tangent to the parabola at P and P' respectively,

and their intersection, T, will be
a point of the directrix. The
lines CM, BN, drawn through
P and P', making, with the axis
of X, angles having for their
common tangent

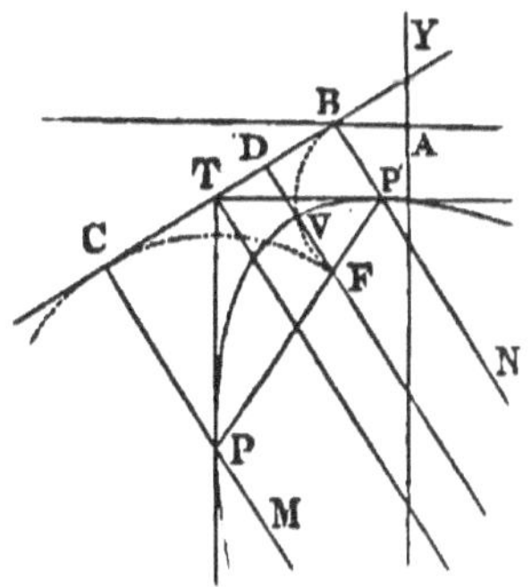

$$-\frac{B}{2A}=-\frac{2C}{B},$$

are diameters of the curve, and
BC drawn through T perpendicular to these diameters,
is the directrix. With P as a center and PC as a radius,
or with P' as a center and $P'B$ as a radius, describe an
arc of a circle. This arc will cut the chord PP' at the
focus F. The perpendicular FD, drawn through F to
the directrix, is the axis, and the middle point, V, of FD,
is the vertex of the parabola.

EXAMPLES.

It will aid in the construction of the curve represented
by any equation to find the points in which it is inter-
sected by the co-ordinate axes. If we make either variable equal to zero in the equation, the values of the other
variable given by the resulting equation will be the distances from the origin to the intersections of the curve,
with axis of the latter variable. When the roots of the
equation which we solve are real and unequal, there will
be two intersections, where real and equal, the axis will
be tangent to the curve at the point thus determined, and
when imaginary, the curve and the axis will have no
common points.

1.—*Construct the curve represented by the equation*
$$y^2+2xy+3x^2-4x=0.$$
Whence $\qquad y=-x\pm\sqrt{-2x(x-2)}.$
Here $A=1$, $B=2$, $C=3$; therefore $B^2-4AC<0$, and

20*

the curve is an ellipse which passes through the origin of co-ordinates, since the equation has no absolute term.

$$y = -x$$

is the equation of a diameter of the curve and the co-ordinates of its vertices are $x' = 0, y' = 0$ and $x'' = 2, y'' = -2$.
By making $x = 1$ in the original equation, we find $y = +. 41+$, or -2.41 for the ordinates of the vertices of the diameter conjugate to the first.

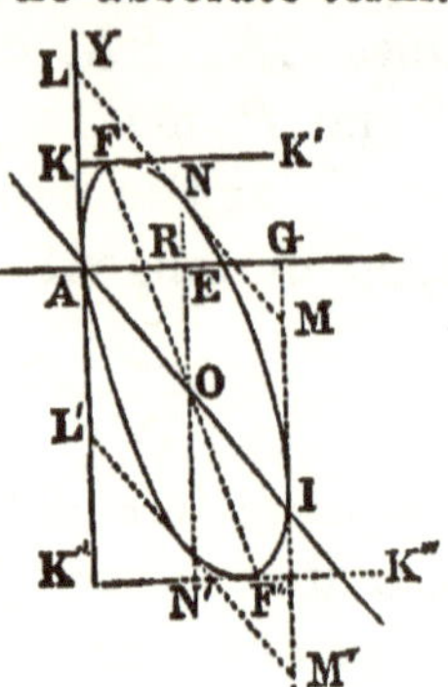

The length of the first diameter is equal to $\sqrt{8} = 2.82+$, and the length of the second is $+.41 + 2.41 = 2.82$.

2.—*Determine the curve that corresponds to the equation*

$$y^2 + 2xy + x^2 - 6y + 9 = 0.$$

Here $A = 1$, $B = 2$, $C = 1$, hence $B^2 - 4AC = 0$, and the curve is a parabola. We find

$$y = -x + 3 \pm \sqrt{-6x},$$

And
$$x = -y \pm \sqrt{6y - 9.}$$

The diameter whose equation is $y = -x + 3$ has $x' = 0$, and $y' = 3$ for the co-ordinates of its vertex. The axis of y is therefore tangent to the curve. The co-ordinates of the vertex of the diameter whose equation is $x = -y$ are, $x'' = -1\frac{1}{2}$, and $y'' = 1\frac{1}{2}$, and a line drawn through this point parallel to the axis of X will be tangent to the curve.

Let P' be the vertex of the first diameter and P that of the second. The chord PP' passes through the focus. $P'S'$, PS making with the axis of X, on the negative side, angles of $45°$ are diameters of the curve, and BT a perpendicular to PS is the directrix.

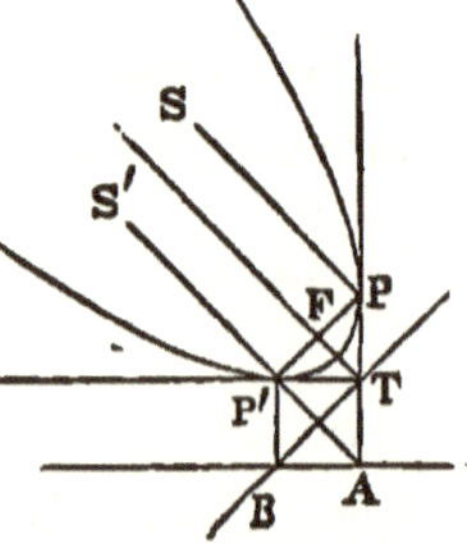

3.—*Determine the curve of which the equation is*

$$y^2 + 2xy - 2x^2 - 4y - x + 10 = 0.$$

In this case $A=1$, $B=2$, $C=-2$; hence $B^2 - 4AC > 0$, and the curve is an hyperbola.   The equation gives

$$y = -x + 2 \pm \sqrt{3x^2 - 3x - 6}.$$

The abscissas of the vertices of the diameter whose equation is

$$y = -x + 2$$

are the roots of the equation

$$3x^2 - 3x - 6 = 0.$$

Whence $x' = -1$, and $x'' = 2$, and the corresponding values of y are $y' = 3$ and $y'' = 0$.

The diameter which is parallel to the axis of y is conjugate to PP', and terminates in the conjugate hyperbola.   The co-ordinates of its vertices are imaginary and may be found by making $x = \frac{1}{2}$ in the original equation.   We would thus find

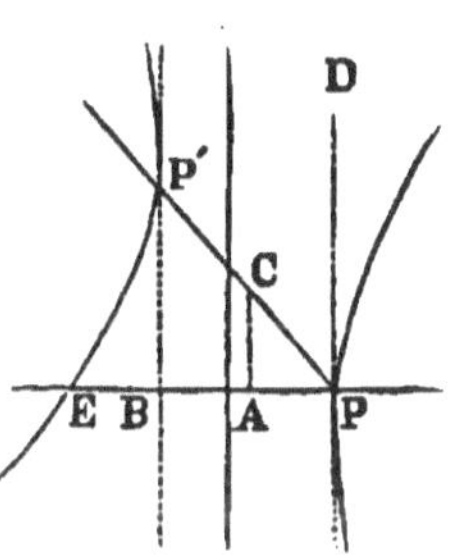

$$y = \tfrac{3}{2} \pm \frac{5.2\sqrt{-1}}{2}$$

The conjugate diameter will therefore be about 5.2. The point E in which the curve intersects the axis of X is on the left of the origin and at a distance from it equal to $2\frac{1}{2}$ units.

4.—*Determine the curve represented by the equation*

$$y^2 + 6xy + 9x^2 - 2y - 6x - 15 = 0.$$

In this, the condition $B^2 - 4AC = 0$ is satisfied, and the curve is the parabola; but it answers to the case in which the parabola reduces to two parallel lines.

In fact the equation may be put under the form

$$(y + 3x)^2 - 2(y + 3x) = 15.$$

Whence $\qquad\quad y + 3x = 1 \pm \sqrt{16},$

Or $\qquad\qquad y + 3x = 5 \text{ or } -3.$

The first member of the equation may therefore be resolved into the factors $y+3x-5$, and $y+3x+3$; which, placed separately equal to zero, give for the parallel lines the equations

$$y=-3x+5,$$
And $$y=-3x-3.$$

5.—Determine the curve of which the equation is

$$y^2-4xy+5x^2-2y+5=0.$$

In this we have $B^2-4AC<0$, and the curve is an ellipse, but it answers to the case in which the curve becomes imaginary. For, resolving the equation in relation to y, we find

$$y=2x+1\pm\sqrt{-(x-2)^2}.$$

The quantity under the radical in this value of y will be negative for every real value of x, hence, all values of y are imaginary; that is, there is no point whose co-ordinates will satisfy the given equation.

By inspection we may also discover that the first member of the equation can be placed under the form

$$(y-2x-1)^2+(x-2)^2,$$

which is the sum of two squares, and must therefore remain positive for all real values of x and y.

6.— What kind of a curve corresponds to the equation

$$y^2-2xy-x^2-2y+2x+3=0\ ?$$

Ans. It is an *hyperbola.* The axis of Y is midway between the two branches. One branch of the curve cuts the axis of X at the point -1; the other branch cuts the same axis at the point $+3$.

7.— Determine the curve represented by the equation

$$y^2-2xy+2x^2-2x+4=0.$$

Resolving, we find

$$(y-x)^2+(x-1)^2+3=0.$$

The condition for the ellipse is satisfied, but the curve is imaginary.

8.—*What kind of a curve corresponds to the equation*

$$y^2 - 2xy + x^2 + x = 0 ?$$

Ans. It is a parabola passing through the origin and extending without limit, in the direction of x and y negative.

9.—*What kind of a curve corresponds to the equation*

$$y^2 - 2xy + x^2 - 2y - 1 = 0 ?$$

Ans. It is a parabola, cutting the axis of X at the distance of —1 and +1 from the origin, and extending indefinitely in the direction of *plus x* and *plus y.*

10.—*What kind of a curve corresponds to the equation*

$$y^2 - 4xy + 4x^2 = 0 ?$$

Ans. It is a straight line passing through the origin, making an angle of $26° \ 34'$ with the axis of Y.

11.—*What kind of a curve corresponds to the equation*

$$y^2 - 2xy + 2x^2 - 2y + 2x = 0 ?$$

Ans. It is an ellipse limited by parallels to the axis of Y drawn through the points —1, and +1, on the axis of X.

CHAPTER VII.

ON THE INTERSECTIONS OF LINES AND THE GEOMETRICAL SOLUTION OF EQUATIONS.

We have seen that the equation of a straight line is

$$y = tx + c,$$

And that the general equation of a circle is

$$(x \pm a)^2 + (y \pm b)^2 = R^2.$$

The first is a simple, the second a quadratic equation,

and if the value of x derived from the first be substituted in the second, we shall have a resulting equation of the second degree, in which y cannot correspond to every point in the straight line, nor to every point in the circumference of the circle, but it will correspond to the two points in which the straight line cuts the circumference, and to those points only.

And if the straight line should not cut the circumference, the values of y in the resulting equation *must necessarily become imaginary.* All this has been shown in the application of the polar equation of the circle, in Chap. 2.

Let us now extend this principle still further. The equation of the parabola is

$$y^2 = 2px,$$

an equation of the second degree, and the equation of a circle is

$$(x \pm a)^2 + (y \pm b)^2 = R^2,$$

also an equation of the second degree. But when two equations of the second degree are combined, they will produce an equation of the fourth degree.

But this resulting equation of the fourth degree cannot correspond to all points in the parabola, nor to all points in the circumference of the circle, but it must correspond equally to both; hence, it will correspond to the points of intersection, and if the two curves do not intersect, the combination of their equations will produce an equation whose roots are *imaginary.*

Let us take the equation $y^2 = 2px$, and take p for the *unit* of measure, (that is, the distance from the directrix to the focus is unity,) then $x = \dfrac{y^2}{2}$, and this value of x substituted in the equation of the circle, will give

$$\left(\frac{y^2}{2} \pm a\right)^2 + (y \pm b)^2 = R^2.$$

Let the vertex of the parabola be the origin of rectangular co-ordinates.

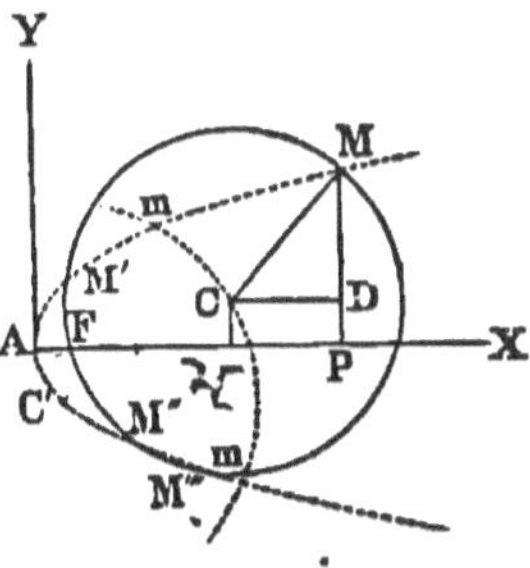

Take $AP=x$, and let it refer to either the parabola or the circle, and let $PM=y$, $AF=\frac{1}{2}$, $AH=a$, $HC=b$, and $CM=R$.

Now in the right angle triangle CMD, we have

$$CD=HP=x-a, \quad MD=y-b,$$

and corresponding to this *particular* figure, **we** shall have in lieu of the proceding equation

$$\left(\frac{y^2}{2}-a\right)^2+(y-b)^2=R^2.$$

Whence $\quad y^4+(4-4a)y^2-8by=4(R^2-a^2-b^2.)$ (F)

This equation is of the fourth degree, hence it must have *four* roots, and this corresponds with the figure, for the circle cuts the parabola in *four* points, M, M', M'', and M'''.

The second term of the equation is wanting, that is, the co-efficient to y^3 is 0, and hence it follows from the theory of equations, that the sum of the *four* roots must be *zero*.

The sum of two of them, which are above the axis of AX, (the two *plus* roots,) must be equal to the sum of the two *minus* roots corresponding to the points M'' and M'''.

The values of a and b and R may be such as to place the center C in such a position that the circumference can cut the parabola in only two points, and then the resulting equation will be such as to give two *real* and two *imaginary* roots.

Indeed, a circumference referred to the same unit of measure and to the same co-ordinates, might not cut the parabola at all, and in that case the resulting equation would have only *imaginary* roots.

In case the circle touches the parabola, the equation will have two equal roots.

Now it is plain that if we can construct a figure that will truly represent any equation in this form, that figure will be a solution to the equation. For instance, a figure correctly drawn will show the magnitude of *PM*, one of ·the roots of the equation.

We will illustrate by the following

EXAMPLES.

1.—*Find the roots of the equation*

$$y^4—11.14y^2—6.74y+9.9225=0.$$

This equation is the same in form as our theoretical equation (F), and therefore we can solve it *geometrically* as follows:

Draw rectangular co-ordinates, as in the figure, and take $AF=\frac{1}{2}$, and construct the *parabola*.

To find the center of the circle and the radius, we put

$$4—4a=—11.14, \quad (1) \qquad —8b=—6.74, \quad (2)$$

and

$$4(R^2—a^2—b^2)=—9.9225. \qquad (3)$$

From eq. (1), $a=3.78$. From eq. (2), $b=0.84$.

And these values of a and b, substituted in eq. (3), give

$$R=3.34, \text{ nearly.}$$

Take from the scale which corresponds to $AF=\frac{1}{2}$, $AH=a=3.78$, $HC=0.84$, and from C as a center, with a radius equal to 3.34, describe the circumference cutting the parabola in the four points, *M*, *M'*, *M''*, and *M'''*. The distance of *M* from the axis of *X* is +3.5, of *M'*

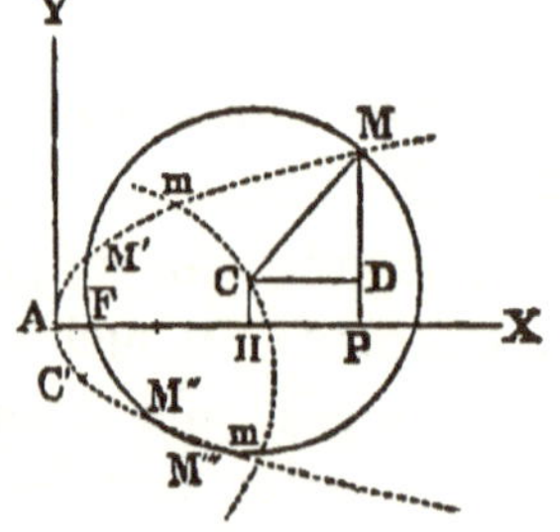

it is +0.7, of *M''* it is —1.5, and of *M'''* it is —2.7, and these are the four roots of the equation.

Their sum is 0, as it ought to be, because the equation contains no third power of y.

2.—*Find the roots of the equation*

$$y^4+y^3+6y^2+12y-72=0.$$

This equation contains the third power of y; therefore this geometrical solution will not apply until that term is removed.

But we can remove that term by putting

$$y=z-\tfrac{1}{4}.$$

(See theory of transforming equations in algebra).

This value of y substituted in the equation, it becomes

$$z^4+5\tfrac{5}{8}z^2+9\tfrac{1}{8}z=74\tfrac{1}{2}\tfrac{6}{5}\tfrac{3}{6},$$

and this equation is in the proper form.

Now put $4-4a=5\tfrac{5}{8}$, $-8b=9\tfrac{1}{8}$, and $4(R^2-a^2-b^2)=74\tfrac{1}{2}\tfrac{6}{5}\tfrac{3}{6}$.
Whence $a=-1\tfrac{3}{32}$, $b=-7\tfrac{3}{64}$, and $R=4.485$.

These values of a and b designate the point C' for the center of the circle. From this center, with a radius $=4.485$, we strike the circumference, cutting the parabola in the two points m and m'. The point m is $2\tfrac{1}{4}$ units above the axis AX, and the point m' is $-2\tfrac{3}{4}$ units from the same line, and these are the two roots of the equation. *The other two roots are imaginary*, shown by the fact that *this* circumference can cut the parabola in two points only.

If we conceive the circumference of a circle to pass through the vertex of the parabola A, then will

$$a^2+b^2=R^2,$$

and this supposition reduces the general equation (F) to

$$y^4+(4-4a)y^2-8by=0.$$

Here $y=\pm0$ will satisfy the equation, and this is as it should be, for the circumference actually touches the parabola on the axis of X.

Now divide this last equation by this value of y, and we have

$$y^3+(4-4a)y=8b. \qquad\text{(G)}$$

Here is an equation of the third degree, referring to a parabola and a circle; the circumference cutting the parabola at its vertex for one point, and if it cuts the parabola in any other point, that other point will designate another root in equation (G).

It is possible for a circle to touch one side of the parabola within, and cut at the vertex A and at some other point. Therefore it is possible for an equation in the form of eq. (G) to have three real roots, and two of them equal.

The circumferences of most circles, however, can cut the parabola in A and in one other point, showing one real root and two *imaginary roots*.

Equation (G) can be used to effect a mechanical solution of all numerical equations of the third degree, in that form.*

We will illustrate this by one or two

EXAMPLES.

1.—*Given* $y^3 + 4y = 39$, *to find the value of* y *by construction.* (See fig. following page)

Put $4 - 4a = 4$, and $8b = 39$. Whence $a = 0$, and $b = 4\frac{7}{8}$.

These values of a and b designate the point C on the axis of Y for the center of the circle, $CA = 4\frac{7}{8}$, the radius.

The circle again cuts the parabola in P, and PQ measures three units, *the only real root of the equation.*

2.—*Given* $y^3 - 75y = 250$, *to find the values of* y *by construction.*

When the co-efficients are large, a large figure is required; but to avoid this inconvenience, we reduce the co-efficients, as shown in Chap. 2.

---

* Observe that the second term, or y^2, in a regular cubic is wanting. Hence, if any example contains that term, it must be removed before a geometrical solution can be given.

Thus put $\quad y=nz.$

Then the equation becomes

$$n^3z^3-75nz=250.$$

$$z^3-\frac{75}{n^2}z=\frac{250}{n^3}.$$

Now take $n=5$, then we have

$$z^3-3z=2.$$

In this last equation the co-efficients are sufficiently small to apply to a construction.

Put $\qquad 4-4a=-3$, and $8b=2.$

Whence $\qquad a=1\frac{3}{4}$, and $b=\frac{1}{4}.$

These values of a and b designate the point D for the center of the circle. DA is the radius.

The circle cuts the parabola in t, and touches it in T, showing that one root of the equation is $+2$, and two others each equal to -1.

But $y=nz.$ That is, $y=5\times2$, or-5, $-5.$

Or the roots of the original equation are $+10$, -5, $-5.$

When an equation contains the second power of the unknown quantity, it must be removed by transformation before this method of solution can be applied.

3.—*Given* $y^3-48y=128$ *to find the values of* y *by construction.* $\qquad$ *Ans.* $+8$, -4, $-4.$

4.—*Given* $y^3-13y=-12$, *to find the values of* y *by construction.* $\qquad$ *Ans.* $+1$, $+3$, and $-4.$

Conversely we can describe a parobola, and take any point, as H, at pleasure, and with HA as a radius, describe a circle and find the equation to which it belongs.

This circle cuts the parabola in the points m, n and o, indicating an equation whose roots are $+1$, $+2.4$, and $-3.4.$

We may also find the particular equation from the general equation

$$y^3+(4-4a)y=8b,$$

observing the locality of H, which corresponds to $a=3\cdot3$ and $b=-1$, and taking these values of a and b, we have

$$y^3-9.2y=-8,$$

for the equation sought.

REMARKS ON THE INTERPRETATION OF EQUATIONS.

In every science it is important to take an occasional retrospective view of first principles, and the conviction that none demand this more imperatively than geometry will excuse us for reconsidering the following truths so often in substance, if not in words, called to mind before.

An equation, geometrically considered, whatever may be its degree, is but the equation of a point, and can only designate a point.

Thus, the equation $y=ax+b$ designates a point, which point is found by measuring any assumed value which may be given to x from the origin of co-ordinates on the axis of X, and from that extremity measuring a distance represented by $(ax+b)$ on a line parallel to the axis of Y.

The extremity of the last measure *is the point designated by the equation.* If we assume another value for x, and measure again in the same way, we shall find the point which now corresponds to the value of x. Again, assume another value for x, and find the designated point.

Lastly, if we connect these several points, we shall find them all in the same *right line,* and in this sense the equation of the first degree, $y=ax+b$, *is the general equation of a right line,* but the right line is found by finding points in the line and connecting them.

In like manner the equation of the second degree

$$y=\pm\sqrt{2Rx-x^2},$$

only designates a point when we assume any value for x, (not inconsistent with the existence of the equation), and take the *plus* sign. It will also designate another point when we take the *minus* sign. Taking another value of x, and thus finding two other points, we shall have four points,—still another value of x and we can find two other points, and so on, we might find any number of points. Lastly, on comparing these points we shall find that *they are all in the circumference of the same circle*, and hence we say that the preceding equation is the equation of a *circle*. Yet it can designate only one, or at most, two points at a time.

If we assume different values for y, and find the corresponding values of x, the result will be the same circle, because the x and y mutually depend upon each other.

Now let us take the last practical example
$$y^3 - 13y = -12,$$
and, for the sake of perspicuity, change y into x, then we shall have
$$x^3 - 13x + 12 = 0.$$

Now we can suppose $y = 0$ to be another equation; then will
$$y = x^3 - 13x + 12 \qquad \text{(A)}$$
be an independent equation between two variables, and of the third degree.

The particular hypothesis that $y = 0$, gives three values to x, $(+1, +3, \text{and} -4)$, that is, *three points* are designated: the first at the distance of one unit to the right of the axis of Y; the second at the distance of three units on the same side of the axis of Y; and the third point four units on the opposite side of the same axis, and *this is all the equation can show until we make another hypothesis.*

Again, let us assume $y = 5$, then equation (A) becomes
$$5 = x^3 - 13x + 12, \text{ or } x^3 - 13x + 7 = 0,$$
and this is, in effect, changing the origin five units on the axis of Y. A solution of this last equation fixes three other points on a line parallel to the axis of X.

Again, let us assume $y = 10$, then equation (A) becomes
$$x^3 - 13x + 2 = 0,$$

21*

and a solution of this equation gives three other points.

And thus we may proceed, assigning different values to y, and deducing the corresponding values of x, as appears in the following table, commencing at the origin of the co-ordinates, where $y=0$, and varying each way.

$$y=30.0388 \quad x=-2.2814 \quad +4.1628 \quad -2.0814$$
$$y=25. \quad x=-1.1 \quad +4.03 \quad -2.91$$
$$y=20. \quad x=-0.40 \quad +3.80 \quad -3.41$$
$$y=15. \quad x=-0.20 \quad +3.70 \quad -3.50$$
$$y=10. \quad x=+0.14 \quad +3.52 \quad -3.66$$
$$y=5. \quad x=+0.55 \quad +3.3 \quad -3.85$$

When $y=0$. then will $x=+1.$ $\quad +3.$ $\quad -4.$
$$y=-5 \quad x=+1.66 \quad +2.477 \quad -4.14$$
$$y=-6.0388 \quad x=+2.0814 \quad +2.0814 \quad -4.1623$$

Taking $y=0$, a solution of the equation $y=x^3-13x+12$, gives the three points a, a, a, on the axis of X.

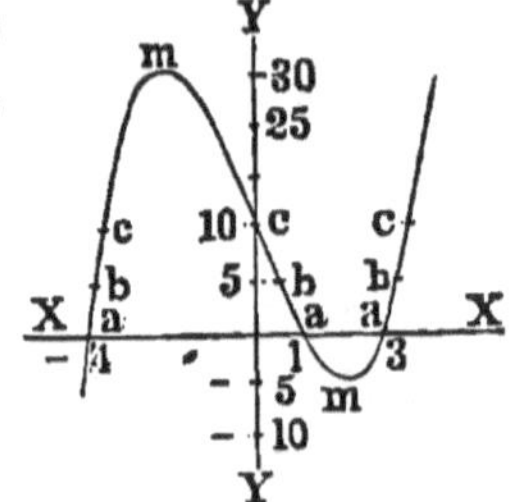

Then taking $y=5$, and a solution gives three points b, b, b, on a line parallel to the axis of X, and at the distance of 5 units above said axis.

Again, taking $y=10$, and another solution gives the three points c, c, c. Now joining the three points $(a, b, c,)$ (a, b, c), and (a, b, c), we shall have apparently *three* curves corresponding to the equation of the *third* degree, and thus, we might hastily conclude that every equation of the third degree would give *three curves*, and every equation of the fourth degree *four curves*, etc., etc., *but this is not true.*

If we continue finding points as before, we shall find that the three curves $(a, b, c,)$ $(a, b, c,)$ and $(a, b, c,)$ are but different portions of the *same curve*, and we can now venture to draw this general conclusion:

That in an equation involving y, the ordinate, to the first power, and the abscissa, x, *to the third power, the axis of* X, *or lines parallel to that axis, may cut the curve in three points.*

From analogy, we also infer that if we have an equation involving x to the *fourth* power, the axis of X, or its parallels, will cut the curve in *four* points; and if we have an equation involving x to the *fifth* power, that axis or its parallels will cut the curve in *five* points, and so on.

In the equation under consideration, $(y=x^3-13x+12)$, if we assume y greater than 30.0388, or less than -6.0388, we shall find that two values of x in each case will become imaginary, and on each side of these limits the parallels to X will cut the curve only *in one point.*

Two points vanish at a time, and this corresponds with the truth demonstrated in algebra, "that *imaginary roots* enter equations in pairs."

The points m, m, the turning points in the curve, are called *maximum* points, and can be found only by approximation, using the ordinary processes of computation, but the peculiar operation of the *calculus* gives these points at once.

To find the points in the *curve* we might have assumed different values of x in succession, and deduced the corresponding values of y, but this would have given but one point for each assumption; and to define the curve with sufficient accuracy, many assumptions must be made with very small variations to x. We solved the equations approximately and with great rapidity by means of the *circle* and *parabola* as previously shown.

We conclude this subject by the following example:

Let the equation of a curve be

$$(a^2-x^2)(x-b)^2=x^2y^2,$$

from which we are required to give a geometrical delineation of the curve. From the equation we have

$$y=\pm\frac{\sqrt{(a^2-x^2)(x-b)^2}}{x}.$$

The following figure represents the curve which will be recognized as corresponding to the equation, after a little explanation.

If $x=0$, then y becomes infinite, and therefore the ordinate at A is an *asymptote* to the curve. If $AB=b$, and P be taken between A and B, then PM and Pm will be equal, and lie on different sides of the abscissa AP. If $x=b$, then the two values of 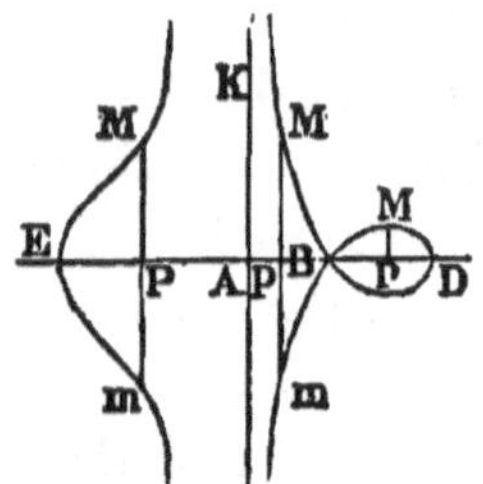 y vanish, because $x-b=0$; and consequently, the curve passes through B, and has there a *duplex* point. If AP be taken greater than AB, then there will be two values of y, as before, having contrary signs, that value which was positive before, now becomes negative, and the negative value becomes positive. But if AD be taken $=a$, and P come to D, then the two values of y vanish, because $\sqrt{a^2-x^2}=0.$ And if AP is taken greater than AD, then a^2-x^2 becomes negative, and the value of y *impossible;* and therefore, the curve does not extend beyond D.

If x now be supposed negative, we shall find

$$y=\pm\sqrt{a^2-x^2}\times(b+x)\div x.$$

If x vanish, both these values of y become infinite, and consequently, the curve has two infinite arcs on each side of the *asymptote* AK. If x increase, it is plain y diminishes, and if x becomes $=-a$, y vanishes, and consequently the curve passes through E, if AE be taken $=AD$, on the opposite side. If x be supposed, numerically, greater than $-a$, then y becomes *impossible;* and no part of the curve can be found beyond E. This curve is the *conchoid* of the ancients.

CHAPTER VIII.

STRAIGHT LINES IN SPACE.

Straight lines in one and the same plane are referred to *two* co-ordinate axes in that plane, —but straight lines in space require *three* co-ordinate axes, made by the intersection of *three planes.*

To take the most simple view of the subject, conceive a *horizontal* plane cut by a *meridian* plane, and by a *perpendicular east* and *west* plane.

The common point of intersection we shall call the origin or *zero point,* and we might conceive this point to be the center of a sphere, and about it will be eight quadrangular spaces corresponding to the eight quadrants of a sphere, which extended, would comprise *all space.*

The horizontal *east and west* line of intersection of these planes, we shall call the axis of X. The horizontal intersection in the direction of the *meridian,* the axis of Y; and that perpendicular to it in the plane of the meridian, the axis of Z. Distances estimated from the zero point horizontally to the right, as we look towards the north, we shall designate as *plus,* to the left *minus.*

Distances measured on the axis of Y and parallel thereto, towards us from the zero point, we shall call *plus;* those in the opposite direction will therefore be *minus.* Perpendicular distances from the horizontal plane upwards are taken as *plus,* downward *minus.*

The horizontal plane is called the plane of xy, the meridian plane is designated as the plane of yz, and the perpendicular east and west plane the plane of xz.

Now let it be observed that x will be *plus* or *minus,* according to its direction from the plane of yz, y will be *plus* or *minus,* according to its direction from the plane

xz, and z will be *plus* or *minus*, according as it is above or below the horizontal place xy.

PROPOSITION I.

To find the equation of a straight line in space.

Conceive a straight line passing in any direction through space, and conceive a plane coinciding with it, and perpendicular to the plane xz. The intersection of this plane with the plane xz, will form a line on the plane xz, and this is said to be the projection of the line on the plane xz, and the equation of this projected line will be in the form

$$x = az + \pi. \quad \text{(Chap. 1, Prop. 1.)}$$

Conceive another plane coinciding with the proposed line, and perpendicular to the plane yz, its intersection with the plane yz is said to be the projection of the line on the plane yx, and the equation of this projected line is in the form

$$y = bz + \beta.$$

These two equations taken together are said to be equations of the line, because the first equation is a general equation for all lines that can be drawn in the first projecting plane, and the second equation is a general equation for all lines that can be drawn in the second projecting plane; therefore taken together, they express the intersection of the two planes, which is the line itself.

For illustration, we give the following example: Construct the line whose equations are

$$\left. \begin{array}{l} x = 2z + 1 \\ y = 3z - 2 \end{array} \right\}$$

Make $z=0$, then $x=1$, and $y=-2$.
Now take $AP=1$, and draw Pm parallel to the axis of Y, making $Pm=-2$; then m is the point in the plane xy, through which the line *must* pass.

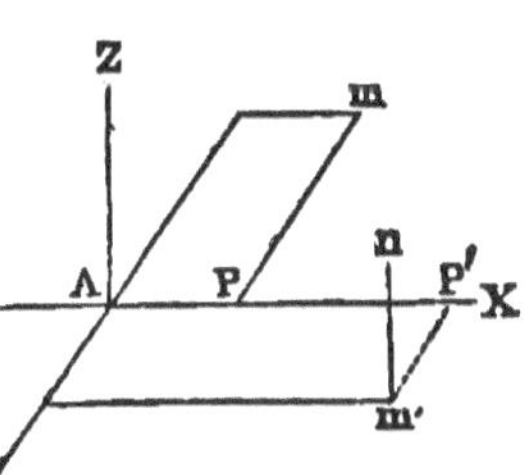

Now take z equal to any number at pleasure, say 1, then we shall have $x=3$ and $y=1$.

Take $AP'=3$, $P'm'=+1$, and from the point m' in the plane xy erect $m'n$ perpendicular to the plane xy, and make it equal to 1, because we took $z=1$, then n is another point in the line. Draw $n\,m$ and produce it, and it will be the line designated by the equations.

PROPOSITION II.

To find the equation of a straight line which shall pass through a given point.

Let the co-ordinates of the given point be represented by x', y', z'.

The equations sought must satisfy the general equations

$$\left.\begin{array}{l} x=az+\pi. \\ y=bz+\beta. \end{array}\right\} \qquad (1)$$

The equations corresponding to the given point are

$$x'=az'+\pi. \qquad y'=bz'+\beta.$$

Subtracting eq. (1) from these, respectively, we have

$$x'-x=a(z'-z), \text{ and } y'-y=b(z'-z),$$

the equations required.

PROPOSITION III.

To find the equations of a straight line which shall pass through two given points.

Let the co-ordinates of the second point be x'', y'', z''. Now by the second proposition, the equations which express the condition that the line passes through the two points, will be

$$x''-x'=a(z''-z'),$$

And
$$y''-y'=b(z''-z').$$

Whence
$$a=\frac{x''-x'}{z''-z'},\ b=\frac{y''-y'}{z''-z'}.$$

Substituting the values of a and b in the equations of a line passing through a single point (Prop. 2,) we have

$$x-x'=\left(\frac{x''-x'}{z''-z'}\right)(z-z').\quad y-y'=\left(\frac{y''-y'}{z''-z''}\right)(z-z'),$$

for the equations required.

PROPOSITION IV.

To find the condition under which two straight lines intersect in space, and the co-ordinates of the point of intersection.

Let the equation of the lines be

$$x=az+\pi.\qquad\qquad y=bz+\beta.$$
$$x=a'z+\pi'.\qquad\qquad y=b'z+\beta'.$$

If the two lines intersect, the co-ordinates of the common point, which may be denoted by x, y, z, will satisfy all of these four equations, therefore by subtraction, we have

$$(a-a')z+\pi-\pi'=0,\qquad\qquad (b-b')z+\beta-\beta'=0.$$

Whence, by eliminating z, we find

$$\frac{\pi-\pi'}{a-a'}=\frac{\beta-\beta'}{b-b'},$$

which is the condition under which two lines intersect.

Now $z=\frac{\pi'-\pi}{a-a'}$, and this value of z being substituted in the first equations, we obtain

$$x=\frac{a\pi'-a'\pi}{a-a'}\ \text{ and }\ y=\frac{b\beta'-b'\beta}{b-b'},$$

for the value of the co-ordinates of the point of intersection.

Cor.—If $a=a'$, the denominators in the second member will become 0, making x and y infinite; that is, the
point of intersection is at an infinite distance from the
origin, and the lines are therefore parallel.

PROPOSITION V.—PROBLEM.

*To express analytically the distance of a given point from
the origin.*

Let P be the given point in
space; it is in the perpendicular
at the point N, which is in the
plane xy.

The angle $AMN=90°$. Also,
the angle $ANP=90°$.

Let $AM=x$, $MN=y$, $NP=z$.

Then $\overline{AN}^2=x^2+y^2$.

But $\overline{AP}^2=\overline{AN}^2+\overline{NP}^2=x^2+y^2+z^2$.

Now if we designate AP by r, we shall **have**
$$r^2=x^2+y^2+z^2$$
for the expression required.

PROPOSITION VI.—PROBLEM

To express analytically the length of a line in space.

Let $PP'=D$ be the line in question.

Let the co-ordinates of the point P
be x, y, z, and of the point P' be x',
y', z'.

Now $\quad MM'=x'-x=NQ.$
$$QN'=y'-y.$$
$$\overline{NN'}^2=(x'-x)^2+(y'-y)^2=\overline{PR}^2$$
$$P'R=z'-z.$$

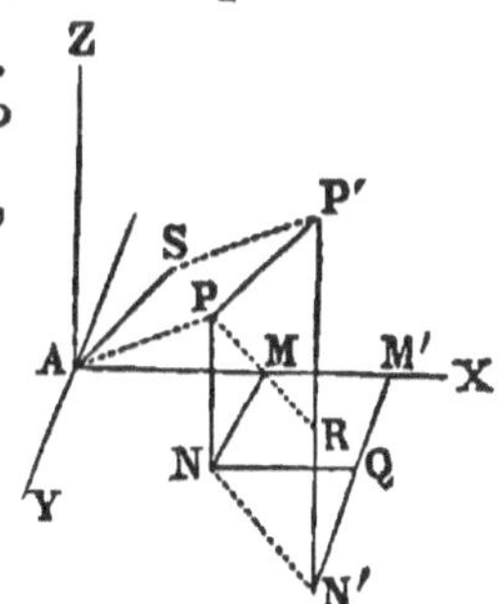

In the triangle PRP' we have

$$\overline{PP'}^2 = \overline{PR}^2 + \overline{P'R}^2 = (x'-x)^2 + (y'-y)^2 + (z'-z)^2.$$

Or $\qquad D^2 = (x'-x)^2 + (y'-y)^2 + (z'-z)^2,$ \qquad (1)

which is the expression required.

Scholium.—If through one extremity of the line, as P, we draw PA to the origin, and from the other extremity P', we draw $P'S$ parallel and equal to PA, and draw AS, it will be parallel to PP', and equal to it, and this virtually reduces this proposition to the previous one. This also may be drawn from the equation, for if A is one extremity of the line, its co-ordinates x, y, and z are each equal to zero, and

$$D^2 = x'^2 + y'^2 + z'^2.$$

PROPOSITION VII.—PROBLEM.

To find the inclination of any line in space to the three axes.

From the origin draw a line parallel to the given line; then the inclination of this line to the axes will be the same as that of the given line.

The equations for the line passing from the origin are

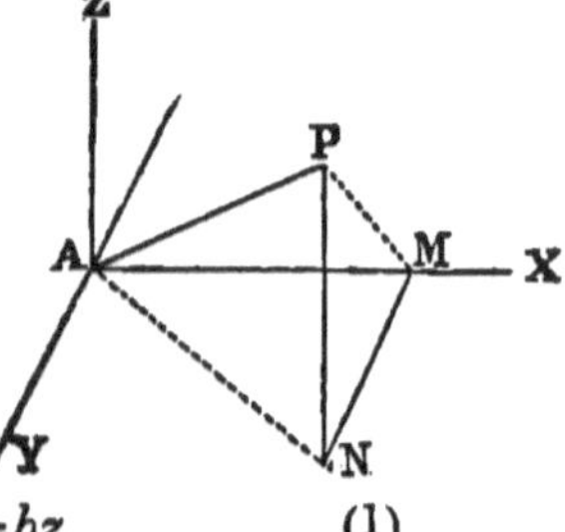

$$x = az, \quad \text{and} \quad y = bz. \qquad (1)$$

Let X represent the inclination of this line with the axis of x, Y its inclination with the axis of y, and Z its inclination with the axis of z.

The three points P, N, M, are in a plane which is parallel to the plane zy, and AM is a perpendicular between the two planes. AMP is a right-angled triangle, the right angle being at M.

Let $AP = r$ and $AM = x$. Then, by trigonometry, we have

As $\qquad r : \sin. 90° : : x : \cos. X.$ \quad Whence $x = r \cos. X.$

Also, as $r : \sin. 90° : : y : \cos. Y.$ \quad Whence $y = r \cos. Y.$

Also, as $r : \sin. 90° :: z : \cos. Z.$ Whence $z = r \cos. Z.$ From Prop. 5 we have

$$r^2 = x^2 + y^2 + z^2. \qquad (2)$$

Substituting the values of x, y, and z, as above, we have

$$r^2 = r^2 \cos.^2 X + r^2 \cos.^2 Y + r^2 \cos.^2 Z.$$

Dividing by r^2 will give

$$\cos.^2 X + \cos.^2 Y + \cos.^2 Z = 1, \qquad (3)$$

an equation which is easily called to mind, and one that is useful in the higher mathematics.

If in eq. (2) we substitute the values of x^2 and y^2 taken from eq. (1), we shall have

$$r^2 = a^2 z^2 + b^2 z^2 + z^2. \qquad (4)$$

But we have three other values of r^2 as follows:

$$r^2 = \frac{x^2}{\cos.^2 X}, \quad r^2 = \frac{y^2}{\cos.^2 Y}, \quad \text{and } r^2 = \frac{z^2}{\cos.^2 Z}.$$

Whence

$$\frac{x}{\cos. X} = \pm z \sqrt{1 + a^2 + b^2}. \qquad (5)$$

$$\frac{y}{\cos. Y} = \pm z \sqrt{1 + a^2 + b^2}. \qquad (6)$$

And

$$\frac{1}{\cos. Z} = \pm \sqrt{1 + a^2 + b^2}. \qquad (7)$$

In eq. (5) put the value of x drawn from eq. (1), and in eq. (6) the value of y from eq. (1), and reduce, and we shall obtain

$$\left. \begin{array}{l} \cos. X = \dfrac{a}{\pm \sqrt{1 + a^2 + b^2}}. \\[2ex] \cos. Y = \dfrac{b}{\pm \sqrt{1 + a^2 + b^2}} \\[2ex] \cos. Z = \dfrac{1}{\pm \sqrt{1 + a^2 + b^2}} \end{array} \right\}$$

The analytical expressions for the inclination of a line in space to the three co-ordinates.

The double sign shows two angles supplemental to each other, the plus sign corresponds to the acute angle, and the minus sign to the obtuse angle.

PROPOSITION VIII.

To find the inclination of two lines in terms of their separate inclinations to the axes.

Through the origin draw two lines respectively parallel to the given lines. An expression for the cosine of the angle between these two lines is the quantity sought.

Let AP be parallel to one of the given lines, and AQ parallel to the other. The angle PAQ is the angle sought.

Let the equations of one of these lines be

$$x=az, \qquad y=bz,$$

and of the other

$$x'=a'z', \qquad y'=b'z'.$$

Let $AP=r$, $AQ=r'$, $PQ=D$, and the angle $PAQ=V$.

Now in plane trigonometry (Prop. 8, p. 260, Geom.,) we have

$$\cos. V=\frac{r^2+r'^2-D^2}{2rr'}. \qquad (1)$$

From Prop. 6 we have

$$D^2=(x'-x)^2+(y'-y)^2+(z'-z)^2.$$

Expanding this, it becomes

$$\begin{cases} D^2=(x'^2+y'^2+z'^2)+(x^2+y^2+z^2) \\ -2x'x-2y'y-2z'z. \end{cases}$$

But by Prop. 5 we have

$$x^2+y^2+z^2=r^2,$$

and

$$x'^2+y'^2+z'^2=r'^2.$$

Whence $\quad 2x'x+2y'y+2z'z=r^2+r'^2-D^2.$

This equation applied to eq. (1) reduces it to

$$\cos. V=\frac{x'x+y'y+z'z}{rr'}.$$

But r and r' may have any values taken at pleasure; their lengths will have no effect on the angle V. Therefore, for convenience, we take each of them equal to unity.

Whence $\qquad \cos. V=x'x+y'y+z'z. \qquad (2)$

But in Prop. 7 we found that $x=r\cos.X$, $y=r\cos.Y$, etc., and that $x'=r'\cos.X'$, $y'=r'\cos.Y'$, etc.; and since we have taken $r=1$ and $r'=1$, $x=\cos.X$, etc., and $x'=\cos.X'$, etc.   Hence

$$\cos.V=\cos.X\cos.X'+\cos.Y\cos.Y'+\cos.Z\cos.Z'. \quad (3)$$

But by Prop. 7 we have

$$\cos.X=\frac{a}{\pm\sqrt{1+a^2+b^2}}, \text{ and } \cos.X=\frac{a'}{\pm\sqrt{1+a'^2+b'^2}}, \text{ etc.}$$

Substituting these values in eq. (3) we have

$$\cos.V=\frac{1+aa'+bb'}{\pm(\sqrt{1+a^2+b^2})(\sqrt{1+a'^2+b'^2})}$$

for the expression required.

The cos. V will be plus or minus, according as we take the signs of the radicals in the denominator alike or unlike.   The plus sign corresponds to an acute angle, the minus sign to its supplement.

Cor. 1.—If we make $V=90°$, then cos. $V=0$, and the equation becomes

$$1+aa'+bb'=0,$$

which is the equation of condition to make two lines at right angles in space.

Cor. 2.—If we make $V=0$, the two straight lines will become parallel, and the equation will become

$$\pm1=\frac{1+aa'+bb'}{\sqrt{1+a^2+b^2}\ \sqrt{1+a'^2+b'^2}}$$

Squaring, clearing of fractions, and reducing, we shall find

$$(a'-a)^2+(b'-b)^2+(ab'-a'b)^2=0.$$

Each term being a square, will be positive, and therefore the equation can only be satisfied by making each term separately equal to 0.

Whence $a'=a$, $b'=b$, and $ab'=a'b$.

The third condition is in consequence of the first two.

22*   R

CHAPTER IX.

ON THE EQUATION OF A PLANE.

An equation which can represent any point in a line is said to be the equation of the line.

Similarly, an equation which can represent or indicate any point in a plane, is, in the language of analytical geometry, the equation of the plane.

PROPOSITION I.

To find the equation of a plane.

Let us suppose that we have a plane which cuts the axes of X, Y and Z at the points B, C and D, respectively; then, if these points be connected by the straight lines BC, CD and DB, it is evi- 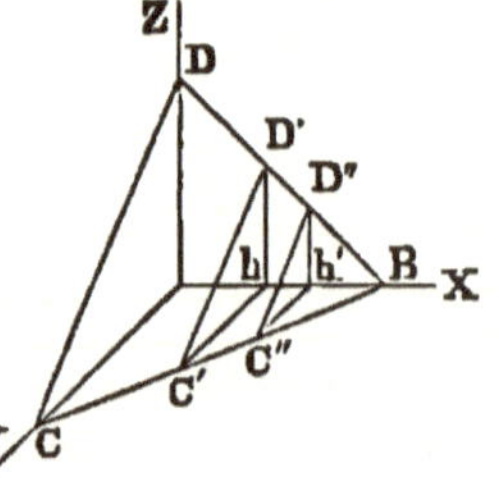
dent that these lines are the inter-
sections of the plane with the
planes of the co-ordinate axes.

Now a plane may be conceived
as a surface generated by moving a
straight line in such a manner that
in all its positions it shall be parallel to its first position and intersect another fixed straight line. Thus the line DC, so moving that in the several positions, $D'C'$, $D''C''$, etc., it remains parallel to DC and constantly intersects DB, will generate the plane determined by the points D, C and B.

The line DB being in the plane xy, its equations are
$$y=0, \quad z=mx+b, \qquad (1)$$
and for the line DC we have
$$x=0, \quad z=ny+b. \qquad (2)$$
The plane passed through the line $D'C'$ parallel to the

plane zy, cuts the axis of X at the point p.   Denoting Ap by c, the equations of the line $D'C'$ become

$$x=c, \quad z=ny+b'. \tag{3}$$

It is obvious that eqs. (3) can be made to represent the moving line in all its positions by giving suitable values to c and b', and that, for any one of its positions, the co-ordinates of its intersection with the line DB must satisfy both eqs. (1) and (3).   That is, c and b', in the first and second of eqs. (3), must be the same as x and z, respectively, in the second of eqs. (1).   Hence

$$b'=z-ny, \text{ and } b'=mx+b.$$

Equating these two values of b', we have

$$z-ny=mx+b,$$

or $$z=mx+ny+b. \tag{4}$$

This equation expresses the relation between the co-ordinates x, y and z for any point whatever in the plane generated by the motion of the line DC, and is, therefore the equation of this plane.

Cor. 1.—Every equation of the first degree between three variables, by transposition and division, may be reduced to the form of eq. (4), and will, therefore, be the equation of a plane.

Cor. 2.—In eq. (4), m is the tangent of the angle which the intersection of the plane with the plane xz makes with the axis of X, n the tangent of the angle that the intersection with the plane yz makes with the axis of Y, and b the distance from the origin to the point in which the plane cuts the axis of Z.

Hence, if any equation of the first degree between three variables be solved with respect to one of the variables, the co-efficient of either of the other variables denotes the tangent of the angle that the intersection of the plane represented by the equation, with the plane of the axes of the first and second variables, makes with the axis of the second variable.

Scholium.—If we assume

$$m=-\frac{A}{C}, \ n=-\frac{B}{C}, \ b=-\frac{D}{C},$$

and substitute these values in eq. (4), it will become, by reduction and transposition,

$$Ax+By+Cz+D=0,$$

which is the form under which the equation of the plane is very often presented.

From this equation we deduce the following general truths:

First.—If we suppose a plane to pass through the origin of the co-ordinates for this point, $x=0$, $y=0$, and $z=0$, and these values substituted in the equation of the plane will give $D=0$ also. Therefore, when a plane passes through the origin of co-ordinates, the general equation for the plane reduces to

$$Ax+By+Cz=0.$$

Second.—To find the points in which the plane cuts the axes, we reason thus:

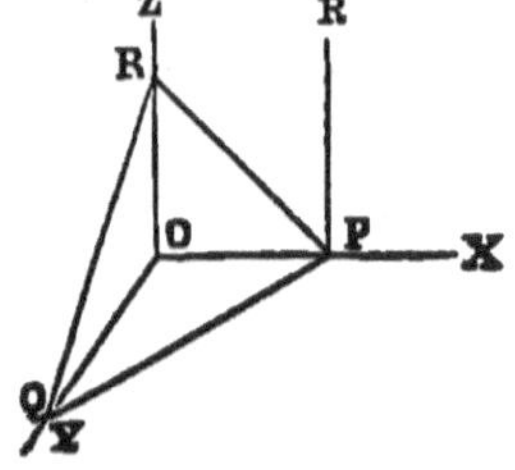

The equation of the plane must respond to each and every point in the plane; the point P, therefore, in which the plane cuts the axis of X, must correspond to $y=0$ and $z=0$, and these values, substituted in the equation, reduces it to

$$Ax+D=0.$$

Or
$$x=-\frac{D}{A}=OP.$$

For the point Q we must take $x=0$ and $z=0$.

And
$$y=-\frac{D}{B}=OQ.$$

For the point R,
$$z=-\frac{D}{C}=OR.$$

Third.—If we suppose the plane to be perpendicular to the plane XY, PR', its intersection with, or *trace* on, the plane XZ, must be drawn parallel to OZ, and the plane will meet the axis of Z at the distance *infinity*. That is, OR, or its equal, $\left(-\frac{D}{C}\right)$, must be infinite, which requires that $C=0$, which reduces the general equation of the plane to

$$Ax + By + D = 0,$$

which is the equation of the *trace* or line PQ on the plane XY. If the plane were perpendicular to the plane ZX, the line OQ, or its equal, $\left(-\dfrac{D}{B}\right)$, must be *infinite*, which requires that $B=0$, and this reduces the general equation to

$$Ax + Cz + D = 0,$$

which is the equation for the *trace PR*, and hence we may conclude in general terms,

That when a plane is perpendicular to any one of the co-ordinate planes, its equation is that of its trace on the same plane.

PROPOSTION II.—PROBLEM.

To find the length of a perpendicular drawn from the origin to a plane, and to find its inclination with the three co-ordinate axes.

Let RPQ be the plane, and from the origin, O, draw Op perpendicular to the plane; this line will be at right-angles to every line drawn in the plane from the point p.

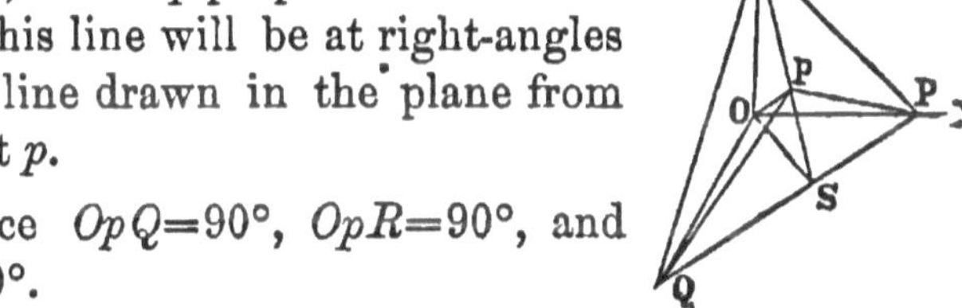

Whence $OpQ=90°$, $OpR=90°$, and $OpP=90°$.

Let $Op=p$.

Designate the angle pOP by X, pOQ by Y, and pOR by Z.

By the *preceding scholium* we learn that

$$OP = -\frac{D}{A}, \; OQ = -\frac{D}{B}, \text{ and } OR = -\frac{D}{C},$$

A, B, C and D being the constants in the equation of a plane.

Now, in the right-angled triangle OpP, we have

$$OP : 1 :: Op : \cos. X.$$

That is, $\qquad -\dfrac{D}{A} : 1 :: p : \cos. X. \qquad\qquad (1)$

The right-angled triangle OpQ gives

$$-\frac{D}{B} : 1 :: p : \cos. Y. \qquad (2)$$

The right-angled triangle OpR gives

$$-\frac{D}{C} : 1 :: p : \cos. Z. \qquad (3)$$

Proportion (1) gives us

$$\cos.^2 X = \frac{p^2}{D^2} A^2, \qquad (4)$$

(2) gives

$$\cos.^2 Y = \frac{p^2}{D^2} B^2, \qquad (5)$$

and (3) gives

$$\cos.^2 Z = \frac{p^2}{D^2} C^2. \qquad (6)$$

Adding these three equations, and observing that the sum of the first members is *unity*, (Prop. 7, Chap. 8), and we have

$$\frac{p^2}{D^2}(A^2 + B^2 + C^2) = 1.$$

Whence

$$p = \pm \frac{D}{\sqrt{A^2 + B^2 + C^2}}. \qquad (7)$$

This value of p placed in eqs. (4), (5) and (6), by reduction, will give

$$\cos. X = \pm \frac{A}{\sqrt{A^2 + B^2 + C^2}}. \qquad (8)$$

$$\cos. Y = \pm \frac{B}{\sqrt{A^2 + B^2 + C^2}}. \qquad (9)$$

$$\cos. Z = \pm \frac{C}{\sqrt{A^2 + B^2 + C^2}}. \qquad (10)$$

Expressions (7), (8), (9) and (10) are those sought.

PROPOSITION III.—PROBLEM.

To find the analytical expressions for the inclination of a plane to the three co-ordinate planes respectively.

Let $Ax+By+Cz+D=0$ be the equation of the plane, and let PQ represent its line of intersection with the co-ordinate plane (xy).

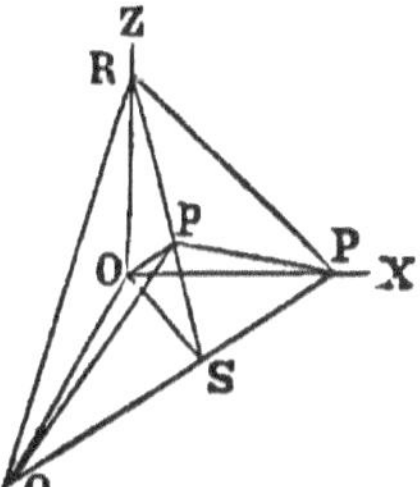

From the origin, O, draw OS perpendicular to the trace PQ. Draw pS. OpS is a right-angled triangle, right-angled at p, and the angle OSp measures the inclination of the plane with the horizontal plane (xy). Our object is to find the angle OSp.

In the right-angled triangle POQ we have found

$$OP=-\frac{D}{A}, \quad OQ=-\frac{D}{B}.$$

Whence $\qquad PQ=\frac{D}{AB}\sqrt{A^2+B^2}.$

Now PS, a segment of the hypothenuse made by the perpendicular OS, is a third proportional to PQ and PO. Therefore

$$\frac{D}{AB}\sqrt{A^2+B^2} : -\frac{D}{A} :: -\frac{D}{A} : PS.$$

Or $\quad \sqrt{A^2+B^2} : -B :: -\frac{D}{A} : PS=-\frac{BD}{A\sqrt{A^2+B^2}}.$

The other segment, QS, is a third proportional to PQ and OQ. Therefore

$$\frac{D}{AB}\sqrt{A^2+B^2} : -\frac{D}{B} :: -\frac{D}{B} : QS.$$

Or $\quad \sqrt{A^2+B^2} : -A :: -\frac{D}{B} : QS=\frac{AD}{B\sqrt{A^2+B^2}}.$

But the perpendicular, OS, is a *mean proportional* between these two segments. Therefore we have

$$OS=\frac{D}{\sqrt{A^2+B^2}}.$$

Now, by simple permutation, we may conclude that the perpendicular from the origin O to the *trace PR* is

$$\frac{D}{\sqrt{A^2+C^2}},$$

and that to the trace QR is

$$\frac{D}{\sqrt{B^2+C^2}}.$$

We shall designate the angle which the plane makes with the plane of (xy) by (xy), and the angle it makes with (xz) by (xz), and that with (yz) by (yz).

Now the triangle OpS gives

$$OS : \sin. 90° :: Op : \sin. OSp.$$

That is, $\dfrac{D}{\sqrt{A^2+B^2}} : 1 :: \dfrac{D}{\sqrt{A^2+B^2+C^2}} : \sin. OSp.$

Whence $\sin.^2 OSp = \sin.^2(xy) = \dfrac{A^2+B^2}{A^2+B^2+C^2}.$

Similarly, $\sin.^2(xz) = \dfrac{A^2+C^2}{A^2+B^2+C^2}.$

And $\sin.^2(yz) = \dfrac{B^2+C^2}{A^2+B^2+C^2}.$

But by trigonometry we know that $\cos.^2 = 1 - \sin.^2$.

Whence $\cos.^2(xy) = 1 - \dfrac{A^2+B^2}{A^2+B^2+C^2} = \dfrac{C^2}{A^2+B^2+C^2}$, etc.

Whence

$$\cos.(xy) = \frac{\pm C}{\sqrt{A^2+B^2+C^2}}$$

$$\cos.(xz) = \frac{\pm B}{\sqrt{A^2+B^2+C^2}}$$

$$\cos.(yz) = \frac{\pm A}{\sqrt{A^2+B^2+C^2}}$$

$\Big\}$ Expressions sought.

Squaring, and adding the last three equations, we find

$$\cos.^2(xy) + \cos.^2(xz) + \cos.^2(yz) = 1.$$

That is, *the sum of the squares of the cosines of the three angles which a plane forms with the three co-ordinate planes, is equal to radius square, or unity.*

PROPOSITION IV.—PROBLEM.

To find the equation of the intersection of two planes.

Let
$$Ax+By+Cz+D=0, \qquad (1)$$
$$A'x+B'y+C'z+D'=0, \qquad (2)$$

be the equations of the two planes.

If the two planes intersect, the values of x, y and z will be the same for any point in the line of intersection. Hence, we may combine the equations for that line.

Multiply eq. (1) by C' and eq. (2) by C, and subtract the products, and we shall have

$$(AC'-A'C)x+(BC'-B'C)y+(DC'-D'C)=0,$$

for the equation of the line of intersection on the plane (xy). If we eliminate y in a similar manner, we shall have the equation of the line of intersection on the plane (xz); and eliminating x will give us the equation of the line of intersection on the plane (yz).

PROPOSITION V.—PROBLEM.

To find the equation to a perpendicular let fall from a given point (x′, y′, z′,) *upon a given plane.*

As the perpendicular is to pass through a given point, its equations must be of the form

$$x-x'=a(z-z'), \qquad (1)$$
$$y-y'=b(z-z'), \qquad (2)$$

in which a and b are to be determined.

The equation of the plane is

$$Ax+By+Cz+D=0.$$

The line and the plane being perpendicular to each other, by hypothesis, the projection of the line and the trace of the plane on any one of the co-ordinate planes will be perpendicular to each other.

For the traces of the given plane on the planes (xz) and (yz), we have $Ax+Cz+D=0$ and $By+Cz+D=0$.

23

From the former $\quad x=-\dfrac{C}{A}z-\dfrac{D}{A}.$ $\qquad$ (3)

From the latter $\quad y=-\dfrac{C}{B}z-\dfrac{D}{B}.$ $\qquad$ (4)

Now eqs. (1) and (3) represent lines which are at right angles with each other.

Also, eqs. (2) and (4) represent lines at right angles with each other.

But when two lines are at right angles, (Prop. 5, Chap. 1), and a and a' are their trigonometrical tangents, we must have $\qquad (aa'+1=0)$.

That is, $\qquad -a\dfrac{C}{A}+1=0,$ or $a=\dfrac{A}{C}.$

Like reasoning gives us $b=\dfrac{B}{C}$, and these values put in eqs. (1) and (2) give

$$\left.\begin{array}{l} x-x'=\dfrac{A}{C}(z-z') \\[2mm] y-y'=\dfrac{B}{C}(z-z') \end{array}\right\} \text{ for the equations sought.}$$

PROPOSITION VI.—PROBLEM.

To find the angle included by two planes given by their equations.

Let $\qquad Ax+By+Cz+D=0,$ $\qquad$ (1)

And $\qquad A'x+By'+C'z+D'=0,$ $\qquad$ (2)

be the equations of the planes.

Conceive lines drawn from the origin perpendicular to each of the planes. Then it is obvious that the angle contained between these two lines is the *supplement* of the inclination of the planes. But an angle and its supplement have numerically the same trigonometrical expression.

Designate the angle between the two planes by V, then Proposition 8, in the last chapter gives

$$\text{cos. } V = \frac{1 + aa' + bb'}{\pm(\sqrt{1 + a^2 + b^2})(\sqrt{1 + a'^2 + b'^2})}. \quad (3)$$

The equations of the two perpendicular lines from the origin must be in the form

$$x = az, \qquad\qquad y = bz,$$
$$x = a'z \qquad\qquad y = b'z.$$

But because the first line is perpendicular to the first plane, we must have

$$a = \frac{A}{C}, \quad \text{and} \quad b = \frac{B}{C}, \quad \text{(Prop. 5.)}$$

And to make the second line perpendicular to the second plane requires that

$$a' = \frac{A'}{C'}, \quad \text{and} \quad b' = \frac{B'}{C'}.$$

These values of a, b, and a', b', substituted in eq. (3) will give, by reduction,

$$\text{cos. } V = \pm \frac{AA' + BB' + CC'}{\sqrt{A^2 + B^2 + C^2}\sqrt{A'^2 + B'^2 + C'^2}},$$

for the equation required.

Cor.—When two planes are at right angles, cos. $V = 0$, which will make

$$AA' + BB' + CC' = 0.$$

PROPOSITION VII.—PROBLEM.

To find the inclination of a line to a plane.

Let MN be the plane given by its equation

$$Ax + By + Cz + D = 0,$$

and let PQ be the line given by its equations

$$x=az+a.$$
$$y=bz+\beta.$$

Take any point P in the given line, and let fall PR, the perpendicular, upon the plane ; RQ is its projection on the plane, and PQR, which we will denote by V, is obviously the least angle included between the line and the plane, and it is the angle sought.

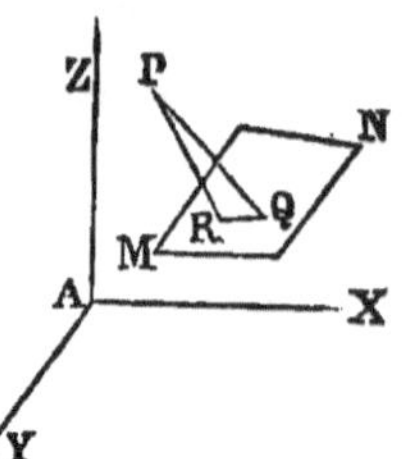

Let $\qquad x=a'z+\pi'$, and $y=b'z+\beta'$,

be the equation of the perpendicular PR, and because it is perpendicular to the plane, we must have (by the last proposition)

$$a'=\frac{A}{C}, \quad \text{and} \quad b'=\frac{B}{C}$$

Because PQ and PR are two lines in space, if we designate the angle included by V, we shall have

$$\cos. V=\pm\frac{1+aa'+bb'}{\sqrt{1+a^2+b^2}\sqrt{1+a'^2+b'^2}}. \quad \text{(Prop. 8, Chap. 8.)}$$

But the cos. V is the same as the sin. PQR, or sin. v, as the two angles are complements of each other.

Making this change, and substituting the values of a' and b', we have

$$\sin. v=\pm\frac{Aa+Bb+C}{\sqrt{1+a^2+b^2}\sqrt{C^2+B^2+A^2}},$$

for the required result.

Cor.—When $v=0$, sin. $v=0$, and this hypothesis gives

$$Aa+Bb+C=0,$$

for the equation expressing the condition that the given line is parallel to the given plane.

We now conclude this branch of our subject with a few practical examples, by which a student can test his knowledge of the two preceding chapters.

EXAMPLES.

1.— *What is the distance between two points in space of which the co-ordinates are*

$$x=3,\ y=5,\ z=-2,\ x'=-2,\ y'=-1,\ z'=6.$$

Ans. 11.180+.

2.— *Of which the co-ordinates are*

$$x=1,\ y=-5,\ z=-3,\ x'=4,\ y'=-4,\ z'=1.$$

Ans. $5\frac{1}{10}$ *nearly.*

3.— *The equations of the projections of a straight line on the co-ordinate planes* (xz), (yz), *are*

$$x=2z+1, \qquad\qquad y=\tfrac{1}{3}z-2,$$

required the equation of projection on the plane (xy).

Ans. $y=\tfrac{1}{6}x-2\tfrac{1}{6}.$

4.— *The equations of the projections of a line on the co-ordinate planes* (xy) *and* (yz) *are*

$$2y=x-5 \qquad and \qquad 2y=z-4,$$

required the equation of the projection on the plane (xz).

Ans. $x=z+1.$

5.— *Required the equations of the three projections of a straight line which passes through two points whose co-ordinates are*

$$x'=2,\ y'=1,\ z'=0,\ and\ x''=-3,\ y''=0,\ z''=-1.$$

What are the projections on the planes (xz) *and* (yz)?

Ans. $x=5z+2,\ y=z+1.$

And from these equations we find the projection on the plane (xy), that is, $5y=x+3$.

(See Prop. 3, Chap. 8.)

6.— *Required the angle included between two lines whose equations are*

$$\left.\begin{array}{l}x=3z+1\\y=2z+6\end{array}\right\}\ of\ the\ 1st,\ and\ \left.\begin{array}{l}x=z+2\\y=-z+1\end{array}\right\}\ of\ the\ 2d.$$

Ans. $V=72°\ 1'\ 28''$

(See Prop. 8, Chap. 8.)

23*

7.—*Find the angles made by the lines designated in the preceding example, with the co-ordinate axes*

(See Prop. 7, Chap. 8.)

$$\text{\textit{Ans.} The 1st line} \begin{cases} 36°\ 42'\ \text{with } X, \\ 57°\ 41'\ 20''\quad Y, \\ 74°\ 29'\ 54''\quad Z, \end{cases} \text{2d line} \begin{cases} 54°44'\ \text{with } X, \\ 125°\ 16'\quad\quad Y, \\ 54°\ 44'\quad\quad Z. \end{cases}$$

8.—*Having given the equation of two straight lines in space,*
as

$$\left.\begin{array}{l} x=3z+1 \\ y=2z+6 \end{array}\right\} \text{ of the 1st, and } \left.\begin{array}{l} x=z+2 \\ y=-z+\beta' \end{array}\right\} \text{ of the 2d,}$$

to find the value of β', so that the lines shall actually intersect, and to find the co-ordinates of the point of intersection.

$$\text{\textit{Ans.} } \begin{cases} \beta'=7\frac{1}{2},\ y=7, \\ x=2\frac{1}{2},\ z=+\frac{1}{2}. \end{cases}$$

(See Prop. 4, Chap. 8.)

9.—*Given the equation of a plane*

$$8x-3y+z-4=0,$$

to find the points in which it cuts the three axes, and the perpendicular distance from the origin to the plane.

(Prop. 2.)

Ans. It cuts the axis of X at the distance of $\frac{1}{2}$ from the origin; the axis of Y at $-1\frac{1}{3}$; and the axis of Z at $+4$.

The origin is .4649+ of unity below the plane.

10.—*Find the equations for the intersections of the two planes*   (Prop. 4.)

$$3x-4y+2z-1=0,$$
$$7x-3y-z+5=0.$$

$$\text{\textit{Ans.} } \begin{cases} \text{On the plane } (xy)\quad 17x-10y+9=0. \\ \text{On the plane } (xz)\quad 19x-10z+23=0. \end{cases}$$

11.—*Find the inclination of these two planes.*
(Prop. 6.)

$$\textit{Ans. } 41°\ 27'\ 41''.$$

12.—*The equations of a line in space are*

$$x=-2z+1, \text{ and } y=3z+2.$$

Find the inclination of this line to the plane represented by the equation (Prop. 7.)

$$8x-3y+z-4=0.$$

Ans. 48° 13′ 13″

13.—*Find the angles made by the plane whose equation is*

$$8x-3y+z-4=0,$$

with the co-ordinate planes.

(Prop. 3.)

$$\text{Ans.} \begin{cases} 83^\circ\ 19'\ 27'' \text{ with } (xy). \\ 110^\circ\ 24'\ 38'' \text{ with } (xz). \\ 21^\circ\ 34'\ 5'' \ \text{ with } (yz). \end{cases}$$

14.—*The equation of a plane being*

$$Ax+By+Cz+D=0,$$

Required the equation of a parallel plane whose perpendicular distance is (a) *from the given plane.*

Ans. Because the planes are to be parallel, their equations must have the same co-efficients, A, B, and C.

In Prop. 2, we learn that the perpendicular distance of the origin from the given plane may be represented by

$$p=\pm\frac{D}{\sqrt{A^2+B^2+C^2}}.$$

Now, as the planes are to be a distance a asunder, the distance of the origin from the required plane must be

$$\frac{D}{\sqrt{A^2+B^2+C^2}}+a \text{ or } \frac{D+a\sqrt{A^2+B^2+C^2}}{\sqrt{A^2+B^2+C^2}}.$$

Whence the equation required is

$$Ax+By+Cz+\left(\frac{D+a\sqrt{A^2+B^2+C^2}}{\sqrt{A^2+B^2+C^2}}\right)=0.$$

15.—*Find the equation of the plane which will cut the axis of* Z *at* 3, *the axis of* X *at* 4, *and the axis of* Y *at* 5.

Ans. $5x+4y+6\tfrac{2}{3}z=20.$

16.—*Find the equation of the plane which will cut the axis of X at 3, the axis of Z at 5, and which will pass at the perpendicular distance 2 from the origin. At what distance from the origin will this plane cut the axis of Y?*

Ans. The equation of the plane is
$$10x+\sqrt{89}\,y+6z-30=0.$$

The plane cuts the axis of Y at $\pm\dfrac{30}{\sqrt{89}}$.

17.—*Find the equations of the intersection of the two planes whose equations are*
$$3x-2y-z-4=0,$$
$$+7x+3y+z-2=0.$$

Ans. $\left\{\begin{array}{l}\text{The equation of the projection of the inter-}\\ \text{section on the plane } (xy) \text{ is}\\ \qquad 10x+y-6=0.\\ \text{On the plane } (xz) \text{ it is}\\ \qquad 23x-z-16=0,\\ \text{and that on the plane } (yz) \text{ is}\\ \qquad 23y+10z+22=0.\end{array}\right.$

18.—*Find the inclination of the planes whose equations are expressed in example 17.*

  Ans. $V=60°\ 50'\ 55''$ or $119°\ 9'\ 5''$.

19.—*A plane intersects the co-ordinate plane (xz) at an inclination of 50°, and the co-ordinate plane (yz) at an inclination of 84°. At what angle will this plane intersect the plane (xy)?*

             Ans. $V=40°\ 38'\ 6''$.

MISCELLANEOUS PROBLEMS.

1. The greatest diameter or major axis of an ellipse is 40 feet, and a line drawn from the center making an angle of 36° with the major axis and terminating in the ellipse is 18 feet long; required the minor axis of this ellipse, its area and excentricity.

Note.—The excentricity of an ellipse is the distance of either focus from the center, when the semi major axis is taken as unity.

$$Ans. \begin{cases} \text{The minor axis is 30.8752.} \\ \text{Area of the ellipse, 969.972 sq. feet.} \\ \text{Excentricity .63575.} \end{cases}$$

2. If equilateral triangles be described as the three sides of any plane triangle and the centers of these equilateral triangles be joined, the triangle so formed will be equilateral; required the proof.

Let ABC represent any plane triangle, A, B and C denoting the angles, and a, b and c the respective sides, the side a being opposite the angle A, and so on.

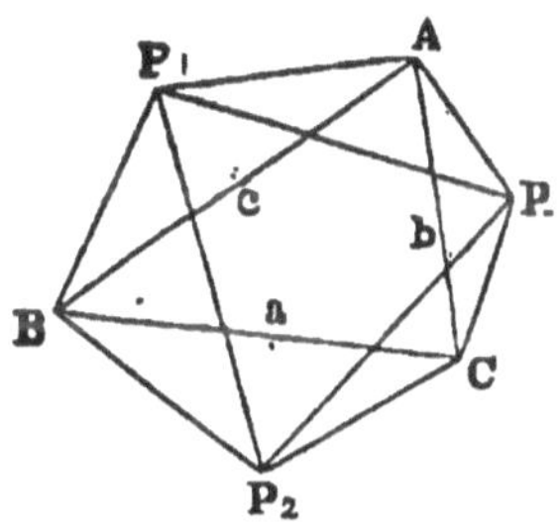

On AC, or b, suppose an equilateral triangle to be drawn, and let P be its center.

Make the same suppositions in regard to the sides c and a, finding P_1 and P_2. Draw PP_1, P_1P_2 and PP_2; then is PP_1P_2 an equilateral triangle, as is to be proved.

We shall assume the principle, which may be easily demonstrated, that *a line drawn from the center of any equilateral triangle to the vertex of either of the angles, is equal to* $\frac{1}{\sqrt{3}}$ *times the side of the triangle.* Hence we have

$$AP=\frac{b}{\sqrt{3}}, \ PC=\frac{b}{\sqrt{3}}, \ AP_1=\frac{c}{\sqrt{3}}, \ P_1B=\frac{c}{\sqrt{3}}, \ BP_2=CP_2=\frac{a}{\sqrt{3}}$$

Also, the angles $PAC=30°$, $P_1AB=30°$, $P_1BA=30°$

and so on. Now it is obvious that the angle PAP_1 is expressed by $(A+60°)$, the angle P_1BP_2 by $(B+60°)$, and PCP_2 by $(C+60°)$. We must now show that the analytical expressions for PP_1 and P_1P_2 are the same. In analytical trigonometry it was found that the cosine of an angle, A, of a plane triangle would be given by the equation

$$\cos. A = \frac{b^2+c^2-a^2}{2bc}$$

Whence, $\quad a^2 = b^2 + c^2 - 2bc \cos. A.$

That is, *The square of one side is equal to the sum of the squares of the other two sides, minus twice the rectangle of the other two sides into the cosine of the opposite angle.*

Applying this to the triangle PAP_1 we have

$$\overline{PP_1}^2 = \frac{b^2}{3} + \frac{c^2}{3} - \frac{2bc}{3} \cos. (A+60°) \qquad (1)$$

Also, $\quad \overline{P_1P_2}^2 = \frac{c^2}{2} + \frac{a^2}{3} - \frac{2ac}{3} \cos. (B+60°) \qquad (2)$

And $\quad \overline{PP_2}^2 = \frac{a}{3} + \frac{b^2}{3} - \frac{2ab}{3} \cos. (C+60°) \qquad (3)$

By trigonometry, $\cos. (A+60) = \cos. A \cos. 60 - \sin. A \sin. 60.$

But $\quad \cos. 60° = \tfrac{1}{2},$ and $\sin. 60 = \tfrac{1}{2}\sqrt{3}$

Whence, $\quad \cos. (A+60) = \tfrac{1}{2} \cos. A - \frac{\sqrt{3}}{2} \sin. A$

This value substituted in eq. (1) that equation becomes

$$\overline{PP_1}^2 = \frac{b^2}{3} + \frac{c^2}{3} - \frac{bc}{3} \cos. A + \frac{bc}{\sqrt{3}} \sin. A \qquad (4)$$

But $\cos. A = \dfrac{b^2+c^2-a^2}{2bc}.$ Whence $\dfrac{bc}{3} \cos. A = \dfrac{b^2+c^2-a^2}{6}$

This value of $\dfrac{bc}{3} \cos. A$ placed in eq. (4), gives

$$\overline{PP_1}^2 = \frac{2b^2}{6} + \frac{2c^2}{6} - \frac{b^2}{6} - \frac{c^2}{6} + \frac{a^2}{6} + \frac{bc}{\sqrt{3}} \sin. A$$

Or, $\quad \overline{PP_1}^2 = \dfrac{a^2+b^2+c^2}{6} + \dfrac{bc}{\sqrt{3}} \sin. A. \qquad (5)$

By a like operation equation (2) becomes

$$\overline{P_1P_2}^2=\frac{a^2+b^2+c^2}{6}+\frac{ac}{\sqrt{3}}\ \text{sin.}\ B \qquad (6)$$

But by the original triangle ABC we have

$$\frac{\text{sin.}\ A}{a}=\frac{\text{sin.}\ B}{b},\ \text{ or sin.}\ A=\frac{a}{b}\ \text{sin.}\ B$$

Placing this value of sin. A in equation (5) that equation becomes

$$\overline{PP_1}^2=\frac{a^2+b^2+c^2}{6}+\frac{ac}{\sqrt{3}}\ \text{sin.}\ B. \qquad (7)$$

We now observe that the second members of (6) and (7) are equal; therefore, $\quad PP_1=P_1P_2$

And in like manner we can prove $PP_1=PP_2$. Therefore the triangle PP_1P_2 has been shown to be equilateral.

PROBLEM.

Given, the excentricity of an Ellipse, to find the difference between the mean and true place of the planet, corresponding to each degree of the mean angle, reckoned from the major axis; the planet describing equal sectors or areas in equal times, about one of the foci, the center of the attractive force.

Let AB be the major axis of an ellipse, of which $CB=CA=A=1$ is the semi-transverse axis, and also let C be the common center of the ellipse and of the circle of which CB is the radius. Then $FC=e$, and F is the focus of the ellipse.

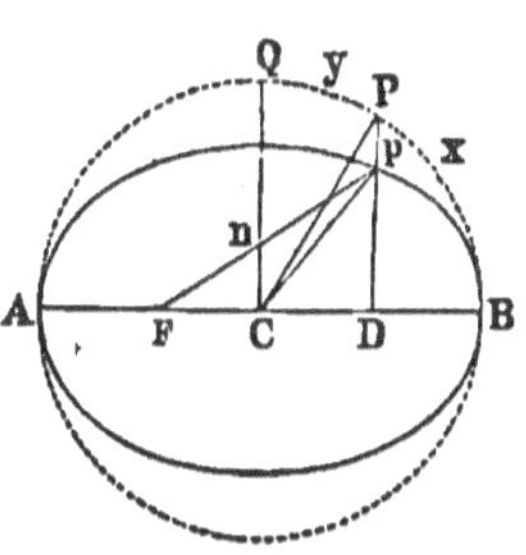

Suppose the planet to be at B, the apogee point of the orbit, (so called in Astronomy). Also, conceive another planet, or material point, to be at B, at the same time. Now, the planet revolves along the ellipse, describing equal areas in equal times, and the hypothetical planet revolves along the circle BPQ, describing, in equal times, equal areas and equal angles about the center C.

It is obvious that the two bodies will arrive at A in the same time. The other halves of the orbits will also be described in the same time, and the two bodies will be together again at the point B.

But at no other points save at A and at B (the apogee and perigee points) will these two bodies be in the same line as seen from F, and the difference of the directions of the two bodies as seen from the focus F is the equation of the center. For instance, suppose the planet to start from B and describe the ellipse as far as p. It has then described the area BFp of the ellipse, about the focus F. In the same time the fictious planet in the circle has moved along the circumference BP to Q, describing the sector BCQ about the center C. Now the areas of these two sectors must be to each other as the area of the ellipse is to the area of the circle. That is,

sector BFp : sector BCQ : : area Ell. : area Cir.

Through p draw PD at right angles to AB, and represent the arc of the circle BP by x.

Then $CD=$cos. x, and $PD=$sin. x. Draw Cp and CP.

But, denoting the semi-conjugate axis by B, we have

area DpB : area DPB : : area Ell. : area Cir.

$$: : \quad B \quad : A$$
$$: : \quad pD \quad : PD$$

Also we have $\triangle CpD : \triangle CPD : : pD : PD$

Hence, area $DpB : \triangle CpD$: : area $DPB : \triangle CPD$

Therefore,

area $DpB + \triangle CpD$: area $DPB + \triangle CPD$: : $B : A$

or,   sector CpB : sector CPB : : $B : A$

$$: : \text{area Ell.} : \text{area Cir.}$$

Hence it follows that

sector FpB : sector CpB : : sector CQB : sector CPB

Whence

sector FpB—sect. CpB : sect. CQB—sect. CPB : : $B : A$

or, $\qquad \triangle FpC$: sector $QCP :: B : A$
$$:: \text{area Ell.} : \text{area. Cir.}$$

But the area of the ellipse is πAB and the area of the circle is $A^2\pi$. But $A=1$ and $B=\sqrt{1-c^2}$.

The area of the triangle FCp is $\tfrac{1}{2}e\,(pD)$, and the area of the sector is $\tfrac{1}{2}y$, representing the arc QP by y.

Whence $\qquad E\,(pD) : y :: \sqrt{1-e^2} : 1.$ $\qquad$ (1)

But we have $\qquad PD : pD :: A : B$
$$:: 1 : \sqrt{1-e^2}, \quad \text{and } PD=\sin. x.$$

Hence, $\quad \sin. x : pD :: 1 : \sqrt{1-e^2}; \quad pD=\sin x\sqrt{1-e^2}$

This value of pD placed in (1) that proportion becomes
$$e\sin. x\sqrt{1-e^2} : y :: \sqrt{1-e^2} : 1$$

Or, $\qquad e\sin. x : y :: 1 : 1. \quad y=e\sin. x.$ $\qquad$ (2)

DEFINITIONS.—1st. The angle x, in astronomy, is called *the excentric anomaly.*

2d. The angle QCB, or $(x+y)$ is called *the mean anomaly.*

3d. The angle pFB is called *the true anomaly.*

4th. The difference between QCB or nCB (of the triangle FnC) and nFC (which is the angle n of the triangle CFn) is *the equation of the center.*

The angle QCB, the *mean* anomaly, is an angle at the center of the ellipse, which is equal to the sum of the angles at n and F; that is, n taken from the angle at the center will give the true angle at the focus, F.

We will designate the angle pFB by t. Now, by the polar equation of an ellipse, we have

$$Fp=\frac{1-e^2}{1-e\cos. t} \qquad A \text{ being } 1.$$

Again, by the triangle FDp, we find,

$$Fp=\sqrt{FD^2+pD^2}$$

But $\quad \overline{FD}^2=(e+\cos. x)^2=e^2+2e\cos. x+\cos.^2 x$

And $\quad \overline{pD}^2=\sin.^2 x\,(1-e^2)=\sin.^2 x-e^2\sin.^2 x$

Therefore, $FD^2+pD^2=e^2+2e\cos. x+1-e^2\sin.^2 x$

But $\qquad e^2\sin.^2 x=e^2-e^2\cos.^2 x.$

24

Substituting this value of e^2 sin.$^2 \cdot x$ in the preceding expression we have

$$\overline{FD}^2 + \overline{pD}^2 = 1 + 2e \cos. x + e^2 \cos.^2 x$$

Whence $\quad Fp = \sqrt{\overline{FD^2 + pD^2}} = 1 + e \cos. x.$

Equating these two values of Fp and we obtain

$$1 - e^2 = (1 + c \cos. x)(1 - e \cos. t)$$

Whence $\quad \cos. t = \dfrac{e + \cos. x}{1 + e \cos. x} \qquad (3)$

Here we have a value of t in terms of x and e, but the equation is not adapted to the use of logarithms.

By equation (27) Plane Trigonometry, we have

$$\tan.^2 \tfrac{1}{2} t = \dfrac{1 - \cos. t}{1 + \cos. t}$$

If the value of cos. t from equation (3) be placed in this we shall have

$$\tan.^2 \tfrac{1}{2} t = \dfrac{1 - \dfrac{e + \cos. x}{1 + e \cos. x}}{1 + \dfrac{e + \cos. x}{1 + e \cos. x}} = \dfrac{1 + e \cos. x - e - \cos. x}{1 + e \cos. x + e + \cos. x}$$

Or, $\tan.^2 \tfrac{1}{2} t = \dfrac{(1 - e) - (1 - e) \cos. x}{(1 + e) + (1 + e) \cos. x} = \dfrac{(1 - e)(1 - \cos. x)}{(1 + e)(1 + \cos. x)}$

That is, $\quad \tan. \tfrac{1}{2} t = \left(\dfrac{1 - e}{1 + e}\right)^{\frac{1}{2}} \tan. \tfrac{1}{2} x. \qquad (4)$

From eq. (2) we obtain

$$\text{Mean Anomaly} = x + e \sin. x. \qquad (5)$$

By assuming x, equation (5) gives the Mean Anomaly. Then equation (4) gives the corresponding *True Anomaly*. To apply these equations to the apparent solar orbit, the value of e is .0167751 the radius of the circle being unity. But $y = e$ sin. x, and as y is a portion of the circumference to the radius unity, we must express e in some known part of the circumference, one degree, for example, as the unit.

Because 180° is equal to 3.14159265, therefore the value of e, in degrees, is found by the following proportion.

$$3.14159265 : 180° : : .0167751 : d \text{ degrees.}$$

By log.,

log. 0167751	—2.2246652
log. 180°	2.2552725
	0.4799377
log. π	0.4971499
Log. e, in degrees, of arc,	—1.9827878
Add log. 60	1.7781513
Log. e, in min. of arc,	1.7609391 constant log.

$$\text{Log. } \sqrt{\frac{1-e}{1+e}} = \log. \left(\frac{0.9832249}{1.0167751}\right)^{\frac{1}{2}} = -1.992714 \text{ cons. log.}$$

We are now prepared to make an application of equations (4) and (5)

For example, we require the equation of the center for the solar orbit, corresponding to 28° of mean anomaly, reckoning from the apogee. The excentric anomaly is less than the mean by about half of the value of the equation of the center at any point; and x must be assumed.

Thus, suppose $x=27° 32'$; then $\frac{1}{2}x=13° 46'$		
sin. $x=$sin. 27° 32'		9.664891
Constant,		1.760939
e sin. $x=$	26' 6518	1.425880
Add x	27° 32'	
Mean Anomaly=	27° 58' 39''1	
Tan. $\frac{1}{2}x$ 13° 46'		9.389178
Const.		—1.992714
tan. $\frac{1}{2}t$ 13° 32' 59''		9.381892
	2	
True anomaly	27° 5' 58''	
Mean Anomaly	27° 58' 39''1	
Equation of center	52' 41''1	

corresponding to the mean anomaly of 27° 58' 39''1, not to 28° as was required.

Now let us take $x=27°$ 40'; then $\frac{1}{2}x=13°$ 50'

sin. x	27° 40	9.666824
	Con.	1.760939
e sin. x	26' 777	1.427763
Add x	27° 40'	

Mean Anomaly, 28° 6' 46''6

tan. $\frac{1}{2}x=$13° 50'		9.391360
Con.		—1.992714
tan. $\frac{1}{2}l$ 13° 36' 43''		9.384074

2
____

$l=$27° 13' 26''

Mean anomaly 28° 6' 46''6

Eq. center,   53' 20''6

          corresponding to 28° 6' 46''6.

Now, we can find the equation corresponding to 28° by the following obvious proportion:

28° 6' 46''6	53' 20''6	28° 00' 00''
27 58 39 1	52 41 1	27 5 39 1
8' 7''5 :	39''5 : :	1' 20''9 : 4''7
Add		52' 41''1

Equation or value sought,   52' 45''1

In like manner we can find the value of the equation of the center of any and every other degree of the mean anomaly in the orbit of the sun, or any other orbit, when the excentricity is known.

LOGARITHMIC TABLES:

ALSO A TABLE OF

NATURAL AND LOGARITHMIC

SINES, COSINES, AND TANGENTS,

TO EVERY MINUTE OF THE QUADRANT.

LOGARITHMS OF NUMBERS

FROM

1 TO 10000.

N.	Log.	N.	Log.	N.	Log.	N.	Log.
1	0 000000	26	1 414973	51	1 707570	76	1 880814
2	0 301030	27	1 431364	52	1 716003	77	1 886491
3	0 477121	28	1 447158	53	1 724276	78	1 892095
4	0 602060	29	1 462398	54	1 732394	79	1 897627
5	0 698970	30	1 477121	55	1 740363	80	1 903090
6	0 778151	31	1 491362	56	1 748188	81	1 908485
7	0 845098	32	1 505150	57	1 755875	82	1 913814
8	0 903090	33	1 518514	58	1 763428	83	1 919078
9	0 954243	34	1 531479	59	1 770852	84	1 924279
10	1 000000	35	1 544068	60	1 778151	85	1 929419
11	1 041393	36	1 556303	61	1 785330	86	1 934498
12	1 079181	37	1 568202	62	1 792392	87	1 939519
13	1 113943	38	1 579784	63	1 799341	88	1 944483
14	1 146128	39	1 591065	64	1 806180	89	1 949390
15	1 176091	40	1 602060	65	1 812913	90	1 954243
16	1 204120	41	1 612784	66	1 819544	91	1 959041
17	1 230449	42	1 623249	67	1 826075	92	1 963788
18	1 255273	43	1 633468	68	1 832509	93	1 968483
19	1 278754	44	1 643453	69	1 838849	94	1 973128
20	1 301030	45	1 653213	70	1 845098	95	1 977724
21	1 322219	46	1 662578	71	1 851258	96	1 982271
22	1 342423	47	1 672098	72	1 857333	97	1 986772
23	1 361728	48	1 681241	73	1 863323	98	1 991226
24	1 380211	49	1 690196	74	1 869232	99	1 995635
25	1 397940	50	1 698970	75	1 875061	100	2 000000

NOTE. In the following table, in the last nine columns of each page, where the first or leading figures change from 9's to 0's, points or dots are now introduced instead of the 0's through the rest of the line, to catch the eye, and to indicate that from thence the corresponding natural number in the first column stands in the *next lower line*, and its annexed first two figures of the Logarithms in the second column.

N.	0	1	2	3	4	5	6	7	8	9
100	000000	0434	0868	1301	1734	2166	2598	3029	3461	3891
101	4321	4750	5181	5609	6038	6466	6894	7321	7748	8174
102	8600	9026	9451	9876	.300	.724	1147	1570	1993	2415
103	012837	3259	3680	4100	4521	4940	5360	5779	6197	6616
104	7033	7451	7868	8284	8700	9116	9532	9947	.361	.775
105	021189	1603	2016	2428	2841	3252	3664	4075	4486	4896
106	5306	5715	6125	6533	6942	7350	7757	8164	8571	8978
107	9384	9789	.195	.600	1004	1408	1812	2216	2619	3021
108	033424	3826	4227	4628	5029	5430	5830	6230	6629	7028
109	7426	7825	8223	8620	9017	9414	9811	.207	.602	.998
110	041393	1787	2182	2576	2969	3362	3755	4148	4540	4932
111	5323	5714	6105	6495	6885	7275	7664	8053	8442	8830
112	9218	9606	9993	.380	.766	1153	1538	1924	2309	2694
113	053078	3463	3846	4230	4613	4996	5378	5760	6142	6524
114	6905	7286	7666	8046	8426	8805	9185	9563	9942	.320
115	060698	1075	1452	1829	2206	2582	2958	3333	3709	4083
116	4458	4832	5206	5580	5953	6326	6699	7071	7443	7815
117	8186	8557	8928	9298	9668	..38	.407	.776	1145	1514
118	071882	2250	2617	2985	3352	3718	4085	4451	4816	5182
119	5547	5912	6276	6640	7004	7368	7731	8094	8457	8819
120	9181	9543	9904	.266	.626	.987	1347	1707	2067	2426
121	082785	3144	3503	3861	4219	4576	4934	5291	5647	6004
122	6360	6716	7071	7426	7781	8136	8490	8845	9198	9552
123	9905	.258	.611	.963	1315	1667	2018	2370	2721	3071
124	093422	3772	4122	4471	4820	5169	5518	5866	6215	6562
125	6910	7257	7604	7951	8298	8644	8990	9335	9681	1026
126	100371	0715	1059	1403	1747	2091	2434	2777	3119	3462
127	3804	4146	4487	4828	5169	5510	5851	6191	6531	6871
128	7210	7549	7888	8227	8565	8903	9241	9579	9916	.253
129	110590	0926	1263	1599	1934	2270	2605	2940	3275	3609
130	3943	4277	4611	4944	5278	5611	5943	6276	6608	6940
131	7271	7603	7934	8265	8595	8926	9256	9586	9915	0245
132	120574	0903	1231	1560	1888	2216	2544	2871	3198	3525
133	3852	4178	4504	4830	5156	5481	5806	6131	6456	6781
134	7105	7429	7753	8076	8399	8722	9045	9368	9690	..12
135	130334	0655	0977	1298	1619	1939	2260	2580	2900	3219
136	3539	3858	4177	4496	4814	5133	5451	5769	6086	6403
137	6721	7037	7354	7671	7987	8303	8618	8934	9249	9564
138	9879	.194	.508	.822	1136	1450	1763	2076	2389	2702
139	143015	3327	3639	3951	4263	4574	4885	5196	5507	5818
140	6128	6438	6748	7058	7367	7676	7985	8294	8603	8911
141	9219	9527	9835	.142	.449	.756	1063	1370	1676	1982
142	152288	2594	2900	3205	3510	3815	4120	4424	4728	5032
143	5336	5640	5943	6246	6549	6852	7154	7457	7759	8061
144	8362	8664	8965	9266	9567	9868	.168	.469	.769	1068
145	161368	1667	1967	2266	2564	2863	3161	3460	3758	4055
146	4353	4650	4947	5244	5541	5838	6134	6430	6726	7022
147	7317	7613	7908	8203	8497	8792	9086	9380	9674	9968
148	170262	0555	0848	1141	1434	1726	2019	2311	2603	2895
149	3186	3478	3769	4060	4351	4641	4932	5222	5512	5802

N.	0	1	2	3	4	5	6	7	8	9
150	176091	6381	6670	6959	7248	7536	7825	8113	8401	8689
151	8977	9264	9552	9839	.126	.413	.699	.985	1272	1558
152	181844	2129	2415	2700	2985	3270	3555	3839	4123	4407
153	4691	4975	5259	5542	5825	6108	6391	6674	6956	7239
154	7521	7803	8084	8366	8647	8928	9209	9490	9771	..51
					281					
155	190332	0612	0892	1171	1451	1730	2010	2289	2567	2846
156	3125	3403	3681	3959	4237	4514	4792	5069	5346	5623
157	5899	6176	6453	6729	7005	7281	7556	7832	8107	8382
158	8657	8932	9206	9481	9755	..29	.303	.577	.850	1124
159	201397	1670	1943	2216	2488	2761	3033	3305	3577	3848
					273					
160	4120	4391	4663	4934	5204	5475	5746	6016	6286	6556
161	6826	7096	7365	7634	7904	8173	8441	8710	8979	9247
162	9515	9783	..51	.319	.586	.853	1121	1388	1654	1921
163	212188	2454	2720	2986	3252	3518	3783	4049	4314	4579
164	4844	5109	5373	5638	5902	6166	6430	6694	6957	7221
					264					
165	7484	7747	8010	8273	8536	8798	9060	9323	9585	9846
166	220108	0370	0631	0892	1153	1414	1675	1936	2196	2456
167	2716	2976	3236	3496	3755	4015	4274	4533	4792	5051
168	5309	5568	5826	6084	6342	6600	6858	7115	7372	7630
169	7887	8144	8400	8657	8913	9170	9426	9682	9938	.193
					257					
170	230449	0704	0960	1215	1470	1724	1979	2234	2488	2742
171	2996	3250	3504	3757	4011	4264	4517	4770	5023	5276
172	5528	5781	6033	6285	6537	6789	7041	7292	7544	7795
173	8046	8297	8548	8799	9049	9299	9550	9800	..50	.300
174	240549	0799	1048	1297	1546	1795	2044	2293	2541	2790
					249					
175	3038	3286	3534	3782	4030	4277	4525	4772	5019	5266
176	5513	5759	6006	6252	6499	6745	6991	7237	7482	7728
177	7973	8219	8464	8709	8954	9198	9443	9687	9932	.176
178	250420	0664	0908	1151	1395	1638	1881	2125	2368	2610
179	2853	3096	3338	3580	3822	4064	4306	4548	4790	5031
					242					
180	5273	5514	5755	5996	6237	6477	6718	6958	7198	7439
181	7679	7918	8158	8398	8637	8877	9116	9355	9594	9833
182	260071	0310	0548	0787	1025	1263	1501	1739	1976	2214
183	2451	2688	2925	3162	3399	3636	3873	4109	4346	4582
184	4818	5054	5290	5525	5761	5996	6232	6467	6702	6937
					235					
185	7172	7406	7641	7875	8110	8344	8578	8812	9046	9279
186	9513	9746	9980	.213	.446	.679	.912	1144	1377	1609
187	271842	2074	2306	2538	2770	3001	3233	3464	3696	3927
188	4158	4389	4620	4850	5081	5311	5542	5772	6002	6232
189	6462	6692	6921	7151	7380	7609	7838	8067	8296	8525
					229					
190	8754	8982	9211	9439	9667	9895	.123	.351	.578	.806
191	281033	1261	1488	1715	1942	2169	2396	2622	2849	3075
192	3301	3527	3753	3979	4205	4431	4656	4882	5107	5332
193	5557	5782	6007	6232	6456	6681	6905	7130	7354	7578
194	7802	8026	8249	8473	8696	8920	9143	9366	9589	9812
					224					
195	290035	0257	0480	0702	0925	1147	1369	1591	1813	2034
196	2256	2478	2699	2920	3141	3363	3584	3804	4025	4246
197	4466	4687	4907	5127	5347	5567	5787	6007	6226	6446
198	6665	6884	7104	7323	7542	7761	7979	8198	8416	8635
199	8853	9071	9289	9507	9725	9943	.161	.378	.595	.813

N.	0	1	2	3	4	5	6	7	8	9
200	301030	1247	1464	1681	1898	2114	2331	2547	2764	2980
201	3196	3412	3628	3844	4059	4275	4491	4706	4921	5136
202	5351	5566	5781	5996	6211	6425	6639	6854	70.8	7282
203	7496	7710	7924	8137	8351	8564	8778	8991	9204	9417
204	9630	9843	..56	.268	.481	.693	.906	1118	1330	1542
					212					
205	311754	1966	2177	2389	2600	2812	3023	3234	3445	3656
206	3867	4078	4289	4499	4710	4920	5130	5340	5551	5760
207	5970	6180	6390	6599	6809	7018	7227	7436	7646	7854
208	8063	8272	8481	8689	8898	9106	9314	9522	9730	9938
209	320146	0354	0562	0769	0977	1184	1391	1598	1805	2012
					207					
210	2219	2426	2633	2839	3046	3252	3458	3665	3871	4077
211	4282	4488	4694	4899	5105	5310	5516	5721	5926	6131
212	6336	6541	6745	6950	7155	7359	7563	7767	7972	8176
213	8380	8583	8787	8991	9194	9398	9601	9805	...8	.211
214	330414	0617	0819	1022	1225	1427	1630	1832	2034	2236
					202					
215	2438	2640	2842	3044	3246	3447	3649	3850	4051	4253
216	4454	4655	4856	5057	5257	5458	5658	5859	6059	6260
217	6460	6660	6860	7060	7260	7459	7659	7858	8058	8257
218	8456	8656	8855	9054	9253	9451	9650	9849	..47	.246
219	340444	0642	0841	1039	1237	1435	1632	1830	2028	2225
					198					
220	2423	2620	2817	3014	3212	3409	3606	3802	3999	4196
221	4392	4589	4785	4981	5178	5374	5570	5766	5962	6157
222	6353	6549	6744	6939	7135	7330	7525	7720	7915	8110
223	8305	8500	8694	8889	9083	9278	9472	9666	9860	..54
224	350248	0442	0636	0829	1023	1216	1410	1603	1796	1989
					193					
225	2183	2375	2568	2761	2954	3147	3339	3532	3724	3916
226	4108	4301	4493	4685	4876	5068	5260	5452	5643	5834
227	6026	6217	6408	6599	6790	6981	7172	7363	7554	7744
228	7935	8125	8316	8506	8696	8886	9076	9266	9456	9646
229	9835	..25	.215	.404	.593	.783	.972	1161	1350	1539
					190					
230	361728	1917	2105	2294	2482	2671	2859	3048	3236	3424
231	3612	3800	3988	4176	4363	4551	4739	4926	5113	5301
232	5488	5675	5862	6049	6236	6423	6610	6796	6983	7169
233	7356	7542	7729	7915	8101	8287	8473	8659	8845	9030
234	9216	9401	9587	9772	9958	.143	.328	.513	.698	.883
					185					
235	371068	1253	1437	1622	1806	1991	2175	2360	2544	2728
236	2912	3096	3280	3464	3647	3831	4015	4198	4382	4565
237	4748	4932	5115	5298	5481	5664	5846	6029	6212	6394
238	6577	6759	6942	7124	7306	7488	7670	7852	8034	8216
239	8398	8580	8761	8943	9124	9306	9487	9668	9849	..30
					182					
240	380211	0392	0573	0754	0934	1115	1296	1476	1656	1837
241	2017	2197	2377	2557	2737	2917	3097	3277	3456	3636
242	3815	3995	4174	4353	4533	4712	4891	5070	5249	5428
243	5606	5785	5964	6142	6321	6499	6677	6856	7034	7212
244	7390	7568	7746	7923	8101	8279	8456	8634	8811	8989
					178					
245	9166	9343	9520	9698	9875	..51	.228	.405	.582	.759
246	390035	1112	1288	1464	1641	1817	1993	2169	2345	2521
247	2697	2873	3048	3224	3400	3575	3751	3926	4101	4277
248	4452	4627	4802	4977	5152	5326	5501	5676	5850	6025
249	6199	6374	6548	6722	6896	7071	7245	7419	7592	7766

N.	0	1	2	3	4	5	6	7	8	9
250	397940	8114	8287	8461	8634	8808	8981	9154	9328	9501
251	9674	9847	..20	.192	.365	.538	.711	.883	1056	1228
252	401401	1573	1745	1917	2089	2261	2433	2605	2777	2949
253	3121	3292	3464	3635	3807	3978	4149	4320	4492	4663
254	4834	5005	5176	5346	5517	5688	5858	6029	6199	6370
					171					
255	6540	6710	6881	7051	7221	7391	7561	7731	7901	8070
256	8240	8410	8579	8749	8918	9087	9257	9426	9595	9764
257	9933	.102	.271	.440	.609	.777	.946	1114	1283	1451
258	411620	1788	1956	2124	2293	2461	2629	2796	2964	3132
259	3300	3467	3635	3803	3970	4137	4305	4472	4639	4806
260	4973	5140	5307	5474	5641	5808	5974	6141	6308	6474
261	6641	6807	6973	7139	7306	7472	7638	7804	7970	8135
262	8301	8467	8633	8798	8964	9129	9295	9460	9625	9791
263	9956	.121	.286	.451	.616	.781	.945	1110	1275	1439
264	421604	1788	1933	2097	2261	2426	2590	2754	2918	3082
265	3246	3410	3574	3737	3901	4065	4228	4392	4555	4718
266	4882	5045	5208	5371	5534	5697	5860	6023	6186	6349
267	6511	6674	6836	6999	7161	7324	7486	7648	7811	7973
268	8135	8297	8459	8621	8783	8944	9106	9268	9429	9591
269	9752	9914	..75	.236	.398	.559	.720	.881	1042	1203
270	431364	1525	1685	1846	2007	2167	2328	2488	2649	2809
271	2969	3130	3290	3450	3610	3770	3930	4090	4249	4409
272	4569	4729	4888	5048	5207	5367	5526	5685	5844	6004
273	6163	6322	6481	6640	6800	6957	7116	7275	7433	7592
274	7751	7909	8067	8226	8384	8542	8701	8859	9017	9175
					158					
275	9333	9491	9648	9806	9964	.122	.279	.437	.594	.752
276	440909	1066	1224	1381	1538	1695	1852	2009	2166	2323
277	2480	2637	2793	2950	3106	3263	3419	3576	3732	3889
278	4045	4201	4357	4513	4669	4825	4981	5137	5293	5449
279	5604	5760	5915	6071	6226	6382	6537	6692	6848	7003
280	7158	7313	7468	7623	7778	7933	8088	8242	8397	8552
281	8706	8861	9015	9170	9324	9478	9633	9787	9941	..95
282	450249	0403	0557	0711	0865	1018	1172	1326	1479	1633
283	1786	1940	2093	2247	2400	2553	2706	2859	3012	3165
284	3318	3471	3624	3777	3930	4082	4235	4387	4540	4692
285	4845	4997	5150	5302	5454	5606	5758	5910	6062	6214
286	6366	6518	6670	6821	6973	7125	7276	7428	7579	7731
287	7882	8033	8184	8336	8487	8638	8789	8940	9091	9242
288	9392	9543	9694	9845	9995	.146	.296	.417	.597	.748
289	460898	1048	1198	1348	1499	1649	1799	1948	2098	2248
290	2398	2548	2697	2847	2997	3146	3296	3445	3594	3744
291	3893	4042	4191	4340	4490	4639	4788	4936	5085	5234
292	5383	5532	5680	5829	5977	6126	6274	6423	6571	6719
293	6868	7016	7164	7312	7460	7608	7756	7904	8052	8200
294	8347	8495	8643	8790	8938	9085	9233	9380	9527	9675
					147					
295	9822	9969	.116	.263	.410	.557	.704	.851	.998	1145
296	471292	1438	1585	1732	1878	2025	2171	2318	2464	2610
297	2756	2903	3049	3195	3341	3487	3633	3779	3925	4071
298	4216	4362	4508	4653	4799	4944	5090	5235	5381	5526
299	5671	5816	5962	6107	6252	6397	6542	6687	6832	6976

N.	0	1	2	3	4	5	6	7	8	9
300	477121	7266	7411	7555	7700	7844	7989	8133	8278	8422
301	8566	8711	8855	8999	9143	9287	9481	9575	9719	9863
302	480007	0151	0294	0438	0582	0725	0869	1012	1156	1299
303	1443	1586	1729	1872	2016	2159	2302	2445	2588	2731
304	2874	3016	3159	3302	3445	3587	3730	3872	4015	4157
					142					
305	4300	4442	4585	4727	4869	5011	5153	5295	5437	5579
306	5721	5863	6005	6147	6289	6430	6572	6714	6855	6997
307	7138	7280	7421	7563	7704	7845	7986	8127	8269	8410
308	8551	8692	8833	8974	9114	9255	9396	9537	9667	9818
309	9959	..99	.239	.380	.520	.661	.801	.941	1081	1222
310	491362	1502	1642	1782	1922	2062	2201	2341	2481	2621
311	2760	2900	3040	3179	3319	3458	3597	3737	3876	4015
312	4155	4294	4433	4572	4711	4850	4989	5128	5267	5406
313	5544	5683	5822	5960	6099	6238	6376	6515	6653	6791
314	6930	7068	7206	7344	7483	7621	7759	7897	8035	8173
315	8311	8448	8586	8724	8862	8999	9137	9275	9412	9550
316	9687	9824	9962	..99	.236	.374	.511	.648	.785	.922
317	501059	1196	1333	1470	1607	1744	1880	2017	2154	2291
318	2427	2564	2700	2837	2973	3109	3246	3382	3518	3655
319	3791	3927	4063	4199	4335	4471	4607	4743	4878	5014
320	5150	5285	5421	5557	5693	5828	5964	6099	6234	6370
321	6505	6640	6776	6911	7046	7181	7316	7451	7586	7721
322	7856	7991	8126	8260	8395	8530	8664	8799	8934	9068
323	9203	9337	9471	9606	9740	9874	...9	.143	.277	.411
324	510345	0679	0813	0947	1081	1215	1349	1482	1616	1750
					134					
325	1883	2017	2151	2284	2418	2551	2684	2818	2951	3084
326	3218	3351	3484	3617	3750	3883	4016	4149	4282	4414
327	4548	4681	4813	4946	5079	5211	5344	5476	5609	5741
328	5874	6006	6139	6271	6403	6535	6668	6800	6932	7064
329	7196	7328	7460	7592	7724	7855	7987	8119	8251	8382
330	8514	8646	8777	8909	9040	9171	9303	9434	9566	9697
331	9828	9959	..90	.221	.353	.484	.615	.745	.876	1007
332	521138	1269	1400	1530	1661	1792	1922	2053	2183	2314
333	2444	2575	2705	2835	2966	3096	3226	3356	3486	3616
334	3746	3876	4006	4136	4266	4396	4526	4656	4785	4915
335	5045	5174	5304	5434	5563	5693	5822	5951	6081	6210
336	6339	6469	6598	6727	6856	6985	7114	7243	7372	7501
337	7630	7759	7888	8016	8145	8274	8402	8531	8660	8788
338	8917	9045	9174	9302	9430	9559	9687	9815	9943	..72
339	530200	0328	0456	0584	0712	0840	0968	1096	1223	1351
340	1479	1607	1734	1862	1990	2117	2245	2372	2500	2627
341	2754	2882	3009	3136	3264	3391	3518	3645	3772	3899
342	4026	4153	4280	4407	4534	4661	4787	4914	5041	5167
343	5294	5421	5547	5674	5800	5927	6053	6180	6306	6432
344	6558	6685	6811	6937	7063	7189	7315	7441	7567	7693
					129					
345	7819	7945	8071	8197	8322	8448	8574	8699	8825	8951
346	9076	9202	9327	9452	9578	9703	9829	9954	..79	.204
347	540329	0455	0580	0705	0830	0955	1080	1205	1330	1454
348	1579	1704	1829	1953	2078	2203	2327	2452	2576	2701
349	2825	2950	3074	3199	3323	3447	3571	3696	3820	3944

N.	0	1	2	3	4	5	6	7	8	9
350	544068	4192	4316	4440	4564	4688	4812	4936	5060	5183
351	5307	5431	5555	5678	5805	5925	6049	6172	6296	6419
352	6543	6666	6789	6913	7036	7159	7282	7405	7529	7652
353	7775	7898	8021	8144	8267	8389	8512	8635	8758	8881
354	9003	9126	9249	9371	9494	9616	9739	9861	9984	.196
					122					
355	550228	0351	0473	0595	0717	0840	0962	1084	1206	1328
356	1450	1572	1694	1816	1938	2060	2181	2303	2425	2547
357	2668	2790	2911	3033	3155	3276	3393	3519	3640	3762
358	3883	4004	4126	4247	4368	4489	4610	4731	4852	4973
359	5094	5215	5346	5457	5578	5699	5820	5940	6061	6182
360	6303	6423	6544	6664	6785	6905	7026	7146	7267	7387
361	7507	7627	7748	7868	7988	8108	8228	8349	8469	8589
362	8709	8829	8948	9068	9188	9308	9428	9548	9667	9787
363	9907	..26	.146	.265	.385	.504	.624	.743	.863	.982
364	561101	1221	1340	1459	1578	1698	1817	1936	2055	2173
365	2293	2412	2531	2650	2769	2887	3006	3125	3244	3362
366	3481	3600	3718	3837	3955	4074	4192	4311	4429	4548
367	4666	4784	4903	5021	5139	5257	5376	5494	5612	5730
368	5848	5966	6084	6202	6320	6437	6555	6673	6791	6909
369	7026	7144	7262	7379	7497	7614	7732	7849	7967	8084
370	8202	8319	8436	8554	8671	8788	8905	9023	9140	9257
371	9374	9491	9608	9725	9882	9959	..76	.193	.309	.426
372	570543	0660	0776	0893	1010	1126	1243	1359	1476	1592
373	1709	1825	1942	2058	2174	2291	2407	2523	2639	2755
374	2872	2988	3104	3220	3336	3452	3568	3684	3800	3915
					116					
375	4031	4147	4263	4379	4494	4610	4726	4841	4957	5072
376	5188	5303	5419	5534	5650	5765	5880	5996	6111	6226
377	6341	6457	6572	6687	6802	6917	7032	7147	7262	7377
378	7492	7607	7722	7836	7951	8066	8181	8295	8410	8525
379	8639	8754	8868	8983	9097	9212	9326	9441	9555	9669
380	9784	9898	..12	.126	.241	.355	.469	.583	.697	.811
381	580925	1039	1153	1267	1381	1495	1608	1722	1836	1950
382	2063	2177	2291	2404	2518	2631	2745	2858	2972	3085
383	3199	3312	3426	3539	3652	3765	3879	3992	4105	4218
384	4331	4444	4557	4670	4783	4896	5009	5122	5235	5348
385	5461	5574	5686	5799	5912	6024	6137	6250	6362	6475
386	6587	6700	6812	6925	7037	7149	7262	7374	7486	7599
387	7711	7823	7935	8047	8160	8272	8384	8496	8608	8720
388	8832	8944	9056	9167	9279	9391	9503	9615	9726	9834
389	9950	..61	.173	.284	.396	.507	.619	.730	.842	.953
390	591065	1176	1287	1399	1510	1621	1732	1843	1955	2066
391	2177	2288	2399	2510	2621	2732	2843	2954	3064	3175
392	3286	3397	3508	3618	3729	3840	3950	4061	4171	4282
393	4393	4503	4614	4724	4834	4945	5055	5165	5276	5386
394	5496	5606	5717	5827	5937	6047	6157	6267	6377	6487
					110					
395	6597	6707	6817	6927	7037	7146	7256	7366	7476	7586
396	7695	7805	7914	8024	8134	8243	8353	8462	8572	8681
397	8791	8900	9009	9119	9228	9337	9446	9556	9666	9774
398	9883	9992	.101	.210	.319	.428	.537	.646	.755	.864
399	600973	1082	1191	1299	1408	1517	1625	1734	1843	1951

N.	0	1	2	3	4	5	6	7	8	9
400	602060	2169	2277	2386	2494	2603	2711	2819	2928	3036
401	3144	3253	3361	3469	3573	3686	3794	3902	4010	4118
402	4226	4334	4442	4550	4658	4766	4874	4982	5089	5197
403	5305	5413	5521	5628	5736	5844	5951	6059	6166	6274
404	6381	6489	6596	6704	6811	6919	7026	7133	7241	7348
					108					
405	7455	7562	7669	7777	7884	7991	8098	8205	8312	8419
406	8526	8633	8740	8847	8954	9061	9167	9274	9381	9488
407	9594	9701	9808	9914	..21	.128	.234	.341	.447	.554
408	610660	0767	0873	0979	1086	1192	1298	1405	1511	1617
409	1723	1829	1936	2042	2148	2254	2360	2466	2572	2678
410	2784	2890	2996	3102	3207	3313	3419	3525	3630	3736
411	3842	3947	4053	4159	4264	4370	4475	4581	4686	4792
412	4897	5003	5108	5213	5319	5424	5529	5634	5740	5845
413	5950	6055	6160	6265	6370	6476	6581	6686	6790	6895
414	7000	7105	7210	7315	7420	7525	7629	7734	7839	7943
415	8048	8153	8257	8362	8466	8571	8676	8780	8884	8989
416	9293	9198	9302	9406	9511	9615	9719	9824	9928	..32
417	620136	0240	0344	0448	0552	0656	0760	0864	0968	1072
418	1176	1280	1384	1488	1592	1695	1799	1903	2007	2110
419	2214	2318	2421	2525	2628	2732	2835	2939	3042	3146
420	3249	3353	3456	3559	3663	3766	3869	3973	4076	4179
421	4282	4385	4488	4591	4695	4798	4901	5004	5107	5210
422	5312	5415	5518	5621	5724	5827	5929	6032	6135	6238
423	6340	6443	6546	6648	6751	6853	6956	7058	7161	7263
424	7366	7468	7571	7673	7775	7878	7980	8082	8185	8287
					103					
425	8389	8491	8593	8695	8797	8900	9002	9104	9206	9308
426	9410	9512	9613	9715	9817	9919	..21	.123	.224	.326
427	630428	0530	0631	0733	0835	0936	1038	1139	1241	1342
428	1444	1545	1647	1748	1849	1951	2052	2153	2255	2356
429	2457	2559	2660	2761	2862	2963	3064	3165	3266	3367
430	3468	3569	3670	3771	3872	3973	4074	4175	4276	4376
431	4477	4578	4679	4779	4880	4981	5081	5182	5283	5383
432	5484	5584	5685	5785	5886	5986	6087	6187	6287	6388
433	6488	6588	6688	6789	6889	6989	7089	7189	7290	7390
434	7490	7590	7690	7790	7890	7990	8090	8190	8290	8389
435	8489	8589	8689	8789	8888	8988	9088	9188	9287	9387
436	9486	9586	9686	9785	9885	9984	..84	.183	.283	.382
437	640481	0581	0680	0779	0879	0978	1077	1177	1276	1375
438	1474	1573	1672	1771	1871	1970	2069	2168	2267	2366
439	2465	2563	2662	2761	2860	2959	3058	3156	3255	3354
440	3453	3551	3650	3749	3847	3946	4044	4143	4242	4340
441	4439	4537	4636	4734	4832	4931	5029	5127	5226	5324
442	5422	5521	5619	5717	5815	5913	6011	6110	6208	6306
443	6404	6502	6600	6698	6796	6894	6992	7039	7187	7285
444	7383	7481	7579	7676	7774	7872	7969	8067	8165	8262
					93					
445	8360	8458	8555	8653	8750	8848	8945	9043	9140	9237
446	9335	9432	9530	9627	9724	9821	9919	..16	.113	.210
447	650308	0405	0502	0599	0696	0793	0890	0987	1084	1181
448	1278	1375	1472	1569	1666	1762	1859	1956	2053	2150
449	2246	2343	2440	2530	2633	2730	2826	2923	3019	3116

N.	0	1	2	3	4	5	6	7	8	9
450	653213	3309	3405	3502	3598	3695	3791	3888	3984	4080
451	4177	4273	4369	4465	4562	4658	4754	4850	4946	5042
452	5138	5235	5331	5427	5526	5619	5715	5810	5906	6002
453	6098	6194	6290	6386	6482	6577	6673	6769	6864	6960
454	7056	7152	7247	7343	7438	7534	7629	7725	7820	7916
					96					
455	8011	8107	8202	8298	8393	8488	8584	8679	8774	8870
456	8965	9060	9155	9250	9346	9441	9536	9631	9726	9821
457	9916	..11	.106	.201	.296	.391	.486	.581	.676	.771
458	660865	0960	1055	1150	1245	1339	1434	1529	1623	1718
459	1813	1907	2002	2096	2191	2286	2380	2475	2569	2663
460	2758	2852	2947	3041	3135	3230	3324	3418	3512	3607
461	3701	3795	3889	3983	4078	4172	4266	4360	4454	4548
462	4642	4736	4830	4924	5018	5112	5206	5299	5393	5487
463	5581	5675	5769	5862	5956	6050	6143	6237	6331	6424
464	6518	6612	6705	6799	6892	6986	7079	7173	7266	7360
465	7453	7546	7640	7733	7826	7920	8013	8106	8199	8293
466	8386	8479	8572	8665	8759	8852	8945	9038	9131	9324
467	9317	9410	9503	9596	9689	9782	9875	9967	..60	.153
468	670241	0339	0431	0524	0617	0710	0802	0895	0988	1080
469	1173	1265	1358	1451	1543	1636	1728	1821	1913	2005
470	2098	2190	2283	2375	2467	2560	2652	2744	2836	2929
471	3021	3113	3205	3297	3390	3482	3574	3666	3758	3850
472	3942	4034	4126	4218	4310	4402	4494	4586	4677	4769
473	4861	4953	5045	5137	5228	5320	5412	5503	5595	5687
474	5778	5870	5962	6053	6145	6236	6328	6419	6511	6602
					91					
475	6694	6785	6876	6968	7059	7151	7242	7333	7424	7516
476	7607	7698	7789	7881	7972	8063	8154	8245	8336	8427
477	8518	8609	8700	8791	8882	8972	9064	9155	9246	9337
478	9428	9519	9610	9700	9791	9882	9973	..63	.154	.245
479	680336	0426	0517	0607	0698	0789	0879	0970	1060	1151
480	1241	1332	1422	1513	1603	1693	1784	1874	1964	2055
481	2145	2235	2326	2416	2506	2596	2686	2777	2867	2957
482	3047	3137	3227	3317	3407	3497	3587	3677	3767	3857
483	3947	4037	4127	4217	4307	4396	4486	4576	4666	4756
484	4854	4935	5025	5114	5204	5294	5383	5473	5563	5652
485	5742	5831	5921	6010	6100	6189	6279	6368	6458	6547
486	6636	6726	6815	6904	6994	7083	7172	7261	7351	7440
487	7529	7618	7707	7796	7886	7975	8064	8153	8242	8331
488	8420	8509	8598	8687	8776	8865	8953	9042	9131	9220
489	9309	9398	9486	9575	9664	9753	9841	9930	..19	.107
490	690196	0285	0373	0362	0550	0639	0728	0816	0905	0993
491	1081	1170	1258	1347	1435	1524	1612	1700	1789	1877
492	1965	2053	2142	2230	2318	2406	2494	2583	2671	2759
493	2847	2935	3023	3111	3199	3287	3375	3463	3551	3639
494	3727	3815	3903	3991	4078	4166	4254	4342	4430	4517
					88					
495	4605	4693	4781	4868	4956	5044	5131	5210	5307	5394
496	5482	5569	5657	5744	5832	5919	6007	6094	6182	6269
497	6356	5444	6531	6618	6706	6793	6880	6968	7055	7142
498	7229	7317	7404	7491	7578	7665	7752	7839	7926	8014
499	8101	8188	8275	8362	8449	8535	8622	8709	8796	8883

N.	0	1	2	3	4	5	6	7	8	9
500	698970	9057	9144	9231	9317	9404	9491	9578	9664	9751
501	9838	9924	..11	..98	.184	.271	.358	.444	.531	.617
502	700704	0790	0877	0963	1050	1136	1222	1309	1395	1482
503	1568	1654	1741	1827	1913	1999	2086	2172	2258	2344
504	2431	2517	2603	2689	2775	2861	2947	3033	3119	3205
					86					
505	3291	3377	3463	3549	3635	3721	3807	3895	3979	4065
506	4151	4236	4322	4408	4494	4579	4665	4751	4837	4922
507	5008	5094	5179	5265	5350	5436	5522	5607	5693	5778
508	5864	5949	6035	6120	6206	6291	6376	6462	6547	6632
509	6718	6803	6888	6974	7059	7144	7229	7315	7400	7485
510	7570	7655	7740	7826	7910	7996	8081	8166	8251	8336
511	8421	8506	8591	8676	8761	8846	8931	9015	9100	9185
512	9270	9355	9440	9524	9609	9694	9779	9863	9948	..33
513	710117	0202	0287	0371	0456	0540	0625	0710	0794	0879
514	0963	1048	1132	1217	1301	1385	1470	1554	1639	1723
515	1807	1892	1976	2060	2144	2229	2313	2397	2481	2566
516	2650	2734	2818	2902	2986	3070	3154	3238	3326	3407
517	3491	3575	3659	3742	3826	3910	3994	4078	4162	4246
518	4330	4414	4497	4581	4665	4749	4833	4916	5000	5084
519	5167	5251	5335	5418	5502	5586	5669	5753	5836	5920
520	6003	6087	6170	6254	6337	6421	6504	6588	6671	6754
521	6838	6921	7004	7088	7171	7254	7338	7421	7504	7587
522	7671	7754	7837	7920	8003	8086	8169	8253	8336	8419
523	8502	8585	8668	8751	8834	8917	9000	9083	9165	9248
524	9331	9414	9497	9580	9663	9745	9828	9911	9994	..77
					82					
525	720159	0242	0325	0407	0490	0573	0655	0738	0821	0903
526	0986	1068	1151	1233	1316	1398	1481	1563	1646	1728
527	1811	1893	.975	2058	2140	2222	2305	2387	2469	2552
528	2634	2716	2798	2881	2963	3045	3127	3209	3291	3374
529	3456	3538	3620	3702	3784	3866	3948	4030	4112	4194
530	4276	4358	4440	4522	4604	4685	4767	4849	4931	5013
531	5095	5176	5258	5340	5422	5503	5585	5667	5748	5830
532	5912	5993	6075	6156	6238	6320	6401	6483	6564	6646
533	6727	6809	6890	6972	7053	7134	7216	7297	7379	7460
534	7541	7623	7704	7785	7866	7948	8029	8110	8191	8273
535	8354	8435	8516	8597	8678	8759	8841	8922	9003	9084
536	9165	9246	9327	9403	9489	9570	9651	9732	9813	9893
537	9974	..55	.136	.217	.298	.378	.459	.440	.621	.702
538	730782	0863	0944	1024	1105	1186	1266	1347	1428	1508
539	1589	1669	1750	1830	1911	1991	2072	2152	2233	2313
540	2394	2474	2555	2635	2715	2796	2876	2956	3037	3117
541	3197	3278	3358	3438	3518	3598	3679	3759	3839	3919
542	3999	4079	4160	4240	4320	4400	4480	4560	4640	4720
543	4800	4880	4960	5040	5120	5200	5279	5359	5439	5519
544	5599	5679	5759	5838	5918	5998	6078	6157	6237	6317
					80					
545	6397	6476	6556	6636	6715	6795	6874	6954	7034	7113
546	7193	7272	7352	7431	7511	7590	7670	7749	7829	7908
547	7987	8067	8146	8225	8305	8384	8463	8543	8622	8701
548	8781	8860	8939	9018	9097	9177	9256	9335	9414	9492
549	9572	9651	9731	9810	9889	9968	..47	.126	.205	.284

N.	0	1	2	3	4	5	6	7	8	9
550	740363	0442	0521	0560	0678	0757	0836	0915	0994	1073
551	1152	1230	1309	1388	1467	1546	1624	1703	1782	1860
552	1939	2018	2096	2175	2254	2332	2411	2489	2568	2646
553	2725	2804	2882	2961	3039	3118	3196	3275	3353	3431
554	3510	3558	3667	3745	3823	3902	3980	4058	4136	4215
					79					
555	4293	4371	4449	4528	4606	4684	4762	4840	4919	4997
556	5075	5153	5231	5309	5387	5465	5543	5621	5699	5777
557	5855	5933	6011	6089	6167	6245	6323	6401	6479	6556
558	6634	6712	6790	6868	6945	7023	7101	7179	7256	7334
559	7412	7489	7567	7645	7722	7800	7878	7955	8033	8110
560	8188	8266	8343	8421	8498	8576	8653	8731	8808	8885
561	8963	9040	9118	9195	9272	9350	9427	9504	9582	9659
562	9736	9814	9891	9968	..45	.123	.200	.277	.354	.431
563	750508	0586	0663	0740	0817	0894	0971	1048	1125	1202
564	1279	1356	1433	1510	1587	1664	1741	1818	1895	1972
565	2048	2125	2202	2279	2356	2433	2509	2586	2663	2740
566	2816	2893	2970	3047	3123	3200	3277	3353	3430	3506
567	3582	3660	3736	3813	3889	3966	4042	4119	4195	4272
568	4348	4425	4501	4578	4654	4730	4807	4883	4960	5036
569	5112	5189	5265	5341	5417	5494	5570	5646	5722	5799
570	5875	5951	6027	6103	6180	6256	6332	6408	6484	6560
571	6636	6712	6788	6864	6940	7016	7092	7168	7244	7320
572	7396	7472	7548	7624	7700	7775	7851	7927	8003	8079
573	8155	8230	8306	8382	8458	8533	8609	8685	8761	8836
574	8912	8988	9068	9139	9214	9290	9366	9441	9517	9592
					74					
575	9668	9743	9819	9894	9970	..45	.121	.196	.272	.347
576	760422	0498	0573	0649	0724	0799	0875	0950	1025	1101
577	1176	1251	1326	1402	1477	1552	1627	1702	1778	1853
578	1928	2003	2078	2153	2228	2303	2378	2453	2529	2604
579	2679	2754	2829	2904	2978	3053	3128	2203	3278	3353
580	3428	3503	3578	3653	3727	3802	3877	3952	4027	4101
581	4176	4251	4326	4400	4475	4550	4624	4699	4774	4848
582	4923	4998	5072	5147	5221	5296	5370	5445	5520	5594
583	5669	5743	5818	5892	5966	6041	6115	6190	6264	6338
584	6413	6487	6562	6636	6710	6785	6859	6933	7007	7082
585	7156	7230	7304	7379	7453	7527	7601	7675	7749	7823
586	7898	7972	8046	8120	8194	8268	8342	8416	8490	8564
587	8638	8712	8786	8860	8934	9008	9082	9156	9230	9303
588	9377	9451	9525	9599	9673	9746	9820	9894	9968	..42
589	770115	0189	0263	0336	0410	0484	0557	0631	0705	0778
590	0852	0926	0999	1073	1146	1220	1293	1367	1440	1514
591	1587	1661	1734	1808	1881	1955	2028	2102	2175	2248
592	2322	2395	2468	3542	2615	2688	2762	2835	2908	2981
593	3055	3128	3201	3274	3348	3421	3494	3567	3640	3713
594	3786	3860	3933	4006	4079	4152	4225	4298	4371	4444
					73					
595	4517	4590	4663	4736	4809	4882	4955	5028	5100	5173
596	5246	5319	5392	5465	5538	5610	5683	5756	5829	5902
597	5974	6047	6120	6193	6265	6338	6411	6483	6556	6629
598	6701	6774	6846	6919	6992	7064	7137	7209	7282	7354
599	7427	7499	7572	7644	7717	7789	7862	7934	8006	8079

N.	0	1	2	3	4	5	6	7	8	9
600	778151	8224	8296	8368	8441	8513	8585	8658	8730	8802
601	8874	8947	9019	9091	9163	9236	9308	9380	9452	9524
602	9596	6669	9741	9813	9885	9957	..29	.101	.173	.245
603	780317	0389	0461	0533	0605	0677	0749	0821	0893	0965
604	1037	1109	1181	1253	1324	1396	1468	1540	1612	1684
					72					
605	1755	1827	1899	1971	2042	2114	2186	2258	2329	2401
606	2473	2544	2616	2688	2759	2831	2902	2974	3046	3117
607	3189	3260	3332	3403	3475	3546	3618	3689	3761	3832
608	3904	3975	4046	4118	4189	4261	4332	4403	4475	4546
609	4617	4689	4760	4831	4902	4974	5045	5116	5187	5259
610	5330	5401	5472	5543	5615	5686	5757	5828	5899	5970
611	6041	6112	6183	6254	6325	6396	6467	6538	6609	6680
612	6751	6822	6893	6964	7035	7106	7177	7248	7319	7390
613	7460	7531	7602	7673	7744	7815	7885	7956	8027	8098
614	8168	8239	8310	8381	8451	8522	8593	8663	8734	8804
615	8875	8946	9016	9087	9157	9228	9299	9369	9440	9510
616	9581	9651	9722	9792	9863	9933	...4	..74	.144	.215
617	790285	0356	0426	0496	0567	0637	0707	0778	0848	0918
618	0988	1059	1129	1199	1269	1340	1410	1480	1550	1620
619	1691	1761	1831	1901	1971	2041	2111	2181	2252	2322
620	2392	2462	2532	2602	2672	2742	2812	2882	2952	3022
621	3092	3162	3231	3301	3371	3441	3511	3581	3651	3721
622	3790	3860	3930	4000	4070	4139	4209	4279	4349	4418
623	4488	4558	4627	4697	4767	4836	4906	4976	5045	5115
624	5185	5254	5324	5393	5463	5532	5602	5672	5741	5811
					69					
625	5880	5949	6019	6088	6158	6227	6297	6366	6436	6505
626	6574	6644	6713	6782	6852	6921	6990	7060	7129	7198
627	7268	7337	7406	7475	7545	7614	7683	7752	7821	7890
628	7960	8029	8098	8167	8236	8305	8374	8443	8513	8582
629	8651	8720	8789	8858	8927	8996	9065	9134	9203	9272
630	9341	9409	9478	9547	9616	9685	9754	9823	9892	9961
631	800026	0098	0167	0236	0305	0373	0442	0511	0580	0648
632	0717	0786	0854	0923	0992	1061	1129	1198	1266	1335
633	1404	1472	1541	1609	1678	1747	1815	1884	1952	2021
634	2089	2158	2226	2295	2363	2432	2500	2568	2637	2705
635	2774	2842	2910	2979	3047	3116	3184	3252	3321	3389
636	3457	3525	3594	3662	3730	3798	3867	3935	4003	4071
637	4139	4208	4276	4354	4412	4480	4548	4616	4685	4753
638	4821	4889	4957	5025	5093	5161	5229	5297	5365	5433
639	5501	5669	5637	5705	5773	5841	5908	5976	6044	6112
640	6180	6248	6316	6384	6451	6519	6587	6655	6723	6790
641	6858	6926	6994	7061	7129	7157	7264	7332	7400	7467
642	7535	7603	7670	7738	7806	7873	7941	8008	8076	8143
643	8211	8279	8346	8414	8481	8549	8616	8684	8751	8818
644	8886	8953	9021	9088	9156	9223	9290	9358	9425	9492
645	9560	9627	9694	9762	9829	9896	9964	..31	..98	.165
646	810233	0300	0367	0434	0501	0596	0636	0703	0770	0837
647	0904	0971	1039	1106	1173	1240	1307	1374	1441	1508
648	1575	1642	1709	1776	1843	1910	1977	2044	2111	2178
649	2245	2312	2379	2445	2512	2579	2646	2713	2780	2847

N.	0	1	2	3	4	5	6	7	8	9
650	812913	2980	3047	3114	3181	3247	3314	3381	3448	3514
651	3581	3648	3714	3781	3848	3914	3981	4048	4114	4181
652	4248	4314	4381	4447	4514	4581	4647	4714	4780	4847
653	4913	4980	5046	5113	5179	5246	5312	5378	5445	5511
654	5578	5644	5711	5777	5843	5910	5976	6042	6109	6175
					67					
655	6241	6308	6374	6440	6506	6573	6639	6705	6771	6838
656	6904	6970	7036	7102	7169	7233	7301	7367	7433	7499
657	7565	7631	7698	7764	7830	7896	7962	8028	8094	8160
658	8226	8292	8358	8424	8490	8556	8622	8688	8754	8820
659	8885	8951	9017	9083	9149	9215	9281	9346	9412	9478
660	9544	9610	9676	9741	9807	9873	9939	...4	..70	.136
661	820201	0267	0333	0399	0464	0530	0595	0661	0727	0792
662	0858	0924	0989	1055	1120	1186	1251	1317	1382	1448
663	1514	1579	1645	1710	1775	1841	1906	1972	2037	2103
664	2168	2233	2299	2364	2430	2495	2560	2626	2691	2756
665	2822	2887	2952	3018	3083	3148	3213	3279	3344	3409
666	3474	3539	3605	3670	3735	3800	3865	3930	3996	4061
667	4126	4191	4256	4321	4386	4451	4516	4581	4646	4711
668	4776	4841	4906	4971	5036	5101	5166	5231	5296	5361
669	5426	5491	5556	5621	5686	5751	5815	5880	5945	6010
670	6075	6140	6204	6269	6334	6399	6464	6528	6593	6658
671	6723	6787	6852	6917	6981	7046	7111	7175	7240	7305
672	7369	7434	7499	7563	7628	7692	7757	7821	7886	7951
673	8015	8080	8144	8209	8273	8338	8402	8467	8531	8595
674	8660	8724	8789	8853	8918	8982	9046	9111	9175	9239
					65					
675	9304	9368	9432	9497	9561	9625	9690	9754	9818	9882
676	9947	..11	..75	.139	.204	.268	.332	.396	.460	.525
677	830589	0653	0717	0781	0845	0909	0973	1037	1102	1166
678	1230	1294	1358	1422	1486	1550	1614	1678	1742	1806
679	1870	1934	1998	2062	2126	2189	2253	2317	2381	2445
680	2509	2573	2637	2700	2764	2828	2892	2956	3020	3083
681	3147	3211	3275	3338	3402	3466	3530	3593	3657	3721
682	3784	3848	3912	3975	4039	4103	4166	4230	4294	4357
683	4421	4484	4548	4611	4675	4739	4802	4866	4929	4993
684	5056	5120	5183	5247	5310	5373	5437	5500	5564	5627
685	5691	5754	5817	5881	5944	6007	6071	6134	6197	6261
686	6324	6387	6451	6514	6577	6641	6704	6767	6830	6894
687	6957	7020	7083	7146	7210	7273	7336	7399	7462	7525
688	7588	7652	7715	7778	7841	7904	7967	8030	8093	8156
689	8219	8282	8345	8408	8471	8534	8597	8660	8723	8786
690	8849	8912	8975	9038	9109	9164	9227	9289	9352	9415
691	9478	9541	9604	9667	9729	9792	9855	9918	9981	..43
692	840106	0169	0232	0294	0357	0420	0482	0545	0608	0671
693	0733	0796	0859	0921	0984	1046	1109	1172	1234	1297
694	1359	1422	1485	1547	1610	1672	1735	1797	1860	1922
					62					
695	1985	2047	2110	2172	2235	2297	2360	2422	2484	2547
696	2609	2672	2734	2796	2859	2921	2983	3046	3108	3170
697	3233	3295	3357	3420	3482	3544	3606	3669	3731	3793
698	3855	3918	3980	4042	4104	4166	4229	4291	4353	4415
699	4477	4539	4601	4664	4726	4788	4850	4912	4974	5036

N.	0	1	2	3	4	5	6	7	8	9
700	845098	5160	5222	5284	5346	5408	5470	5532	5594	5656
701	5718	5780	5842	5904	5966	6028	6090	6151	6213	6275
702	6337	6399	6461	6523	6585	6646	6708	6770	6832	6894
703	6955	7017	7079	7141	7202	7264	7326	7388	7449	7511
704	7573	7634	7676	7758	7819	7881	7943	8004	8066	8128
					62					
705	8189	8251	8312	8374	8435	8497	8559	8620	8682	8743
706	8805	8866	8928	8989	9051	9112	9174	9235	9297	9358
707	9419	9481	9542	9604	9665	9726	9788	9849	9911	9972
708	850033	0095	0156	0217	0279	0340	0401	0462	0524	0585
709	0646	0707	0769	0830	0891	0952	1014	1075	1136	1197
710	1258	1320	1381	1442	1503	1564	1625	1686	1747	1809
711	1870	1931	1992	2053	2114	2175	2236	2297	2358	2419
712	2480	2541	2602	2663	2724	2785	2846	2907	2968	3029
713	3090	3150	3211	3272	3333	3394	3455	3516	3577	3637
714	3698	3759	3820	3881	3941	4002	4063	4124	4185	4245
715	4306	4367	4428	4488	4549	4610	4670	4731	4792	4852
716	4913	4974	5034	5095	5156	5216	5277	5337	5398	5459
717	5519	5580	5640	5701	5761	5822	5882	5943	6003	6064
718	6124	6185	6245	6306	6366	6427	6487	6548	6608	6668
719	6729	6789	6850	6910	6970	7031	7091	7152	7212	7272
720	7332	7393	7453	7513	7574	7634	7694	7755	7815	7875
721	7935	7995	8056	8116	8176	8236	8297	8357	8417	8477
722	8537	8597	8657	8718	8778	8838	8898	8958	9018	9078
723	9138	9198	9258	9318	9379	9439	9499	9559	9619	9679
724	9739	9799	9859	9918	9978	..38	..98	.158	.218	.278
					60					
725	860338	0398	0458	0518	0578	0637	0697	0757	0817	0877
726	0937	0996	1056	1116	1176	1236	1295	1355	1415	1475
727	1534	1594	1654	1714	1773	1833	1893	1952	2012	2072
728	2131	2191	2251	2310	2370	2430	2489	2549	2608	2668
729	2728	2787	2847	2906	2966	3025	3085	3144	3204	3263
730	3323	3382	3442	3501	3561	3620	3680	3739	3799	3858
731	3917	3977	4036	4096	4155	4214	4274	4333	4392	4452
732	4511	4570	4630	4689	4748	4808	4867	4926	4985	5045
733	5104	5163	5222	5282	5341	5400	5459	5519	5578	5637
734	5696	5755	5814	5874	5933	5992	6051	6110	6169	6228
735	6287	6346	6405	6465	6524	6583	6642	6701	6760	6819
736	6878	6937	6996	7055	7114	7173	7232	7291	7350	7409
737	7467	7526	7585	7644	7703	7762	7821	7880	7939	7998
738	8056	8115	8174	8233	8292	8350	8409	8468	8527	8586
739	8644	8703	8762	8821	8879	8938	8997	9056	9114	9173
740	9232	9290	9349	9408	9466	9525	9584	9642	9701	9760
741	9818	9877	9935	9994	..53	.111	.170	.228	.287	.345
742	870404	0462	0521	0579	0638	0696	0755	0813	0872	0930
743	0989	1047	1106	1164	1223	1281	1339	1398	1456	1515
744	1573	1631	1690	1748	1806	1865	1923	1981	2040	2098
					59					
745	2156	2215	2273	2331	2389	2448	2506	2564	2622	2681
746	2739	2797	2855	2913	2972	3030	3088	3146	3204	3262
747	3321	3379	3437	3495	3553	3611	3669	3727	3785	3844
748	3902	3960	4018	4076	4134	4192	4250	4308	4360	4424
749	4482	4540	4598	4656	4714	4772	4830	4888	4945	5003

N.	0	1	2	3	4	5	6	7	8	9
750	875061	5119	5177	5235	5293	5351	5409	5466	5524	5582
751	5640	5698	5756	5813	5871	5929	5987	6045	6102	6160
752	6218	6276	6333	6391	6449	6507	6564	6622	6680	6737
753	6795	6853	6910	6968	7026	7083	7141	7199	7256	7314
754	7371	7429	7487	7544	7602	7659	7717	7774	7832	7889
					57					
755	7947	8004	8062	8119	8177	8234	8292	8349	8407	8464
756	8522	8579	8637	8694	8752	8809	8866	8924	8981	9039
757	9096	9153	9211	9268	9325	9383	9440	9497	9555	9612
758	9669	9726	9784	9841	9898	9956	..13	..70	.127	.185
759	880242	0299	0356	0413	0471	0528	0580	0642	0699	0756
760	0814	0871	0928	0985	1042	1099	1156	1213	1271	1328
761	1385	1442	1499	1556	1613	1670	1727	1784	1841	1898
762	1955	2012	2069	2126	2183	2240	2297	2354	2411	2468
763	2525	2581	2638	2695	2752	2809	2866	2923	2980	3037
764	3093	3150	3207	3264	3321	3377	3434	3491	3548	3605
765	3661	3718	3775	3832	3888	3945	4002	4059	4115	4172
766	4229	4285	4342	4399	4455	4512	4569	4625	4682	4739
767	4795	4852	4909	4965	5022	5078	5135	5192	5248	5305
768	5361	5418	5474	5531	5587	5644	5700	5757	5813	5870
769	5926	5983	6039	6096	6152	6209	6265	6321	6378	6434
770	6491	6547	6604	6660	6716	6773	6829	6885	6942	6998
771	7054	7111	7167	7223	7280	7336	7392	7449	7505	7561
772	7617	7674	7730	7786	7842	7898	7955	8011	8067	8123
773	8179	8236	8292	8348	8404	8460	8516	8573	8629	8655
774	8741	8797	8853	8909	8965	9021	9077	9134	9190	9246
					56					
775	9302	9358	9414	9470	9526	9582	9638	9694	9750	9806
776	9862	9918	0974	..30	..86	.141	.197	.253	.309	.365
777	890421	0477	0533	0589	0645	0700	0756	0812	0868	0924
778	0980	1035	1091	1147	1203	1259	1314	1370	1426	1482
779	1537	1593	1649	1705	1760	1816	1872	1928	1983	2039
780	2095	2150	2206	2262	2317	2373	2429	2484	2540	2595
781	2651	2707	2762	2818	2873	2929	2985	3040	3096	3151
782	3207	3262	3318	3373	3429	3484	3540	3595	3651	3706
783	3762	3817	3873	3928	3984	4039	4094	4150	4205	4261
784	4316	4371	4427	4482	4538	4593	4648	4704	4759	4814
785	4870	4925	4980	5036	5091	5146	5201	5257	5312	5367
786	5423	5478	5533	5588	5644	5699	5754	5809	5864	5920
787	5975	6030	6085	6140	6195	6251	6306	6361	6416	6471
788	6526	6581	6636	6692	6747	6802	6857	6912	6967	7022
789	7077	7132	7187	7242	7297	7352	7407	7462	7517	7572
790	7627	7683	7737	7792	7847	7902	7957	8012	8067	8122
791	8176	8231	8286	8341	8396	8451	8506	8561	8615	8670
792	8725	8780	8835	8890	8944	8999	9054	9109	9164	9218
793	9273	9328	9383	9437	9492	9547	9602	9656	9711	9766
794	9821	9875	9930	9985	..39	..94	.149	.203	.258	.312
					55					
795	900367	0422	0476	0531	0586	0640	0695	0749	0804	0859
796	0913	0968	1022	1077	1131	1186	1240	1295	1349	1404
797	1458	1513	1567	1622	1676	1736	1785	1840	1894	1948
798	2003	2057	2112	2166	2221	2275	2329	2384	2438	2492
799	2547	2601	2655	2710	2764	2818	2873	2927	2981	3036

N.	0	1	2	3	4	5	6	7	8	9
800	903090	3144	3199	3253	3307	3361	3416	3470	3524	3578
801	3633	3687	3741	3795	3849	3904	3958	4012	4066	4120
802	4174	4229	4283	4337	4391	4445	4499	4553	4607	4661
803	4716	4770	4824	4878	4932	4986	5040	5094	5148	5202
804	5256	5310	5364	5418	5472	5526	5580	5634	5688	5742
					54					
805	5796	5850	5904	5958	6012	6066	6119	6173	6227	6281
806	6335	6389	6443	6497	6551	6604	6658	6712	6766	6820
807	6874	6927	6981	7035	7089	7143	7196	7250	7304	7358
808	7411	7465	7519	7573	7626	7680	7734	7787	7841	7895
809	7949	8002	8056	8110	8163	8217	8270	8324	8378	8431
810	8485	8539	8592	8646	8699	8753	8807	8860	8914	8967
811	9021	9074	9128	9181	9235	9289	9342	9396	9449	9503
812	9556	9610	9663	9716	9770	9823	9877	9930	9984	..37
813	910091	0144	0197	0251	0304	0358	0411	0464	0518	0571
814	0624	0678	0731	0784	0838	0891	0944	0998	1051	1104
815	1158	1211	1264	1317	1371	1424	1477	1530	1584	1637
816	1690	1743	1797	1850	1903	1956	2009	2063	2115	2169
817	2222	2275	2323	2381	2435	2488	2541	2594	2645	2700
818	2753	2806	2859	2913	2966	3019	3072	3125	3178	3231
819	3284	3337	3390	3443	3496	3549	3602	3655	3708	3761
820	3814	3867	3920	3973	4026	4079	4132	4184	4237	4290
821	4343	4396	4449	4502	4555	4608	4660	4713	4766	4819
822	4872	4925	4977	5030	5083	5136	5189	5241	5594	5347
823	5400	5453	5505	5558	5611	5664	5716	5769	5822	5875
824	5927	5980	6033	6085	6138	6191	6243	6296	6349	6401
825	6454	6507	6559	6612	6664	6717	6770	6822	6875	6927
826	6980	7033	7035	7138	7190	7243	7295	7348	7400	7453
827	7506	7558	7611	7663	7716	7768	7820	7873	7925	7978
828	8030	8083	8185	8188	8240	8293	8345	8397	8450	8502
829	8555	8607	8659	8712	8764	8816	8869	8921	8973	9026
830	9078	9130	9183	9235	9287	9340	9392	9444	9496	9549
831	9601	9653	9706	9758	9810	9862	9914	9967	..19	..71
832	920123	0176	0228	0280	0332	0384	0436	0489	0541	0593
833	0645	0697	0749	0801	0853	0906	0958	1010	1062	1114
834	1166	1218	1270	1322	1374	1426	1478	1530	1582	1634
835	1686	1738	1790	1842	1894	1946	1998	2050	2102	2154
836	2206	2258	2310	2362	2414	2466	2518	2570	2622	2674
837	2725	2777	2829	2881	2933	2985	3037	3089	3140	3192
838	3244	3296	3348	3399	3451	3503	3555	3607	3658	3710
839	3762	3814	3865	3917	3969	4021	4072	4124	4147	4228
840	4279	4331	4383	4434	4486	4538	4589	4641	4693	4744
841	4796	4848	4899	4951	5003	5054	5106	5157	5209	5261
842	5312	5364	5415	5467	5518	5570	5621	5673	5725	5776
843	5828	5874	5931	5982	6034	6085	6137	6188	6240	6291
844	6342	6394	6445	6497	6548	6600	6651	6702	6754	6805
					52					
845	6857	6908	6959	7011	7062	7114	7165	7216	7268	7319
846	7370	7422	7473	7524	7576	7627	7678	7730	7783	7832
847	7883	7935	7986	8037	8088	8140	8191	8242	8293	8345
848	8396	8447	8498	8549	8601	8652	8703	8754	8805	8857
849	8908	8959	9010	9061	9112	9163	9216	9266	9317	9368

N.	0	1	2	3	4	5	6	7	8	9
850	929419	9473	9521	9572	9623	9674	9725	9776	9827	9879
851	9930	9981	..32	..83	.134	.185	.236	.287	.338	.389
852	930440	0491	0542	0592	0643	0694	0745	0796	0847	0898
853	0949	1000	1051	1102	1153	1204	1254	1305	1356	1407
854	1458	1509	1560	1610	1661	1712	1763	1814	1865	1915
					51					
855	1966	2017	2068	2118	2169	2220	2271	2322	2372	2423
856	2474	2524	2575	2626	2677	2727	2778	2829	2879	2930
857	2981	3031	3082	3133	3183	3234	3285	3335	3386	3437
858	3487	3538	3589	3639	3690	3740	3791	3841	3892	3943
859	3993	4044	4094	4145	4195	4246	4269	4347	4397	4448
860	4493	4549	4599	4650	4700	4751	4801	4852	4902	4953
861	5003	5054	5104	5154	5205	5255	5303	5356	5406	5457
862	5507	5558	5608	5658	5709	5759	5809	5860	5910	5960
863	6011	6061	6111	6162	6212	6262	6313	6363	6413	6463
864	6514	6564	6614	6665	6715	6765	6815	6865	6916	6966
865	7016	7066	7117	7167	7217	7267	7317	7367	7418	7468
866	7518	7568	7618	7668	7718	7769	7819	7869	7919	7969
867	8019	8069	8119	8169	8219	8269	8320	8370	8420	8470
868	8520	8570	8620	8670	8720	8770	8820	8870	8919	8970
869	9020	9070	9120	9170	9220	9270	9320	9369	9419	9469
870	9519	9569	9616	9669	9719	9769	9819	9869	9918	9968
871	940018	0068	0118	0168	0218	0267	0317	0367	0417	0467
872	0516	0566	0616	0666	0716	0765	0815	0865	0915	0964
873	1014	1064	1114	1163	1213	1263	1313	1362	1412	1462
874	1511	1561	1611	1660	1710	1760	1809	1859	1909	1958
875	2008	2058	2107	2157	2207	2256	2306	2355	2405	2455
876	2504	2554	2603	2653	2702	2752	2801	2851	2901	2950
877	3000	3049	3099	3148	3198	3247	3297	3346	3396	3445
878	3495	3544	3593	3643	3692	3742	3791	3841	3890	3939
879	3989	4038	4088	4137	4186	4236	4285	4335	4384	4433
880	4483	4532	4581	4631	4680	4729	4779	4828	4877	4927
881	4976	5025	5074	5124	5173	5222	5272	5321	5370	5419
882	5469	5518	5567	5616	5665	5715	5764	5813	5862	5912
883	5961	6010	6059	6108	6157	6207	6256	6305	6354	6403
884	6452	6501	6551	6600	6649	6698	6747	6796	6845	6894
885	6943	6992	7041	7090	7140	7189	7238	7287	7336	7385
886	7434	7483	7532	7581	7630	7679	7728	7777	7826	7875
887	7924	7973	8022	8070	8119	8168	8217	8266	8315	8365
888	8413	8462	8511	8560	8609	8657	8706	8755	8804	8853
889	8902	8951	8999	9048	9097	9146	9195	9244	9292	9341
890	9390	9439	9488	9536	9585	9634	9683	9731	9780	9829
891	9878	9926	9975	..24	..73	.121	.170	.219	.267	.316
892	950365	0414	0462	0511	0560	0608	0657	0706	0754	0803
893	0851	0900	0949	0997	1046	1095	1143	1192	1240	1289
894	1338	1386	1435	1483	1532	1580	1629	1677	1726	1775
					48					
895	1823	1872	1920	1969	2017	2066	2114	2163	2211	2260
896	2308	2356	2405	2453	2502	2550	2599	2647	5696	2744
897	2792	2841	2889	2938	2986	3034	3083	3131	3180	3228
898	3276	3325	3373	3421	3470	3518	3566	3615	3663	3711
899	3760	3808	3856	3905	3953	4001	4019	4098	4146	4194

N.	0	1	2	3	4	5	6	7	8	9
900	954243	4291	4339	4387	4435	4484	4532	4580	4628	4677
901	4725	4773	4821	4869	4918	4966	5014	5062	5110	5158
902	5207	5255	5303	5351	5399	5447	5495	5543	5592	5640
903	5688	5736	5784	5832	5880	5928	5976	6024	6072	6120
904	6168	6216	6265	6313	6361	6409	6457	6505	6553	6601
					48					
905	6649	6697	6745	6793	6840	6888	6936	6984	7032	7080
906	7128	7176	7224	7272	7320	7368	7416	7464	7512	7559
907	7607	7655	7703	7751	7799	7847	7894	7942	7990	8038
908	8086	8134	8181	8229	8277	8325	8373	8421	8468	8516
909	8564	8612	8659	8707	8755	8803	8850	8898	8946	8994
910	9041	9089	9137	9185	9232	9280	9328	9375	9423	9471
911	9518	9566	9614	9661	9709	9757	9804	9852	9900	9947
912	9995	..42	..90	.138	.185	.233	.280	.328	.376	.423
913	960471	0518	0566	0613	0661	0709	0756	0804	0851	0899
914	0946	0994	1041	1089	1136	1184	1231	1279	1326	1374
915	1421	1469	1516	1563	1611	1658	1706	1753	1801	1848
916	1895	1943	1990	2038	2085	2132	2180	2227	2275	2322
917	2369	2417	2464	2511	2559	2606	2653	2701	2748	2795
918	2843	2890	2937	2985	3032	3079	3126	3174	3221	3268
919	3316	3363	3410	3457	3504	3552	3599	3646	3693	3741
920	3788	3835	3882	3929	3977	4024	4071	4118	4165	4212
921	4260	4307	4354	4401	4448	4495	4542	4590	4637	4684
922	4731	4778	4825	4872	4919	4966	5013	5061	5108	5155
923	5202	5249	5296	5343	5390	5437	5484	5531	5578	5625
924	5672	5719	5766	5813	5860	5907	5954	6001	6048	6095
925	6142	6189	6236	6283	6329	6376	6423	6470	6517	6564
926	6611	6658	6705	6752	6799	6845	6892	6939	6986	7033
927	7080	7127	7173	7220	7267	7314	7361	7408	7454	7501
928	7548	7595	7642	7688	7735	7782	7829	7875	7922	7969
929	8016	8062	8109	8156	8203	8249	8296	8343	8390	8436
930	8483	8530	8576	8623	8670	8716	8763	8810	8856	8903
931	8950	8996	9043	9090	9136	9183	9229	9276	9323	9369
932	9416	9463	9509	9556	9602	9649	9695	9742	9789	9835
933	9882	9928	9975	..21	..68	.114	.161	.207	.254	.300
934	970347	0393	0440	0486	0533	0579	0626	0672	0719	0765
935	0812	0858	0904	0951	0997	1044	1090	1137	1183	1229
936	1276	1322	1369	1415	1461	1508	1554	1601	1647	1693
937	1740	1786	1832	1879	1925	1971	2018	2064	2110	2157
938	2203	2249	2295	2342	2388	2434	2481	2527	2573	2619
939	2666	2712	2758	2804	2851	2897	2943	2989	3035	3082
940	3128	3174	3220	3266	3313	3359	3405	3451	3497	3543
941	3590	3636	3682	3728	3774	3820	3866	3913	3959	4005
942	4051	4097	4143	4189	4235	4281	4327	4374	4420	4466
943	4512	4558	4604	4650	4696	4742	4788	4834	4880	4926
944	4972	5018	5064	5110	5156	5202	5248	5294	5340	5386
					46					
945	5432	5478	5524	5570	5616	5662	5707	5753	5799	5845
946	5891	5937	5983	6029	6075	6121	6167	6212	6258	6304
947	6350	6396	6442	6488	6533	6579	6625	6671	6717	6763
948	6808	6854	6900	6946	6992	7037	7083	7129	7175	7220
949	7266	7312	7358	7403	7449	7495	7541	7586	7632	7678

N.	0	1	2	3	4	5	6	7	8	9
950	977724	7769	7815	7861	7906	7952	7998	8043	8089	8135
951	8181	8226	8272	8317	8363	8409	8454	8500	8546	8591
952	8637	8683	8728	8774	8819	8865	8911	8956	9002	9047
953	9093	9138	9184	9230	9275	9321	9366	9412	9457	9503
954	9548	9594	9639	9685	9730	9776	9821	9867	9912	9958
					46					
955	980003	0049	0094	0140	0185	0231	0276	0322	0367	0412
956	0458	0503	0549	0594	0640	0685	0730	0776	0821	0867
957	0912	0957	1003	1048	1093	1139	1184	1229	1275	1320
958	1366	1411	1456	1501	1547	1592	1637	1683	1728	1773
959	1819	1864	1909	1954	2000	2045	2090	2135	2181	2226
960	2271	2316	2362	2407	2452	2497	2543	2588	2633	2678
961	2723	2769	2814	2859	2904	2949	2994	3040	3085	3130
962	3175	3220	3265	3310	3356	3401	3446	3491	3536	3581
963	3626	3671	3716	3762	3807	3852	3897	3942	3987	4032
964	4077	4122	4167	4212	4257	4302	4347	4392	4437	4482
965	4527	4572	4617	4662	4707	4752	4797	4842	4887	4932
966	4977	5022	5067	5112	5157	5202	5247	5292	5337	5382
967	5426	5471	5516	5561	5606	5651	5699	5741	5786	5830
968	5875	5920	5965	6010	6055	6100	6144	6189	6234	6279
969	6324	6369	6413	6458	6503	6548	6593	6637	6682	6727
970	6772	6817	6861	6906	6951	6996	7040	7035	7130	7175
971	7219	7264	7309	7353	7398	7443	7488	7532	7577	7622
972	7666	7711	7756	7800	7845	7890	7934	7979	8024	8068
973	8113	8157	8202	8247	8291	8336	8381	8425	8470	8514
974	8559	8604	8648	8693	8737	8782	8826	8871	8916	8960
975	9005	9049	9093	9138	9183	9227	9272	9316	9361	9405
976	9450	9494	9539	9583	9628	9672	9717	9761	9806	9850
977	9895	9939	9983	..28	..72	.117	.161	.206	.250	.294
978	990339	0383	0428	0472	0516	0561	0605	0650	0694	0738
979	0783	0827	0871	0916	0960	1004	1049	1093	1137	1182
980	1226	1270	1315	1359	1403	1448	1492	1536	1580	1625
981	1669	1713	1758	1802	1846	1890	1935	1979	2023	2067
982	2111	2156	2200	2244	2288	2333	2377	2421	2465	2509
983	2554	2598	2642	2686	2730	2774	2819	2863	2907	2951
984	2995	3039	3083	3127	3172	3216	3260	3304	3348	3392
985	3436	3480	3524	3568	3613	3657	3701	3745	3789	3833
986	3877	3921	3965	4009	4053	4097	4141	4185	4229	4273
987	4317	4361	4405	4449	4493	4537	4581	4625	4669	4713
988	4757	4801	4845	4886	4933	4977	5021	5065	5108	5152
989	5196	5240	5284	5328	5372	5416	5460	5504	5547	5591
990	5635	5679	5723	5767	5811	5854	5898	5942	5986	6030
991	6074	6117	6161	6205	6249	6293	6337	6380	6424	6468
992	6512	6555	6599	6643	6687	6731	6774	6818	6862	6906
993	6949	6993	7037	7030	7124	7168	7212	7255	7299	7343
994	7386	7430	7474	7517	7561	7605	7648	7692	7736	7779
					44					
995	7823	7867	7910	7954	7998	8041	8085	8129	8172	8216
996	8259	8303	8347	8390	8434	8477	8521	8564	8608	8652
997	8695	8739	8792	8826	8869	8913	8956	9000	9043	9087
998	9131	9174	9218	9261	9305	9348	9392	9435	9479	9522
999	9565	9609	9652	9696	9739	9783	9826	9870	9913	9957

′	Sine.	D 10″	Cosine.	D.10″	Tang.	D.10″	Cotang.	N.sine.	N. cos.	
0	0.000000		10.000000		0.000000		Infinite.	00000	100000	60
1	6.463726		000000		6.463726		13.536274	00029	100000	59
2	764753		000000		764756		235244	00058	100000	58
3	940347		000000		940847		059153	00087	100000	57
4	7.065786		000000		7.065786		12.934214	00116	100000	56
5	162696		000000		162696		837304	00145	100000	55
6	241877		9.999999		241878		758122	00175	100000	54
7	308824		999999		308825		691175	00204	100000	53
8	366816		999999		366817		633183	00233	100000	52
9	417968		999999		417970		582030	00262	100000	51
10	463725		999998		463727		536273	00291	100000	50
11	7.505118		9.999998		7.505120		12.494880	00320	99999	49
12	542906		999997		542909		457091	00349	99999	48
13	577668		999997		577672		422328	00378	99999	47
14	609853		999996		609857		390143	00407	99999	46
15	639816		999996		639820		360180	00436	99999	45
16	667845		999995		667849		332151	00465	99999	44
17	694173		999995		694179		305821	00495	99999	43
18	718997		999994		719003		280997	00524	99999	42
19	742477		999993		742484		257516	00553	99998	41
20	764754		999993		764761		235239	00582	99998	40
21	7.785943		9.999992		7.785951		12.214049	00611	99998	39
22	806146		999991		806155		193845	00640	99998	38
23	825451		999990		825460		174540	00669	99998	37
24	843934		999989		843944		156056	00698	99998	36
25	861663		999988		861674		138326	00727	99997	35
26	878695		999988		878708		121292	00756	99997	34
27	895085		999987		895099		104901	00785	99997	33
28	910879		999986		910894		089106	00814	99997	32
29	926119		999985		926134		073866	00844	99996	31
30	940842		999983		940858		059142	00873	99996	30
31	7.955082		9.999982		7.955100		12.044900	00902	99996	29
32	968870	2298	999981	0.2	968889	2298	031111	00931	99996	28
33	982233	2227	999980	0.2	982253	2227	017747	00960	99995	27
34	995198	2161	999979	0.2	995219	2161	004781	00989	99995	26
35	8.007787	2098	999977	0.2	8.007809	2098	11.992191	01018	99995	25
36	020021	2039	999976	0.2	020045	2039	979955	01047	99995	24
37	031919	1983	999975	0.2	031945	1983	968055	01076	99994	23
38	043501	1930	999973	0.2	043527	1930	956473	01105	99994	22
39	054781	1880	999972	0.2	054809	1880	945191	01134	99994	21
40	065776	1832	999971	0.2	065806	1833	934194	01164	99993	20
41	8.076500	1787	9.999969	0.2	8.076531	1787	11.923469	01193	99993	19
42	086965	1744	999968	0.2	086997	1744	913003	01222	99993	18
43	097183	1703	999966	0.2	097217	1703	902783	01251	99992	17
44	107167	1664	999964	0.2	107202	1664	892797	01280	99992	16
45	116926	1626	999963	0.3	116963	1627	883037	01309	99991	15
46	126471	1591	999961	0.3	126510	1591	873490	01338	99991	14
47	135810	1557	999959	0.3	135851	1557	864149	01367	99991	13
48	144953	1524	999958	0.3	144996	1524	855004	01396	99990	12
49	153907	1492	999956	0.3	153952	1493	846048	01425	99990	11
50	162681	1462	999954	0.3	162727	1463	837273	01454	99989	10
51	8.171280	1433	9.999952	0.3	8.171328	1434	11.828672	01483	99989	9
52	179713	1405	999950	0.3	179763	1406	820237	01513	99989	8
53	187985	1379	999948	0.3	188036	1379	811964	01542	99988	7
54	196102	1353	999946	0.3	196156	1353	803844	01571	99988	6
55	204070	1328	999944	0.3	204126	1328	795874	01600	99987	5
56	211895	1304	999942	0.3	211953	1304	788047	01629	99987	4
57	219581	1281	999940	0.4	219641	1281	780359	01658	99986	3
58	227134	1259	999938	0.4	227195	1259	772805	01687	99986	2
59	234557	1237	999936	0.4	234621	1238	765379	01716	99985	1
60	241855	1216	999934	0.4	241921	1217	758079	01745	99985	0
	Cosine.		Sine.		Cotang.		Tang.	N. cos.	N. sine	′

89 Degrees.

'	Sine.	D. 10"	Cosine.	D. 10"	Tang.	D. 10"	Cotang.	N. sine.	N. cos.	'
0	8.241855	1196	9.999934	0.4	8.241921	1197	11.758079	01742	99985	60
1	249033	1177	999932	0.4	249102	1177	750898	01774	99984	59
2	256094	1158	999929	0.4	256165	1158	743835	01803	99984	58
3	263042	1140	999927	0.4	263115	1140	736885	01832	99983	57
4	269881	1122	999925	0.4	269956	1122	730044	01862	99983	56
5	276614	1105	999922	0.4	276691	1105	723309	01891	99982	55
6	283243	1088	999920	0.4	283323	1089	716677	01920	99982	54
7	289773	1072	999918	0.4	289856	1073	710144	01949	99981	53
8	296207	1056	999915	0.4	296292	1057	703708	01978	99980	52
9	302546	1041	999913	0.4	302634	1042	697366	02007	99980	51
10	308794	1027	999910	0.4	308884	1037	691116	02036	99979	50
11	8.314954	1012	9.999907	0.4	8.315046	1013	11.684954	02065	99979	49
12	321027	998	999905	0.4	321122	999	678878	02094	99978	48
13	327016	985	999902	0.4	327114	985	672886	02123	99977	47
14	332924	971	999899	0.5	333025	972	666975	02152	99977	46
15	338753	959	999897	0.5	338856	959	661144	02181	99976	45
16	344504	946	999894	0.5	344610	946	655390	02211	99976	44
17	350181	934	999891	0.5	350289	934	649711	02240	99975	43
18	355783	922	999888	0.5	355895	922	644105	02269	99974	42
19	361315	910	999885	0.5	361430	911	638570	02298	99974	41
20	366777	899	999882	0.5	366895	899	633105	02327	99973	40
21	8.372171	888	9.999879	0.5	8.372292	888	11.627708	02356	99972	39
22	377499	877	999876	0.5	377622	879	622378	02385	99972	38
23	382762	867	999873	0.5	382889	867	617111	02414	99971	37
24	387962	856	999870	0.5	388092	857	611908	02443	99970	36
25	393101	846	999867	0.5	393234	847	606766	02472	99969	35
26	398179	837	999864	0.5	398315	837	601685	02501	99969	34
27	403199	827	999861	0.5	403338	828	596662	02530	99968	33
28	408161	818	999858	0.5	408304	818	591696	02560	99967	32
29	413068	809	999854	0.5	413213	809	586787	02589	99966	31
30	417919	800	999851	0.5	418068	800	581932	02618	99966	30
31	8.422717	791	9.999848	0.6	8.422869	791	11.577131	02647	99965	29
32	427462	782	999844	0.6	427618	783	572382	02676	99964	28
33	432156	774	999841	0.6	432315	774	567685	02705	99963	27
34	436800	766	999838	0.6	436962	766	563038	02734	99963	26
35	441394	758	999834	0.6	441560	758	558440	02763	99962	25
36	445941	750	999831	0.6	446110	750	553890	02792	99961	24
37	450440	742	999827	0.6	450613	743	549387	02821	99960	23
38	454893	735	999823	0.6	455070	735	544930	02850	99959	22
39	459301	727	999820	0.6	459481	728	540519	02879	99959	21
40	463665	720	999816	0.6	463849	720	536151	02908	99958	20
41	8.467985	712	9.999812	0.6	8.468172	713	11.531828	02938	99957	19
42	472263	706	999809	0.6	472454	707	527546	02967	99956	18
43	476498	699	999805	0.6	476693	700	523307	02996	99955	17
44	480693	692	999801	0.6	480892	693	519108	03025	99954	16
45	484848	686	999797	0.6	485050	686	514950	03054	99953	15
46	488963	679	999793	0.7	489170	680	510830	03083	99952	14
47	493040	673	999790	0.7	493250	674	506750	03112	99952	13
48	497078	667	999786	0.7	497293	668	502707	03141	99951	12
49	501080	661	999782	0.7	501298	661	498702	03170	99950	11
50	505045	655	999778	0.7	505267	655	494733	03199	99949	10
51	8.508974	649	9.999773	0.7	8.509200	650	11.490800	03228	99948	9
52	512867	643	999769	0.7	513098	644	486902	03257	99947	8
53	516726	637	999765	0.7	516961	638	483039	03286	99946	7
54	520551	632	999761	0.7	520790	633	479210	03316	99945	6
55	524343	626	999757	0.7	524586	627	475414	03345	99944	5
56	528102	621	999753	0.7	528349	622	471651	03374	99943	4
57	531828	616	999748	0.7	532080	616	467920	03403	99942	3
58	535523	611	999744	0.7	535779	611	464221	03432	99941	2
59	539186	605	999740	0.7	539447	606	460553	03461	99940	1
60	542819		999735		543084		456916	03490	99939	0
	Cosine.		Sine.		Cotang.		Tang.	N. cos.	N. sinc.	'

88 Degrees.

′	Sine.	D. 10″	Cosine.	D. 10″	Tang.	D. 10″	Cotang.	N. sine.	N. cos.	′
0	8.542819	600	9.999735	0.7	8.543084	602	11.456916	03490	99939	60
1	546422	595	999731	0.7	546691	593	453309	03519	99938	59
2	549995	591	999726	0.7	550268	591	449732	03548	99937	58
3	553539	586	999722	0.7	553817	587	446183	03577	99936	57
4	557054	581	999717	0.8	557336	582	442664	03606	99935	56
5	560540	576	999713	0.8	560828	577	439172	03635	99934	55
6	563999	572	999708	0.8	564291	573	435709	03664	99933	54
7	567431	567	999704	0.8	567727	568	432273	03693	99932	53
8	570836	563	999699	0.8	571137	564	428863	03723	99931	52
9	574214	559	999694	0.8	574520	559	425480	03752	99930	51
10	577566	554	999689	0.8	577877	555	422123	03781	99929	50
11	8.580892	550	9.999685	0.8	8.581208	551	11.418792	03810	99927	49
12	584193	546	999680	0.8	584514	547	415486	03839	99926	48
13	587469	542	999675	0.8	587795	543	412205	03868	99925	47
14	590721	538	999670	0.8	591051	539	408949	03897	99924	46
15	593948	534	999665	0.8	594283	535	405717	03926	99923	45
16	597152	530	999660	0.8	597492	531	402508	03955	99922	44
17	600332	526	999655	0.8	600677	527	399323	03984	99921	43
18	603469	522	999650	0.8	603839	523	396161	04013	99919	42
19	606623	519	999645	0.8	606978	519	393022	04042	99918	41
20	609734	515	999640	0.8	610094	516	389906	04071	99917	40
21	8.612823	511	9.999635	0.9	8.613189	512	11.386811	04100	99916	39
22	615891	508	999629	0.9	616262	508	383738	03129	99915	38
23	618937	504	999324	0.9	619313	505	380687	04159	99913	37
24	621962	501	999619	0.9	622343	501	377657	04188	99912	36
25	624965	497	999614	0.9	625352	498	374648	04217	99911	35
26	627948	494	999608	0.9	628340	495	371660	04246	99910	34
27	630911	490	999603	0.9	631308	491	368692	04275	99909	33
28	633854	487	999597	0.9	634256	488	365744	04304	99907	32
29	636776	484	999592	0.9	637184	485	362816	04333	99906	31
30	639680	481	999586	0.9	640093	482	359907	04362	99905	30
31	8.642563	477	0.999581	0.9	8.642982	478	11.357018	04391	99904	29
32	645498	474	999575	0.9	645853	475	354147	04420	99902	28
33	648274	471	999570	0.9	648704	472	351296	04449	99901	27
34	651102	468	999564	0.9	651537	469	348463	04478	99900	26
35	653911	465	.999556	1.0	654352	466	345648	04507	99898	25
36	656702	462	999553	1.0	657149	463	342851	04536	99897	24
37	659475	459	999547	1.0	659928	460	340072	04565	99896	23
38	662230	456	999541	1.0	662689	457	337311	04594	99894	22
39	664968	453	999535	1.0	665433	454	334567	04623	99893	21
40	667689	451	999529	1.0	668160	453	331840	04653	99892	20
41	8.670393	448	9.999524	1.0	8.670870	449	11.329130	04682	99890	19
42	673080	445	999518	1.0	673563	446	326437	04711	99889	18
43	675751	442	999512	1.0	676239	443	323761	04740	99888	17
44	678405	440	999506	1.0	678900	442	321100	04769	99886	16
45	681043	437	999500	1.0	681544	438	318456	04798	99885	15
46	683665	434	999493	1.0	684172	435	315828	04827	99883	14
47	686272	432	999487	1.0	6 6784	433	313216	04856	99882	13
48	688863	429	999481	1.0	689381	430	310619	04885	99881	12
49	691438	427	999475	1.0	691963	428	308037	04914	99879	11
50	693998	424	999469	1.0	694529	425	305471	04943	99878	10
51	8.696543	422	0.999463	1.0	8.697081	423	11.302919	04972	99876	9
52	699073	419	999456	1.1	699617	420	300383	05001	99875	8
53	701589	417	999450	1.1	702139	418	297861	05030	99873	7
54	704090	414	999443	1.1	704246	415	295354	05059	99872	6
55	706577	412	999437	1.1	707140	413	292860	05088	99870	5
56	709049	410	999431	1.1	709618	411	290382	05117	99869	4
57	711507	407	999424	1.1	702083	408	287917	05146	99867	3
58	713952	405	999418	1.1	714534	406	285465	05175	99866	2
59	716383	403	999411	1.1	716972	404	283028	05205	99864	1
60	718800		999404		719396		280604	05234	99863	0
	Cosine.		Sine.		Cotang.		Tang.	N. cos.	N. sine.	′

′	Sine.	D. 10″	Cosine.	D. 10″	Tang.	D. 10″	Cotang.	N. sine.	N. cos.	
0	8.718800	401	9.909404	1.1	8.719396	402	11.280604	05234	99863	60
1	721204	398	999398	1.1	721806	399	278194	05263	99861	59
2	723595	396	999391	1.1	724204	397	275796	05292	99860	58
3	725972	394	999384	1.1	726588	395	273412	05321	99858	57
4	728337	392	999378	1.1	728959	393	271041	05350	99857	56
5	730688	390	999371	1.1	731317	391	268683	05379	99855	55
6	733027	388	999364	1.1	733663	389	266337	05408	99854	54
7	735354	386	999357	1.2	735996	387	264004	05437	99852	53
8	737667	384	999350	1.2	738317	385	261683	05466	99851	52
9	739969	382	999343	1.2	740626	383	259374	05495	99849	51
10	742259	380	999336	1.2	742922	381	257078	05524	99847	50
11	8.744536	378	9.999329	1.2	8.745207	379	11.254793	05553	99846	49
12	746802	376	999322	1.2	747479	377	252521	05582	99844	48
13	749055	374	999315	1.2	749740	375	250260	05611	99842	47
14	751297	372	999308	1.2	751989	373	248011	05640	99841	46
15	753528	370	999301	1.2	754227	371	245773	05669	99839	45
16	755747	368	999294	1.2	756453	369	243547	05698	99838	44
17	757935	366	999286	1.2	758668	367	241332	05727	99836	43
18	760151	364	999279	1.2	760872	365	239128	05756	99834	42
19	762337	362	999272	1.2	763065	364	236935	05785	99833	41
20	764511	361	999265	1.2	765246	362	234754	05814	99831	40
21	8.766675	359	9.999257	1.2	8.767417	360	11.232583	05844	99829	39
22	768828	357	999250	1.3	769578	358	230422	05873	99827	38
23	770970	355	999242	1.3	771727	356	228273	05902	99826	37
24	773101	353	999235	1.3	773866	355	226134	05931	99824	36
25	775223	352	999227	1.3	775995	353	224005	05960	99822	35
26	777333	350	999220	1.3	778114	351	221886	05989	99821	34
27	779434	348	999212	1.3	780222	350	219778	06018	99819	33
28	781524	347	999205	1.3	782320	348	217680	06047	99817	32
29	783605	345	999197	1.3	784408	346	215592	06076	99815	31
30	785675	343	999189	1.3	786486	345	213514	06105	99813	30
31	8.787736	342	9.999181	1.3	8.788554	343	11.211446	06134	99812	29
32	789787	340	999174	1.3	790613	341	209387	06163	99810	28
33	791828	339	999166	1.3	792662	340	207338	06192	99808	27
34	793859	337	999158	1.3	794701	338	205299	06221	99806	26
35	795881	335	999150	1.3	796731	337	203269	06250	99804	25
36	797894	334	999142	1.3	798752	335	201248	06279	99803	24
37	799897	332	999134	1.3	800763	334	199237	06308	99801	23
38	801892	331	999126	1.3	802765	332	197235	06337	99799	22
39	803876	329	999118	1.3	804858	331	195242	06366	99797	21
40	805852	328	999110	1.3	806742	329	193258	06395	99795	20
41	8.807819	326	9.999102	1.3	8.808717	328	11.191283	06424	99793	19
42	809777	325	999094	1.4	810683	326	189317	06453	99792	18
43	811726	323	999086	1.4	812641	325	187359	06482	99790	17
44	813667	322	999077	1.4	814589	323	185411	06511	99788	16
45	815599	320	999069	1.4	816529	322	183471	06540	99786	15
46	817522	319	999061	1.4	818461	320	181539	06569	99784	14
47	819436	318	999053	1.4	820384	319	179616	06598	99782	13
48	821343	316	999044	1.4	822298	318	177702	06627	99780	12
49	823240	315	999036	1.4	824205	316	175795	06656	99778	11
50	825130	313	999027	1.4	826103	315	173897	06685	99776	10
51	8.827011	312	9.999019	1.4	8.827992	314	11.172008	06714	99774	9
52	828884	311	999010	1.4	829874	312	170126	06743	99772	8
53	830749	309	999002	1.4	831748	311	168252	06773	99770	7
54	832607	308	998993	1.4	833613	310	166387	06802	99768	6
55	834456	307	998984	1.4	835471	308	164529	06831	99766	5
56	836297	306	998976	1.4	837321	307	162679	06860	99764	4
57	838130	304	998967	1.4	839163	306	160837	06889	99762	3
58	839956	303	998958	1.5	840998	304	159002	06918	99760	2
59	841774	302	998950	1.5	842825	303	157175	06947	99758	1
60	843585		998941	1.5	844644		155356	06976	99756	0
	Cosine.		Sine.		Cotang.		Tang.	N. cos.	N. sine.	′

'	Sine.	D. 10"	Cosine.	D. 10"	Tang.	D. 10"	Cotang.	N. sine.	N. cos.	'
0	8.843585	300	9.998941	1.5	8.844644	302	11.155356	06976	99756	60
1	845387	299	998932	1.5	846455	301	153545	07005	99754	59
2	847183	298	998923	1.5	848260	299	151740	07034	99752	58
3	848971	297	998914	1.5	850057	298	149943	07063	99750	57
4	850751	295	998905	1.5	851846	297	148154	07092	99748	56
5	852525	294	998896	1.5	853628	296	146372	07121	99746	55
6	854291	293	998887	1.5	855403	295	144597	07150	99744	54
7	856049	292	998878	1.5	857171	293	142829	07179	99742	53
8	857801	291	998869	1.5	858932	292	141068	07208	99740	52
9	859546	290	998860	1.5	860686	291	139314	07237	99738	51
10	861283	288	998851	1.5	862433	290	137567	07266	99736	50
11	8.863014	287	9.998841	1.5	8.864173	289	11.135827	07295	99734	49
12	864738	286	998832	1.5	865906	288	134094	07324	99731	48
13	866455	285	998823	1.6	867632	287	132368	07353	99729	47
14	868165	284	998813	1.6	869351	285	130649	07382	99727	46
15	869868	283	998804	1.6	871064	284	128936	07411	99725	45
16	871565	282	998795	1.6	872770	283	127230	07440	99723	44
17	873255	281	998785	1.6	874469	282	125531	07469	99721	43
18	874938	279	998776	1.6	876162	281	123838	07498	99719	42
19	876615	279	998766	1.6	877849	280	122151	07527	99716	41
20	878285	277	998757	1.6	879529	279	120471	07556	99714	40
21	8.879949	276	9.998747	1.6	8.881202	278	11.118798	07585	99712	39
22	881607	275	998738	1.6	882869	277	117131	07614	99710	38
23	883258	274	998728	1.6	884530	276	115470	07643	99708	37
24	·884903	273	998718	1.6	886185	275	113815	07672	99705	36
25	886542	272	998708	1.6	887833	274	112167	07701	99703	35
26	888174	271	998699	1.6	889476	273	110524	07730	99701	34
27	889801	270	998689	1.6	891112	272	108888	07759	99699	33
28	891421	269	998679	1.6	892742	271	107258	07788	99696	32
29	893035	268	998669	1.6	894366	270	105634	07817	99694	31
30	894643	267	998659	1.7	895984	269	104016	07846	99692	30
31	8.896246	266	9.998649	1.7	8.897596	268	11.102404	07875	99689	29
32	897842	265	998639	1.7	899203	267	100797	07904	99687	28
33	899432	264	998629	1.7	900803	266	099197	07933	99685	27
34	901017	263	998619	1.7	902398	265	097602	07962	99683	26
35	902596	262	998609	1.7	903987	264	096013	07991	99680	25
36	904169	261	998599	1.7	905570	263	094430	08020	99678	24
37	905736	260	998589	1.7	907147	262	092853	08049	99676	23
38	907297	259	998578	1.7	908719	261	091281	08078	99673	22
39	908853	258	998568	1.7	910285	260	089715	08107	99671	21
40	910404	257	998558	1.7	911846	259	088154	08136	99668	20
41	8.911949	257	9.998548	1.7	8.913401	258	11.086599	08165	99666	19
42	913488	256	998537	1.7	914951	257	085049	08194	99664	18
43	915022	255	998527	1.7	916495	256	083505	08223	99661	17
44	916550	254	998516	1.7	918034	256	081966	08252	99659	16
45	918073	253	998506	1.8	919568	255	080432	08281	99657	15
46	919591	252	998495	1.8	921096	254	078904	08310	99654	14
47	921103	251	998485	1.8	922619	253	077381	08339	99652	13
48	922610	250	998474	1.8	924136	252	075864	08368	99649	12
49	924112	249	998464	1.8	925649	251	074351	08397	99647	11
50	925609	249	998453	1.8	927156	250	072844	08426	99644	10
51	8.927100	248	9.998442	1.8	8.928658	249	11.071342	08455	99642	9
52	928587	247	998431	1.8	930155	249	069845	08484	99639	8
53	930068	246	998421	1.8	931647	248	068353	08513	99637	7
54	931544	245	998410	1.8	933131	247	066866	08542	99635	6
55	933015	244	998399	1.8	934616	246	065384	08571	99632	5
56	934481	213	998388	1.8	936093	245	063907	08600	99630	4
57	935943	243	998377	1.8	937565	244	062435	08629	99627	3
58	937398	242	998366	1.8	939032	244	060968	08658	99625	2
59	938850	241	998355	1.8	940494	243	059506	08687	99622	1
60	940296		998344		941952		058048	08716	99619	0
	Cosine.		Sine.		Cotang.		Tang.	N. cos.	N. sine.	'

85 Degrees.

'	Sine.	D. 10"	Cosine.	D. 10"	Tang.	D. 10"	Cotang.	N. sine.	N. cos.	'
0	8.940296	240	9.998344	1.9	8.941952	242	11.058048	08716	99619	60
1	941738	239	998333	1.9	943404	241	056596	08745	99617	59
2	943174	239	998322	1.9	944852	240	055148	08774	99614	58
3	944606	238	998311	1.9	946295	240	053705	08803	99612	57
4	946034	237	998300	1.9	947734	239	052266	08831	99609	56
5	947456	236	998289	1.9	949168	238	050832	08860	99607	55
6	948874	235	998277	1.9	950597	237	049403	08889	99604	54
7	950287	235	998266	1.9	952021	237	047979	08918	99602	53
8	951696	234	998255	1.9	953441	236	046559	08947	99599	52
9	953100	233	998243	1.9	954856	235	045144	08976	99596	51
10	954499	232	998232	1.9	956267	234	043733	09005	99594	50
11	8.955894	232	9.998220	1.9	8.957674	234	11.042326	09034	99591	49
12	957284	231	998209	1.9	959075	233	040925	09063	99588	48
13	958670	230	998197	1.9	960473	232	039527	09092	99586	47
14	960052	229	998186	1.9	961866	231	038134	09121	99583	46
15	961429	229	998174	1.9	963255	231	036745	09150	99580	45
16	962801	228	998163	1.9	964639	230	035361	09179	99578	44
17	964170	227	998151	1.9	966019	229	033981	09208	99575	43
18	965534	227	998139	1.9	967394	229	032606	09237	99572	42
19	966893	226	998128	2.0	968766	228	031234	09266	99570	41
20	968249	225	998116	2.0	970133	227	029867	09295	99567	40
21	8.969600	224	9.998104	2.0	8.971496	226	11.028504	09324	99564	39
22	970947	224	998092	2.0	972855	226	027145	09353	99562	38
23	972289	223	998080	2.0	974209	225	025791	09382	99559	37
24	973628	222	998068	2.0	975560	224	024440	09411	99556	36
25	974962	222	998056	2.0	976906	224	023094	09440	99553	35
26	976293	221	998044	2.0	978248	223	021752	09469	99551	34
27	977619	220	998032	2.0	979586	222	020414	09498	99548	33
28	978941	220	998020	2.0	980921	222	019079	09527	99545	32
29	980259	219	998008	2.0	982251	221	017749	09556	99542	31
30	981573	218	997996	2.0	983577	220	016423	09585	99540	30
31	8.982883	218	9.997984	2.0	8.984899	220	11.015101	09614	99537	29
32	984189	217	997972	2.0	986217	219	013783	09642	99534	28
33	985491	216	997959	2.0	987532	218	012468	09671	99531	27
34	986789	216	997947	2.0	988842	218	011158	09700	99528	26
35	988083	215	997935	2.0	990149	217	009851	09729	99526	25
36	989374	214	997922	2.1	991451	216	008549	06758	99523	24
37	990660	214	997910	2.1	992750	216	007250	09787	99520	23
38	991943	213	997897	2.1	994045	215	005955	09816	99517	22
39	993222	212	997885	2.1	995337	215	004663	09845	99514	21
40	994497	212	997872	2.1	996624	214	003376	09874	99511	20
41	8.995768	211	9.997860	2.1	8.997908	213	11.002092	09903	99508	19
42	997036	211	997847	2.1	999188	213	000812	09932	99506	18
43	998299	210	997835	2.1	9.000465	212	10.999535	09961	99503	17
44	999560	209	997822	2.1	001738	211	998262	09990	99500	16
45	9.000816	209	997809	2.1	003007	211	996993	10019	99497	15
46	002069	208	997797	2.1	004272	210	995728	10048	99494	14
47	003318	208	997784	2.1	005534	210	994466	10077	99491	13
48	004563	207	997771	2.1	006792	209	993208	10106	99488	12
49	005805	206	997758	2.1	008047	208	991953	10135	99485	11
50	007044	206	997745	2.1	009298	208	990702	10164	99482	10
51	9.008278	205	9.997732	2.1	9.010546	207	10.989454	10192	99479	9
52	009510	205	997719	2.1	011790	207	988210	10221	99476	8
53	010737	204	997706	2.1	013031	206	686969	10250	99473	7
54	011962	203	997693	2.1	014268	206	985732	10279	99470	6
55	013182	203	997680	2.2	015502	205	984498	10308	99467	5
56	014400	202	997667	2.2	016732	204	983268	10337	99464	4
57	015613	202	997654	2.2	017959	204	983041	10366	99461	3
58	016824	201	997641	2.2	019183	203	980817	10395	99458	2
59	018031	201	997628	2.2	020403	203	979597	10424	99455	1
60	019235		997614		021620		978380	10453	99452	0
	Cosine.		Sine.		Cotang.		Tang.	N. cos.	N. sine.	'

84 Degrees.

′	Sine.	D. 10″	Cosine.	D. 10″	Tang.	D. 10″	Cotang.	N. sine.	N. cos.	′
0	9.019235	200	9.997614	2.2	9.021620	202	10.978380	10453	99452	60
1	020435	199	997601	2.2	022834	202	977166	10482	99449	59
2	021632	199	997588	2.2	024044	201	975956	10511	99446	58
3	022825	198	997574	2.2	025251	201	974749	10546	99443	57
4	024016	198	997561	2.2	026455	200	973545	10569	99440	56
5	025203	197	997547	2.2	027655	199	972345	10597	99437	55
6	026386	197	997534	2.2	028852	199	971148	10626	99434	54
7	027567	196	997520	2.3	030046	198	969951	10655	99431	53
8	028744	196	997507	2.3	031237	198	968763	10684	99428	52
9	029918	195	997493	2.3	032425	197	967575	10713	99424	51
10	031089	195	997480	2.3	033609	197	966391	10742	99421	50
11	9.032257	194	9.997466	2.3	9.034791	196	10.965209	10771	99418	49
12	033421	194	997452	2.3	035969	196	964031	10800	99415	48
13	034582	193	997439	2.3	037144	195	962856	10829	99412	47
14	035741	192	997425	2.3	038316	195	961684	10858	99409	46
15	036896	192	997411	2.3	039485	194	960515	10887	99406	45
16	038048	191	997397	2.3	040651	194	959349	10916	99402	44
17	039197	191	997383	2.3	041813	193	958187	10945	99399	43
18	040342	190	997369	2.3	042973	193	957027	10973	99396	42
19	041485	190	997355	2.3	044130	192	955870	11002	99393	41
20	042625	189	997341	2.3	045284	192	954716	11031	99390	40
21	9.043762	189	9.997327	2.3	9.046434	191	10.953566	11060	99386	39
22	044895	180	997313	2.4	047582	191	952418	11089	99383	38
23	046026	188	997299	2.4	048727	190	951273	11118	99380	37
24	047154	187	997285	2.4	049869	190	950131	11147	99377	36
25	048279	187	997271	2.4	051008	189	948992	11176	99374	35
26	049400	186	997257	2.4	052144	189	947856	11205	99370	34
27	050519	186	997242	2.4	053277	188	946723	11234	99367	33
28	051635	185	997228	2.4	054407	188	945593	11263	99364	32
29	052749	185	997214	2.4	055535	187	944465	11291	99360	31
30	053859	184	997199	2.4	056659	187	943341	11320	99357	30
31	9.054966	184	9.997185	2.4	9.057781	186	10.942219	11349	99354	29
32	056071	184	997170	2.4	058900	186	941100	11378	99351	28
33	057172	184	997156	2.4	060016	185	939984	11407	99347	27
34	058271	183	997141	2.4	061130	185	938870	11436	99344	26
35	059367	183	997127	2.4	062240	185	937760	11465	99341	25
36	060460	182	997112	2.4	063348	184	936652	11494	99337	24
37	061551	182	997098	2.4	064453	184	935547	11523	99334	23
38	062639	181	997083	2.5	065556	183	934444	11552	99331	22
39	063724	181	997068	2.5	066655	183	933345	11580	99327	21
40	064806	180	997053	2.5	067752	182	932248	11609	99324	20
41	9.065885	180	9.997039	2.5	9.068846	182	10.931154	11638	99320	19
42	066962	179	997024	2.5	069038	181	930062	11667	99317	18
43	068036	179	997009	2.5	071027	181	928973	11696	99314	17
44	069107	179	996994	2.5	072113	181	927887	11725	99310	16
45	070176	178	996979	2.5	073197	180	926803	11754	99307	15
46	071242	178	996964	2.5	074278	180	925722	11783	99303	14
47	072306	177	996949	2.5	075356	179	924644	11812	99300	13
48	073366	177	996934	2.5	076432	179	923568	11840	99297	12
49	074424	176	996919	2.5	077505	178	922495	11869	99293	11
50	075480	176	996904	2.5	078576	178	921424	11898	99290	10
51	9.076533	175	9.996889	2.5	9.079644	178	10.920356	11927	99286	9
52	077583	175	996874	2.5	080710	177	919290	11956	99283	8
53	078631	175	996858	2.5	081773	177	918227	11985	99279	7
54	079676	174	996843	2.5	082833	176	917167	12014	99276	6
55	080719	174	996828	2.5	083891	176	916109	12043	99272	5
56	081759	173	996812	2.6	084947	175	915053	12071	99269	4
57	082797	173	996797	2.6	086000	175	914000	12100	99265	3
58	083832	172	996782	2.6	087050	175	912950	12129	99262	2
59	084864	172	996766	2.6	088098	174	911902	12158	99258	1
60	085894		996751		089144		910856	12187	99255	0
	Cosine.		Sine.		Cotang.		Tang.	N. cos.	N. sine.	′

Log. Sines and Tangents. (7°) Natural Sines. TABLE II.

′	Sine.	D. 10″	Cosine.	D. 10″	Tang.	D. 10″	Cotang.	N. sine.	N. cos.	′
0	9.085894	171	9.996751	2.6	9.089144	174	10.910856	12187	99255	60
1	086922	171	996735	2.6	090187	173	909813	12216	99251	59
2	087947	170	996720	2.6	091228	173	908772	12245	99248	58
3	088970	170	996704	2.6	092266	173	907734	12274	99244	57
4	089990	170	996688	2.6	093302	172	906698	12302	99240	56
5	091008	169	996673	2.6	094336	172	905664	12331	99237	55
6	092024	169	996657	2.6	095367	172	904633	12360	99233	54
7	093037	168	996641	2.6	096395	171	903605	12389	99230	53
8	094047	168	996625	2.6	097422	171	902578	12418	99226	52
9	095056	168	996610	2.6	098446	171	901554	12447	99222	51
10	096062	167	996594	2.6	099468	170	900532	12476	99219	50
11	9.097065	167	9.996578	2.6	).100487	170	10.899513	12504	99215	49
12	098066	167	996562	2.7	101504	169	898496	12533	99211	48
13	099065	166	996546	2.7	102519	169	897481	12562	99208	47
14	100062	166	996530	2.7	103532	169	896468	12591	99204	46
15	101056	166	996514	2.7	104542	168	895458	12620	99200	45
16	102048	165	996498	2.7	105550	168	894450	12649	99197	44
17	103037	165	996482	2.7	106556	168	893444	12678	99193	43
18	104025	164	996465	2.7	107559	167	892441	12706	99189	42
19	105010	164	996449	2.7	108560	167	891440	12735	99186	41
20	105992	164	996433	2.7	109559	166	890441	12764	99182	40
21	9.106973	163	9.996417	2.7	9.110556	166	10.889444	12793	99178	39
22	107951	163	996400	2.7	111551	166	888449	12822	99175	38
23	108927	163	996384	2.7	112543	165	887457	12851	99171	37
24	109901	162	996368	2.7	113533	165	886467	12880	99167	36
25	110873	162	996351	2.7	114521	165	885479	12908	99163	35
26	111842	162	996335	2.7	115507	164	884493	12937	99160	34
27	112809	161	996318	2.7	116491	164	883509	12966	99156	33
28	113774	161	996302	2.8	117472	164	882528	12995	99152	32
29	114737	160	996285	2.8	118452	163	881548	13024	99148	31
30	115698	160	996269	2.8	119429	163	880571	13053	99144	30
31	9.116656	160	9.996252	2.8	9.120404	162	10.879596	13081	99141	29
32	117613	159	996235	2.8	121377	162	878623	13110	99137	28
33	118567	159	996219	2.8	122348	162	877652	13139	99133	27
34	119519	159	996202	2.8	123317	161	876683	13168	99129	26
35	120469	158	996185	2.8	124284	161	875716	13197	99125	25
36	121417	158	996168	2.8	125249	161	874751	13226	99122	24
37	122362	158	996151	2.8	126211	160	873789	13254	99118	23
38	123306	157	996134	2.8	127172	160	872828	13283	99114	22
39	124248	157	996117	2.8	128130	160	871870	13312	99110	21
40	125187	157	996100	2.8	129087	159	870913	13341	99106	20
41	9.126125	156	9.996083	2.8	9.130041	159	10.869959	13370	99102	19
42	127060	156	996066	2.9	130994	159	869006	13399	99098	18
43	127993	156	996049	2.9	131944	158	868056	13427	99094	17
44	128925	155	996032	2.9	132893	158	867107	13456	99091	16
45	129854	155	996015	2.9	133839	158	866161	13485	99087	15
46	130781	154	995998	2.9	134784	157	865216	13514	99083	14
47	131706	154	995980	2.9	135726	157	864274	13543	99079	13
48	132630	154	995963	2.9	136367	157	863333	13572	99075	12
49	133551	153	995946	2.9	137605	156	862395	13600	99071	11
50	134470	153	995928	2.9	138542	156	861458	13629	99067	10
51	9.135387	153	9.995911	2.9	9.139476	156	10.860524	13658	99063	9
52	136303	152	995894	2.9	140409	155	859591	13687	99059	8
53	137216	152	995876	2.9	141340	155	858660	13716	99055	7
54	138128	152	995859	2.9	142269	155	857731	13744	99051	6
55	139037	152	995841	2.9	143196	154	856804	13773	99047	5
56	139944	151	995823	2.9	144121	154	855879	13802	99043	4
57	140850	151	995806	2.9	145044	154	854956	13831	99039	3
58	141754	151	995.88	2.9	145966	153	854034	13860	99035	2
59	142655	150	995771	2.9	146885	153	853115	13889	99031	1
60	143555	150	995753		147803	153	852197	13917	99027	0
	Cosine.		Sine.		Cotang.		Tang.	N. cos.	N. sine.	′

82 Degrees.

′	Sine.	D. 10″	Cosine.	D. 10″	Tang.	D. 10″	Cotang.	N. sine	N. cos.	
0	9.143555	150	9.995753	3.0	9.147803	153	10.852197	13917	99027	60
1	144453	149	995735	3.0	148718	152	851282	13946	99023	59
2	145349	149	995717	3.0	149632	152	850368	13975	99019	58
3	146243	149	995699	3.0	150544	152	849456	14004	99015	57
4	147136	148	995681	3.0	151454	152	848546	14033	99011	56
5	148026	148	995664	3.0	152363	151	847637	14061	99006	55
6	148915	148	995646	3.0	153269	151	846731	14090	99002	54
7	149802	148	995628	3.0	154174	151	845826	14119	98998	53
8	150686	147	995610	3.0	155077	150	844923	14148	98994	52
9	151569	147	995591	3.0	155978	150	844022	14177	98990	51
10	152451	147	995573	3.0	156877	150	843123	14205	98986	50
11	9.153330	147	9.995555	3.0	9.157775	150	10.842225	14234	98982	49
12	154208	146	995537	3.0	158671	149	841329	14263	98978	48
13	155083	146	995519	3.0	159565	149	840435	14292	98973	47
14	155957	146	995501	3.0	160457	149	839543	14320	98969	46
15	156830	146	995482	3.1	161347	148	838653	14349	98965	45
16	157700	145	995464	3.1	162236	148	837764	14378	98961	44
17	158569	145	995446	3.1	163123	148	836877	14407	98957	43
18	159435	145	995427	3.1	164008	148	835992	14436	98953	42
19	160301	144	995409	3.1	164892	147	835108	14464	98948	41
20	161164	144	995390	3.1	165774	147	834226	14493	98944	40
21	9.162025	144	9.995372	3.1	9.166654	147	10.833346	14522	98940	39
22	162885	144	995353	3.1	167532	146	832468	14551	98936	38
23	163743	143	995334	3.1	168409	146	831591	14580	98931	37
24	164600	143	995316	3.1	169284	146	830716	14608	98927	36
25	165454	143	995297	3.1	170157	145	829843	14637	98923	35
26	166307	142	995278	3.1	171029	145	828971	14666	98919	34
27	167159	142	995260	3.1	171899	145	828101	14695	98914	33
28	168008	142	995241	3.1	172767	145	827233	14723	98910	32
29	168856	142	995222	3.2	173634	144	826366	14752	98906	31
30	169702	141	995203	3.2	174499	144	825501	14781	98902	30
31	9.170547	141	9.995184	3.2	9.175362	144	10.824638	14810	98897	29
32	171389	141	995165	3.2	176224	144	823776	14838	98893	28
33	172230	140	995146	3.2	177084	143	822916	14867	98889	27
34	173070	140	995127	3.2	177942	143	822058	14896	98884	26
35	173908	140	995108	3.2	178799	143	821201	14925	98880	25
36	174744	140	995089	3.2	179655	142	820345	14954	98876	24
37	175578	139	995070	3.2	180508	142	819492	14982	98871	23
38	176411	139	995051	3.2	181360	142	818640	15011	98867	22
39	177242	139	995032	3.2	182211	142	817789	15040	98863	21
40	178072	139	995013	3.2	183059	141	816941	15069	98858	20
41	9.178900	138	9.994993	3.2	9.183907	141	10.816093	15097	98854	19
42	179726	138	994974	3.2	184752	141	815248	15126	98849	18
43	180551	138	994955	3.2	185597	141	814403	15155	98845	17
44	181374	137	994935	3.2	186439	140	813561	15184	98841	16
45	182196	137	994916	3.2	187280	140	812720	15212	98836	15
46	183016	137	994896	3.3	188120	140	811880	15241	98832	14
47	183834	137	994877	3.3	188958	140	811042	15270	98827	13
48	184651	136	994857	3.3	189794	139	810206	15299	98823	12
49	185466	136	994838	3.3	190629	139	809371	15327	98818	11
50	186280	136	994818	3.3	191462	139	808538	15356	98814	10
51	9.187092	136	9.994798	3.3	9.192294	139	10.807706	15385	98809	9
52	187903	135	994779	3.3	193124	138	806876	15414	98805	8
53	188712	135	994759	3.3	193953	138	806047	15442	98800	7
54	189519	135	994739	3.3	194780	138	805220	15471	98796	6
55	190325	135	994719	3.3	195606	138	804394	15500	98791	5
56	191130	134	994700	3.3	196430	137	803570	15529	98787	4
57	191933	134	994680	3.3	197253	137	802747	15557	98782	3
58	192734	134	994660	3.3	198074	137	801926	15586	98778	2
59	193534	134	994640	3.3	198894	137	801106	15615	98773	1
60	194332	133	994620		199713	136	800287	15643	98769	0
	Cosine.		Sine.		Cotang.		Tang.	N. cos.	N. sine.	′

′	Sine.	D. 10″	Cosine.	D. 10″	Tang.	D. 10″	Cotang.	N. sine.	N. cos.	
0	9.194332	133	9.994620	3.3	9.199713	136	10.800287	15643	98769	60
1	195129	133	994600	3.3	200529	136	799471	15672	98764	59
2	195925	132	994580	3.3	201345	136	798655	15701	98760	58
3	196719	132	994560	3.4	202159	135	797841	15730	98755	57
4	197511	132	994540	3.4	202971	135	797029	15758	98751	56
5	198302	132	994519	3.4	203782	135	796218	15787	98746	55
6	199091	131	994499	3.4	204592	135	795408	15816	98741	54
7	199879	131	994479	3.4	205400	134	794600	15845	98737	53
8	200666	131	994459	3.4	206207	134	793793	15873	98732	52
9	201451	131	994438	3.4	207013	134	792987	15902	98728	51
10	202234	130	994418	3.4	207817	134	792183	15931	98723	50
11	9.203017	130	9.994397	3.4	9.208619	133	10.791381	15959	98718	49
12	203797	130	994377	3.4	209420	133	790580	15988	98714	48
13	204577	130	994357	3.4	210220	133	789780	16017	98709	47
14	205354	129	994336	3.4	211018	133	788982	16046	98704	46
15	206131	129	994316	3.4	211815	133	788185	16074	98700	45
16	206906	129	994295	3.4	212611	132	787389	16103	98695	44
17	207679	129	994274	3.5	213405	132	786595	16132	98690	43
18	208452	128	994254	3.5	214198	132	785802	16160	98686	42
19	209222	128	994233	3.5	214989	132	785011	16189	98681	41
20	209992	128	994212	3.5	215780	131	784220	16218	98676	40
21	9.210760	128	9.994191	3.5	9.216568	131	10.783432	16246	98671	39
22	211526	127	994171	3.5	217356	131	782644	16275	98667	38
23	212291	127	994150	3.5	218142	131	781858	16304	98662	37
24	213055	127	994129	3.5	218926	130	781074	16333	98657	36
25	213818	127	994108	3.5	219710	130	780290	16361	98652	35
26	214579	127	994087	3.5	220492	130	779508	16390	98648	34
27	215338	126	994066	3.5	221272	130	778728	16419	98643	33
28	216097	126	994045	3.5	222052	130	777948	16447	98638	32
29	216854	126	994024	3.5	222830	129	777170	16476	98633	31
30	217609	126	994003	3.5	223606	129	776394	16505	98629	30
31	9.218363	125	9.993981	3.5	9.224382	129	10.775618	16533	98624	29
32	219116	125	993960	3.5	225156	129	774844	16562	98619	28
33	219868	125	993939	3.5	225929	129	774071	16591	98614	27
34	220618	125	993918	3.5	226700	129	773300	16620	98609	26
35	221367	125	993896	3.6	227471	128	772529	16648	98604	25
36	222115	124	993875	3.6	228239	128	771761	16677	98600	24
37	222861	124	993854	3.6	229007	128	770993	16706	98595	23
38	223606	124	993832	3.6	229773	128	770227	16734	98590	22
39	224349	124	993811	3.6	230539	127	769461	16763	98585	21
40	225092	123	993789	3.6	231302	127	768698	16792	98580	20
41	9.225833	123	9.993768	3.6	9.232065	127	10.767935	16820	98575	19
42	226573	123	993746	3.6	232826	127	767174	16849	98570	18
43	227311	123	993725	3.6	233586	127	766414	16878	98565	17
44	228048	123	993703	3.6	234345	126	765655	16906	98561	16
45	228784	122	993681	3.6	235103	126	764897	16935	98556	15
46	229518	122	993660	3.6	235859	126	764141	16964	98551	14
47	230252	122	993638	3.6	236614	126	763386	16992	98546	13
48	230984	122	993616	3.6	237368	126	762632	17021	98541	12
49	231714	122	993594	3.7	238120	125	761880	17050	98536	11
50	232444	121	993572	3.7	238872	125	761128	17078	98531	10
51	9.233172	121	9.993550	3.7	9.239622	125	10.760378	17107	98526	9
52	233899	121	994528	3.7	240371	125	759629	17136	98521	8
53	234625	121	993506	3.7	241118	124	758882	17164	98516	7
54	235349	120	993484	3.7	241865	124	758135	17193	98511	6
55	236073	120	993462	3.7	242610	124	757390	17222	98506	5
56	236795	120	9.3440	3.7	243354	124	756646	17250	98501	4
57	237515	120	993418	3.7	244097	124	755903	17279	98496	3
58	238235	120	993396	3.7	244839	123	755161	17308	98491	2
59	238953	119	993374	3.7	245579	123	754421	17336	98486	1
60	239570		993351		246319		753681	17365	98481	0
	Cosine.		Sine.		Cotang.		Tang.	N. cos.	N. sine.	′

′	Sine.	D. 10″	Cosine.	D. 10″	Tang.	D. 10″	Cotang.	N.sine	N. cos.	
0	9.239670	119	9.993351	3.7	9.246319	123	10.753681	17305	98481	60
1	240386	119	993329	3.7	247057	123	752943	17393	98476	59
2	241101	119	993307	3.7	247794	123	752206	17422	98471	58
3	241814	119	993285	3.7	248530	122	751470	17451	98466	57
4	242526	119	993262	3.7	249264	122	750736	17479	98461	56
5	243237	118	993240	3.7	249998	122	750002	17508	98455	55
6	243947	118	993217	3.7	250730	122	749270	17537	98450	54
7	244656	118	993195	3.8	251461	122	748539	17565	98445	53
8	245363	118	993172	3.8	252191	122	747809	17594	98440	52
9	246069	118	993149	3.8	252920	121	747080	17623	98435	51
10	246775	117	993127	3.8	253648	121	746352	17651	98430	50
11	9.247478	117	9.993104	3.8	9.254374	121	10.745626	17680	98425	49
12	248181	117	993081	3.8	255100	121	744900	17708	98420	48
13	248883	117	993059	3.8	255824	121	744176	17737	98414	47
14	249583	117	993036	3.8	256547	120	743453	17766	98409	46
15	250282	116	993013	3.8	257269	120	742731	17794	98404	45
16	250980	116	992990	3.8	257990	120	742010	17823	98399	44
17	251677	116	992967	3.8	258710	120	741290	17852	98394	43
18	252373	116	992944	3.8	259429	120	740571	17880	98389	42
19	253067	116	992921	3.8	260146	120	739854	17909	98383	41
20	253761	115	992898	3.8	260863	119	739137	17937	98378	40
21	9.254453	115	9.992875	3.8	9.261578	119	10.738422	17966	98373	39
22	255144	115	992852	3.8	262292	119	737708	17995	98368	38
23	255834	115	992829	3.8	263005	119	736995	18023	98362	37
24	256523	115	992806	3.9	263717	119	736283	18052	98357	36
25	257211	115	992783	3.9	264428	118	735572	18081	98352	35
26	257898	114	992759	3.9	265138	118	734862	18109	98347	34
27	258583	114	992736	3.9	265847	118	734153	18138	98341	33
28	259268	114	992713	3.9	266555	118	733445	18166	98336	32
29	259951	114	992690	3.9	267261	118	732739	18195	98331	31
30	260633	114	992666	3.9	267967	117	732033	18224	98325	30
31	9.261314	113	9.992643	3.9	9.268671	117	10.731329	18252	98320	29
32	261994	113	992619	3.9	269375	117	730625	18281	98315	28
33	262673	113	992596	3.9	270077	117	729923	18309	98310	27
34	263351	113	992572	3.9	270779	117	729221	18338	98304	26
35	264027	113	992549	3.9	271479	117	728521	18367	98299	25
36	264703	113	992525	3.9	272178	116	727822	18395	98294	24
37	265377	112	992501	3.9	272876	116	727124	18424	98288	23
38	266051	112	992478	3.9	273573	116	726427	18452	98283	22
39	266723	112	992454	4.0	274269	116	725731	18481	98277	21
40	267395	112	992430	4.0	274964	116	725036	18509	98272	20
41	9.268065	112	9.992406	4.0	9.275658	115	10.724342	18538	98267	19
42	268734	111	992382	4.0	276351	115	723649	18567	98261	18
43	269402	111	992359	4.0	277043	115	722957	18595	98256	17
44	270069	111	992335	4.0	277734	115	722266	18624	98250	16
45	270735	111	992311	4.0	278424	115	721576	18652	98245	15
46	271400	111	992287	4.0	279113	115	720887	18681	98240	14
47	272064	111	992263	4.0	279801	115	720199	18710	98234	13
48	272726	110	992239	4.0	280488	114	719512	18738	98229	12
49	273388	110	992214	4.0	281174	114	718826	18767	98223	11
50	274049	110	992190	4.0	281858	114	718142	18795	98218	10
51	9.274708	110	9.992166	4.0	9.282542	114	10.717458	18824	98212	9
52	275367	110	992142	4.0	283225	114	716775	18852	98207	8
53	276024	109	992117	4.1	283907	113	716093	18881	98201	7
54	276681	109	992093	4.1	284588	113	715412	18910	98196	6
55	277337	109	992069	4.1	285268	113	714732	18938	98190	5
56	277991	109	992044	4.1	285947	113	714053	18967	98185	4
57	278644	109	992020	4.1	286624	113	713376	18995	98179	3
58	279297	109	991996	4.1	287301	113	712699	19024	98174	2
59	279948	108	991971	4.1	287977	112	712023	19052	98168	1
60	280599		991947		288652		711348	19081	98163	0
	Cosine.		Sine.		Cotang.		Tang.	N. cos.	N. sine.	′

79 Degrees.

′	Sine.	D. 10″	Cosine.	D. 10″	Tang.	D. 10	Cotang.	N. sine.	N. cos.	
0	9.280590	103	9.991947	4.1	9.288652	112	10.711348	19081	98163	60
1	281248	108	991922	4.1	289326	112	710674	19109	98157	59
2	281897	108	991897	4.1	289999	112	710001	19138	98152	58
3	282544	108	991873	4.1	290671	112	709329	19167	98146	57
4	283190	108	991848	4.1	291342	112	708658	19195	98140	56
5	283836	108	991823	4.1	292013	112	707987	19224	98135	55
6	284480	107	991799	4.1	292682	111	707318	19252	98129	54
7	285124	107	991774	4.1	293350	111	706650	19281	98124	53
8	285766	107	991749	4.2	294017	111	705983	19309	98118	52
9	286408	107	991724	4.2	294684	111	705316	19338	98112	51
10	287048	107	991699	4.2	295349	111	704651	19366	98107	50
11	9.287687	107	9.991674	4.2	9.296013	111	10.703987	19395	98101	49
12	288326	106	991649	4.2	296677	111	703323	19423	98096	48
13	288964	106	991624	4.2	297339	110	702661	19452	98090	47
14	289600	106	991599	4.2	298001	110	701999	19481	98084	46
15	290236	106	991574	4.2	298662	110	701338	19509	98079	45
16	290870	106	991549	4.2	299322	110	700678	19538	98073	44
17	291504	106	991524	4.2	299980	110	700020	19566	98067	43
18	292137	105	991498	4.2	300638	110	699362	19595	98061	42
19	292768	105	991473	4.2	301295	109	698705	19623	98056	41
20	293399	105	991448	4.2	301951	109	698049	19652	98050	40
21	9.294029	105	9.991422	4.2	9.302607	109	10.697393	19680	98044	39
22	294658	105	991397	4.2	303261	109	696739	19709	98039	38
23	295286	105	991372	4.2	303914	109	696086	19737	98033	37
24	295913	104	991346	4.3	304567	109	695433	19766	98027	36
25	296539	104	991321	4.3	305218	109	694782	19794	98021	35
26	297164	104	991295	4.3	305869	108	694131	19823	98016	34
27	297788	104	991270	4.3	306519	108	693481	19851	98010	33
28	298412	104	991244	4.3	307168	108	692832	19880	98004	32
29	299034	104	991218	4.3	307815	108	692185	19908	97998	31
30	299655	104	991193	4.3	308463	108	691537	19937	97992	30
31	9.300276	103	9.991167	4.3	9.309109	108	10.690891	19965	97987	29
32	300895	103	991141	4.3	309754	107	690246	19994	97981	28
33	301514	103	991115	4.3	310398	107	689602	20022	97975	27
34	302132	103	991090	4.3	311042	107	688958	20051	97969	26
35	302748	103	991064	4.3	311685	107	688315	20079	97963	25
36	303364	103	991038	4.3	312327	107	687673	20108	97958	24
37	303979	102	991012	4.3	312967	107	687033	20136	97952	23
38	304593	102	990986	4.3	313608	106	686392	20165	97946	22
39	305207	102	990960	4.3	314247	106	685753	20193	97940	21
40	305819	102	990934	4.3	314885	106	685115	20222	97934	20
41	9.306430	102	9.990908	4.4	9.315523	106	10.684477	20250	97928	19
42	307041	102	990882	4.4	316159	106	683841	20279	97922	18
43	307650	102	990855	4.4	316795	106	683205	20307	97916	17
44	308259	101	990829	4.4	317430	106	682570	20336	97910	16
45	308867	101	990803	4.4	318064	105	681936	20364	97905	15
46	309474	101	990777	4.4	318697	105	681303	20393	97899	14
47	310080	101	990750	4.4	319329	105	680671	20421	97893	13
48	310685	101	990724	4.4	319961	105	680039	20450	97887	12
49	311289	100	990697	4.4	320592	105	679408	20478	97881	11
50	311893	100	990671	4.4	321222	105	678778	20507	97875	10
51	9.312495	100	9.990644	4.4	9.321851	105	10.678149	20535	97869	9
52	313097	100	990618	4.4	322479	104	677521	20563	97863	8
53	313698	100	990591	4.4	323106	104	676894	20592	97857	7
54	314297	100	990565	4.4	323733	104	676267	20620	97851	6
55	314897	100	990538	4.4	324358	104	675642	20649	97845	5
56	315495	100	990511	4.5	324983	104	675017	20677	97839	4
57	316092	99	990485	4.5	325607	104	674393	20706	97833	3
58	316689	99	990458	4.5	326231	104	673769	20734	97827	2
59	317284	99	990431	4.5	326853	104	673147	20763	97821	1
60	317879		990404		327475		672525	20791	97815	0
′	Cosine.		Sine.		Cotang.		Tang.	N. cos.	N. sine.	′

′	Sine.	D. 10″	Cosine.	D. 10″	Tang.	D. 10″	Cotang.	N. sine.	N. cos.	
0	9.317879	99.0	9.990404	4.5	9.327474	103	10.672526	20791	97815	60
1	318473	98.8	990378	4.5	328035	103	671905	20820	97809	59
2	319066	98.7	990351	4.5	328715	103	671285	20848	97803	58
3	319658	98.6	990324	4.5	329334	103	670666	20877	97797	57
4	320249	98.4	990297	4.5	329953	103	670047	20905	97791	56
5	320840	98.3	990270	4.5	330570	103	669430	20933	97784	55
6	321430	98.2	990243	4.5	331187	103	668813	20962	97778	54
7	322019	98.0	990215	4.5	331803	103	668197	20990	97772	53
8	322607	97.9	990188	4.5	332418	102	667582	21019	97766	52
9	323194	97.7	990161	4.5	333033	102	666967	21047	97760	51
10	323780	97.6	990134	4.5	333646	102	666354	21076	97754	50
11	9.324366	97.5	9.990107	4.5	9.334259	102	10.665741	21104	97745	49
12	324950	97.3	990079	4.6	334871	102	665129	21132	97742	48
13	325534	97.2	990052	4.6	335482	102	664518	21161	97735	47
14	326117	97.0	990025	4.6	336093	102	663907	21189	97729	46
15	326700	96.9	989997	4.6	336702	102	663298	21218	97723	45
16	327281	96.8	989970	4.6	337311	101	662689	21246	97717	44
17	327862	96.6	989942	4.6	337919	101	662081	21275	97711	43
18	328442	96.5	989915	4.6	338527	101	661473	21303	97705	42
19	329021	96.4	989887	4.6	339133	101	660867	21331	97698	41
20	329599	96.2	989860	4.6	339739	101	660261	21360	97692	40
21	9.330176	96.1	9.989832	4.6	9.340344	101	10.659656	21388	97686	39
22	330753	96.0	989804	4.6	340948	101	659052	21417	97680	38
23	331329	95.8	989777	4.6	341552	101	658448	21445	97673	37
24	331903	95.7	989749	4.7	342155	100	657845	21474	97667	36
25	332478	95.6	989721	4.7	342757	100	657243	21502	97661	35
26	333051	95.4	989693	4.7	343358	100	656642	21530	97655	34
27	333624	95.3	989665	4.7	343958	100	656042	21559	97648	33
28	334195	95.2	989637	4.7	344558	100	655442	21587	97642	32
29	334766	95.0	989609	4.7	345157	100	654843	21616	97636	31
30	335337	94.9	989582	4.7	345755	100	654245	21644	97630	30
31	9.335906	94.8	9.989553	4.7	9.346353	99.4	10.653647	21672	97623	29
32	336475	94.6	989525	4.7	346949	99.3	653051	21701	97617	28
33	337043	94.5	989497	4.7	347545	99.2	652455	21729	97611	27
34	337610	94.4	989469	4.7	348141	99.1	651859	21758	97604	26
35	338176	94.3	989441	4.7	348735	99.0	651265	21786	97598	25
36	338742	94.1	989413	4.7	349329	98.8	650671	21814	97592	24
37	339306	94.0	989384	4.7	349922	98.7	650078	21843	97585	23
38	339871	93.9	989356	4.7	350514	98.6	649486	21871	97579	22
39	340434	93.7	989328	4.7	351106	98.5	648894	21899	97573	21
40	340996	93.6	989300	4.7	351697	98.3	648303	21928	97566	20
41	9.341558	93.5	9.989271	4.7	9.352287	98.2	10.647713	21956	97560	19
42	342119	93.4	989243	4.7	352876	98.1	647124	21985	97553	18
43	342679	93.2	989214	4.7	353465	98.0	646535	22013	97547	17
44	343239	93.1	989186	4.7	354053	97.9	645947	22041	97541	16
45	343797	93.0	989157	4.7	354640	97.7	645360	22070	97534	15
46	344355	92.9	989128	4.8	355227	97.6	644773	22098	97528	14
47	344912	92.7	989100	4.8	355813	97.5	644187	22126	97521	13
48	345469	92.6	989071	4.8	356398	97.4	643602	22155	97515	12
49	346024	92.5	989042	4.8	356982	97.3	643018	22183	97508	11
50	346579	92.4	989014	4.8	357566	97.1	642434	22212	97502	10
51	9.347134	92.2	9.988985	4.8	9.358149	97.0	10.641851	22240	97496	9
52	347687	92.1	988956	4.8	358731	96.9	641269	22268	97489	8
53	348240	92.0	988927	4.8	359313	96.8	640687	22297	97483	7
54	348792	91.9	988898	4.8	359893	96.7	640107	22325	97476	6
55	349343	91.7	988869	4.8	360474	96.6	639526	22353	97470	5
56	349893	91.6	988840	4.8	361053	96.5	638947	22382	97463	4
57	350443	91.5	988811	4.9	361632	96.3	638368	22410	97457	3
58	350992	91.4	988782	4.9	362210	96.2	637790	22438	97450	2
59	351540	91.3	988753	4.9	362787	96.1	637213	22467	97444	1
60	352088		988724		363364		636636	22495	97437	0
	Cosine.		Sine.		Cotang.		Tang.	N. cos.	N.sine.	

′	Sine.	D. 10″	Cosine.	D. 10″	Tang.	D. 10″	Cotang.	N. sine	N. cos.	
0	9.352088	91.1	9.988724	4.9	9.363364	96.0	10.636636	22495	97437	60
1	352635	91.0	988695	4.9	363940	95.9	636060	22523	97430	59
2	353181	90.9	988666	4.9	364515	95.8	635485	22552	97424	58
3	353726	90.8	988636	4.9	365090	95.7	634910	22580	97417	57
4	354271	90.7	988607	4.9	365664	95.5	634336	22608	97411	56
5	354815	90.5	988578	4.9	366237	95.4	633763	22637	97404	55
6	355358	90.4	988548	4.9	366810	95.3	633190	22665	97398	54
7	355901	90.3	988519	4.9	367382	95.2	632618	22693	97391	53
8	356443	90.2	988489	4.9	367953	95.1	632047	22722	97384	52
9	356984	90.1	988460	4.9	368524	95.0	631476	22750	97378	51
10	357524	89.9	988430	4.9	369094	94.9	630906	22778	97371	50
11	9.358064	89.8	9.988401	4.9	9.369663	94.8	10.630337	22807	97365	49
12	358603	89.7	988371	4.9	370232	94.6	629768	22835	97358	48
13	359141	89.6	988342	4.9	370799	94.5	629201	22863	97351	47
14	359678	89.5	988312	4.9	371367	94.4	628633	22892	97345	46
15	360215	89.3	988282	5.0	371933	94.3	628067	22920	97338	45
16	360752	89.2	988252	5.0	372499	94.2	627501	22948	97331	44
17	361287	89.1	988223	5.0	373064	94.1	626936	22977	97325	43
18	361822	89.0	988193	5.0	373629	94.0	626371	23005	97318	42
19	362356	88.9	988163	5.0	374193	93.9	625807	23033	97311	41
20	362889	88.8	988133	5.0	374756	93.8	625244	23062	97304	40
21	9.363422	88.7	9.988103	5.0	9.375319	93.7	10.624681	23090	97298	39
22	363954	88.5	988073	5.0	375881	93.5	624119	23118	97291	38
23	364485	88.4	988043	5.0	376442	93.4	623558	23146	97284	37
24	365016	88.3	988013	5.0	377003	93.3	622997	23175	97278	36
25	365546	88.2	987983	5.0	377563	93.2	622437	23203	97271	35
26	366075	88.1	987953	5.0	378122	93.1	621878	23231	97264	34
27	366604	88.0	987922	5.0	378681	93.0	621319	23260	97257	33
28	367131	87.9	987892	5.0	379239	92.9	620761	23288	97251	32
29	367659	87.7	987862	5.0	379797	92.8	620203	23316	97244	31
30	368185	87.6	987832	5.1	380354	92.7	619646	23345	97237	30
31	9.368711	87.5	9.987801	5.1	9.380910	92.6	10.619090	23373	97230	29
32	369236	87.4	987771	5.1	381466	92.5	618534	23401	97223	28
33	369761	87.3	987740	5.1	382020	92.4	617980	23429	97217	27
34	370285	87.2	987710	5.1	382575	92.3	617425	23458	97210	26
35	370808	87.1	987679	5.1	383129	92.2	616871	23486	97203	25
36	371330	87.0	987649	5.1	383682	93.1	616318	23514	97196	24
37	371852	86.9	987618	5.1	384234	92.0	615766	23542	97189	23
38	372373	86.7	987588	5.1	384786	91.9	615214	23571	97182	22
39	372894	86.6	987557	5.1	385337	91.8	614663	23599	97176	21
40	373414	86.5	987526	5.1	385888	91.7	614112	23627	97169	20
41	9.373933	86.4	9.987496	5.1	9.386438	91.5	10.613562	23656	97162	19
42	374452	86.3	987465	5.1	386987	91.4	613013	23684	97155	18
43	374970	86.2	987434	5.1	387536	91.3	612464	23712	97148	17
44	375487	86.1	987403	5.2	388084	91.2	611916	23740	97141	16
45	376003	86.0	987372	5.2	388631	91.1	611369	23769	97134	15
46	376519	85.9	987341	5.2	389178	91.0	610822	23797	97127	14
47	377035	85.8	987310	5.2	389724	90.9	610276	23825	97120	13
48	377549	85.7	987279	5.2	390270	90.8	609730	23853	97113	12
49	378063	85.6	987248	5.2	390815	90.7	609185	23882	97106	11
50	378577	85.4	987217	5.2	391360	90.7	608640	23910	97100	10
51	9.379089	85.3	9.987186	5.2	9.391903	90.6	10.608097	23938	97093	9
52	379601	85.2	987155	5.2	392447	90.5	607553	23966	97086	8
53	380113	85.1	987124	5.2	392989	90.4	607011	23995	97079	7
54	380624	85.0	987092	5.2	393531	90.3	606469	24023	97072	6
55	381134	84.9	987061	5.2	394073	90.2	605927	24051	97065	5
56	381643	84.8	987030	5.2	394614	90.1	605386	24079	97053	4
57	382152	84.7	986998	5.2	395154	90.0	604846	24108	97051	3
58	382661	84.6	986967	5.2	395694	89.9	604306	24136	97044	2
59	383168	84.5	986936	5.2	396233	89.8	603767	24164	97037	1
60	383675		986904		396771	89.7	603229	24192	97030	0
	Cosine.		Sine.		Cotang.		Tang.	N. cos.	N. sine.	′

76 Degrees.

′	Sine.	D. 10″	Cosine.	D. 10″	Tang.	D. 10″	Cotang.	N. sine.	N. cos.	′
0	9.383675	84.4	9.986904	5.2	9.396771	89.6	10.603229	24192	97030	60
1	384182	84.3	986873	5.3	397309	89.6	602691	24220	97023	59
2	384687	84.2	986841	5.3	397846	89.5	602154	24249	97015	58
3	385192	84.1	986809	5.3	398383	89.4	601617	24277	97008	57
4	385697	84.0	986778	5.3	398919	89.3	601081	24305	97001	56
5	386201	83.9	986746	5.3	399455	89.2	600545	24333	96994	55
6	386704	83.8	986714	5.3	399990	89.1	600010	24362	96987	54
7	387207	83.7	986683	5.3	400524	89.0	599476	24390	96980	53
8	387709	83.6	986651	5.3	401058	88.9	598942	24418	96973	52
9	388210	83.5	986619	5.3	401591	88.8	598409	24446	96966	51
10	388711	83.4	986587	5.3	402124	88.7	597876	24474	96959	50
11	9.389211	83.3	9.986555	5.3	9.402656	88.6	10.597344	24503	96952	49
12	389711	83.2	986523	5.3	403187	88.5	596813	24531	96945	48
13	390210	83.1	986491	5.3	403718	88.4	596282	24559	96937	47
14	390708	83.0	986459	5.3	404249	88.3	595751	24587	96930	46
15	391206	82.8	986427	5.3	404778	88.2	595222	24615	96923	45
16	391703	82.7	986395	5.3	405308	88.1	594692	24644	96916	44
17	392199	82.6	986363	5.3	405836	88.0	594164	24672	96909	43
18	392695	82.5	986331	5.4	406364	87.9	593636	24700	96902	42
19	393191	82.4	986299	5.4	406892	87.8	593108	24728	96894	41
20	393685	82.3	986266	5.4	407419	87.7	592581	24756	96887	40
21	9.394179	82.2	9.986234	5.4	9.407945	87.6	10.592055	24784	96880	39
22	394673	82.1	986202	5.4	408471	87.5	591529	24813	96873	38
23	395166	82.0	986169	5.4	408997	87.4	591003	24841	96866	37
24	395658	81.9	986137	5.4	409521	87.4	590479	24869	96858	36
25	396150	81.8	986104	5.4	410045	87.3	589955	24897	96851	35
26	396641	81.7	986072	5.4	410569	87.2	589431	24925	96844	34
27	397132	81.7	986039	5.4	411092	87.1	588908	24954	96837	33
28	397621	81.6	986007	5.4	411615	87.0	588385	24982	96829	32
29	398111	81.5	985974	5.4	412137	86.9	587863	25010	96822	31
30	398600	81.4	985942	5.4	412658	86.8	587342	25038	96815	30
31	9.399088	81.3	9.985909	5.4	9.413179	86.7	10.586821	25066	96807	29
32	399575	81.2	985876	5.5	413699	86.6	586301	25094	96800	28
33	400062	81.1	985843	5.5	414219	86.5	585781	25122	96793	27
34	400549	81.0	985811	5.5	414738	86.4	585262	25151	96786	26
35	401035	80.9	985778	5.5	415257	86.4	584743	25179	96778	25
36	401520	80.8	985745	5.5	415775	86.3	584225	25207	96771	24
37	402005	80.7	985712	5.5	416293	86.2	583707	25235	96764	23
38	402489	80.6	985679	5.5	416810	86.1	583190	25263	96756	22
39	402972	80.5	985646	5.5	417326	86.0	582674	25291	96749	21
40	403455	80.4	985613	5.5	417842	85.9	582158	25320	96742	20
41	9.403938	80.3	9.985580	5.5	9.418358	85.8	10.581642	25348	96734	19
42	404420	80.2	985547	5.5	418873	85.7	581127	25376	96727	18
43	404901	80.1	985514	5.5	419387	85.6	580613	25404	96719	17
44	405382	80.0	985480	5.5	419901	85.5	580099	25432	96712	16
45	405862	79.9	985447	5.5	420415	85.5	579585	25460	96705	15
46	406341	79.8	985414	5.6	420927	85.4	579073	25488	96697	14
47	406820	79.7	985380	5.6	421440	85.3	578560	25516	96690	13
48	407299	79.6	985347	5.6	421952	85.2	578048	25545	96682	12
49	407777	79.5	985314	5.6	422463	85.1	577537	25573	96675	11
50	403254	79.4	985280	5.6	422974	85.0	577026	25601	96667	10
51	9.408731	79.4	9.985247	5.6	9.423484	84.9	10.576516	25629	96660	9
52	409207	79.3	985213	5.6	423993	84.8	576007	25657	96653	8
53	409682	79.2	985180	5.6	424503	84.8	575497	25685	96645	7
54	410157	79.1	985146	5.6	425011	84.7	574989	25713	96638	6
55	410632	79.0	985113	5.6	425519	84.6	574481	25741	96630	5
56	411106	78.9	985079	5.6	426027	84.5	573973	25766	96623	4
57	411579	78.8	985045	5.6	426534	84.4	573466	25798	96615	3
58	412052	78.7	985011	5.6	427041	84.3	572959	25826	96608	2
59	412524	78.6	984978	5.6	427547	84.3	572453	25854	96600	1
60	412996		984944		428052		571948	25882	96593	0
	Cosine.		Sine.		Cotang.		Tang.	N. cos.	N. sine.	′

75 Degrees.

′	Sine.	D. 10″	Cosine.	D. 10″	Tang.	D. 10″	Cotang.	N. sine.	N. cos.	
0	9.412996	78.5	9.984944	5.7	9.428052	84.2	10.571948	25882	96593	60
1	413467	78.4	984910	5.7	428557	84.1	571443	25910	96585	59
2	413938	78.3	984876	5.7	429062	84.0	570938	25938	96578	58
3	414408	78.3	984842	5.7	429566	83.9	570434	25966	96570	57
4	414878	78.2	984808	5.7	430070	83.8	569930	25994	96562	56
5	415347	78.1	984774	5.7	430573	83.8	569427	26022	96555	55
6	415815	78.0	984740	5.7	431075	83.7	568925	26050	96547	54
7	416283	77.9	984706	5.7	431577	83.6	568423	26079	96540	53
8	416751	77.8	984672	5.7	432079	83.5	567921	26107	96532	52
9	417217	77.7	984637	5.7	432580	83.4	567420	26135	96524	51
10	417684	77.6	984603	5.7	433080	83.3	566920	26163	96517	50
11	9.418150	77.5	9.984569	5.7	9.433580	83.2	10.566420	26191	96509	49
12	418615	77.4	984535	5.7	434080	83.2	565920	26219	96502	48
13	419079	77.3	984500	5.7	434579	83.1	565421	26247	96494	47
14	419544	77.3	984466	5.7	435078	83.0	564922	26275	96486	46
15	420007	77.2	984432	5.7	435576	82.9	564424	26303	96479	45
16	420470	77.1	984397	5.8	436073	82.8	563927	26331	96471	44
17	420933	77.0	984363	5.8	436570	82.8	563430	26359	96463	43
18	421395	76.9	984328	5.8	437067	82.8	562933	26387	96456	42
19	421857	76.8	984294	5.8	437563	82.7	562437	26415	96448	41
20	422318	76.7	984259	5.8	438059	82.6	561941	26443	96440	40
21	9.422778	76.7	9.984224	5.8	9.438554	82.5	10.561446	26471	96433	39
22	423238	76.6	984190	5.8	439048	82.4	560952	26500	96425	38
23	423697	76.5	984155	5.8	439543	82.3	560457	26528	96417	37
24	424156	76.4	984120	5.8	440036	82.3	559964	26556	96410	36
25	424615	76.3	984085	5.8	440529	82.2	559471	26584	96402	35
26	425073	76.2	984050	5.8	441022	82.1	558978	26612	96394	34
27	425530	76.1	984015	5.8	441514	82.0	558486	26640	96386	33
28	425987	76.0	983981	5.8	442006	81.9	557994	26668	96379	32
29	426443	76.0	983946	5.8	442497	81.9	557503	26696	96371	31
30	426899	75.9	983911	5.8	442988	81.8	557012	26724	96363	30
31	9.427354	75.8	9.983875	5.8	9.443479	81.7	10.556521	26752	96355	29
32	427809	75.7	983840	5.8	443968	81.6	556032	26780	96347	28
33	428263	75.6	983805	5.9	444458	81.6	555542	26808	96340	27
34	428717	75.5	983770	5.9	444947	81.5	555053	26836	96332	26
35	429170	75.4	983735	5.9	445435	81.4	554565	26864	96324	25
36	429623	75.3	983700	5.9	445923	81.3	554077	26892	96316	24
37	430075	75.2	983664	5.9	446411	81.2	553589	26920	96308	23
38	430527	75.2	983629	5.9	446898	81.2	553102	26948	96301	22
39	430978	75.1	983594	5.9	447384	81.1	552616	26976	96293	21
40	431429	75.0	983558	5.9	447870	81.0	552130	27004	96285	20
41	9.431879	75.0	9.983523	5.9	9.448356	80.9	10.551644	27032	96277	19
42	432329	74.9	983487	5.9	448841	80.9	551159	27060	96269	18
43	432778	74.9	983452	5.9	449326	80.8	550674	27088	96261	17
44	433226	74.8	983416	5.9	449810	80.7	550190	27116	96253	16
45	433675	74.7	983381	5.9	450294	80.6	549706	27144	96246	15
46	434122	74.6	983345	5.9	450777	80.6	549223	27172	96238	14
47	434569	74.5	983309	5.9	451260	80.5	548740	27200	96230	13
48	435016	74.4	983273	5.9	451743	80.4	548257	27228	96222	12
49	435462	74.4	983238	6.0	452225	80.3	547775	27256	96214	11
50	435908	74.3	983202	6.0	452706	80.2	547294	27284	96206	10
51	9.436353	74.2	9.983166	6.0	9.453187	80.2	10.546813	27312	96198	9
52	436798	74.1	983130	6.0	453668	80.1	546332	27340	96190	8
53	437242	74.0	983094	6.0	454148	80.0	545852	27368	96182	7
54	437686	74.0	983058	6.0	454628	79.9	545372	27396	96174	6
55	438129	73.9	983022	6.0	455107	79.9	544893	27424	96166	5
56	438572	73.8	982986	6.0	455586	79.8	544414	27452	96158	4
57	439014	73.7	982950	6.0	456064	79.7	543936	27480	96150	3
58	439456	73.6	982914	6.0	456542	79.6	543458	27508	96142	2
59	439897	73.6	982878	6.0	457019	79.6	542981	27536	96134	1
60	440338	73.5	982842	6.0	457496	79.5	542504	27564	96126	0
	Cosine.		Sine.		Cotang.		Tang.	N. cos.	N. sine.	′

′	Sine.	D. 10″	Cosine.	D. 10″	Tang.	D. 10″	Cotang.	N. sine.	N. cos.	
0	9.410338	73.4	9.982842	6.0	9.457496	79.4	10.542504	27564	96126	60
1	440778	73.3	982805	6.0	457973	79.3	542027	27592	96118	59
2	441218	73.2	982769	6.1	458449	79.3	541551	27620	96110	58
3	441658	73·1	982733	6.1	458925	79.2	541075	27648	96102	57
4	442096	73.1	982696	6.1	459400	79.1	540600	27676	96094	56
5	442535	73.0	982660	6.1	459875	79.0	540125	27704	96086	55
6	442973	72.9	982624	6.1	460349	79.0	539651	27731	96078	54
7	443410	72.8	982587	6.1	460823	78.9	539177	27759	96070	53
8	443847	72.7	982551	6.1	461297	78.8	538703	27787	96062	52
9	444284	72.7	982514	6.1	461770	78.9	538230	27815	96054	51
10	444720	72.6	982477	6.1	462242	78.7	537758	27843	96046	50
11	9.445155	72.5	9.982441	6.1	9.462714	78.6	10.537286	27871	96037	49
12	445590	72.4	982404	6.1	463186	78.5	536814	27899	96029	48
13	446025	72.3	982367	6.1	463658	78.5	536342	27927	96021	47
14	446459	72.3	982331	6.1	464129	78.4	535871	27955	96013	46
15	446893	72.2	982294	6.1	464599	78.3	535401	27983	96005	45
16	447326	72.1	982257	6.1	465069	78.3	534931	28011	95997	44
17	447759	72.0	982220	6.2	465539	78.2	534461	28039	95989	43
18	448191	72.0	982183	6.2	466008	78.1	533992	28067	95981	42
19	448623	71.9	982146	6.2	466476	78.0	533524	28095	95972	41
20	449054	71.8	982109	6.2	466945	78.0	533055	28123	95964	40
21	9.449485	71.7	9.982072	6.2	9.467413	77.9	10.532587	28150	95956	39
22	449915	71.6	982035	6.2	467880	77.8	532120	28178	95948	38
23	450345	71.6	981998	6.2	468347	77.8	531653	28206	95940	37
24	450775	71.5	981961	6.2	468814	77.7	531186	28234	95931	36
25	451204	71.4	981924	6.2	469280	77.6	530720	28262	95923	35
26	451632	71.3	981886	6.2	469746	77.5	530254	28290	95915	34
27	452060	71.3	981849	6.2	470211	77.5	529789	28318	95907	33
28	452488	71.2	981812	6.2	470676	77.4	529324	28346	95898	32
29	452915	71.1	981774	6.2	471141	77.3	528859	28374	95890	31
30	453342	71.0	981737	6.2	471605	77.3	528395	28402	95882	30
31	9.453768	71.0	9.981699	6.3	9.472068	77.2	10.527932	28429	95874	29
32	454194	70.9	981662	6.3	472532	77.1	527468	28457	95865	28
33	454619	70.8	981625	6.3	472995	77.1	527005	28485	95857	27
34	455044	70.7	981587	6.3	473457	77.0	526543	28513	95849	26
35	455469	70.7	981549	6.3	473919	76.9	526081	28541	95841	25
36	455893	70.6	981512	6.3	474381	76.9	525619	28569	95832	24
37	456316	70.5	981474	6.3	474842	76.8	525158	28597	95824	23
38	456739	70.4	981436	.6.3	475303	76.7	524697	28625	95816	22
39	457162	70.4	981399	6.3	475763	76.7	524237	28652	95807	21
40	457584	70.3	981361	6.3	476223	76.6	523777	28680	95799	20
41	9.458006	70.2	9.981323	6.3	9.476683	76.5	10.523317	28708	95791	19
42	458427	70.1	981285	6.3	477142	76.5	522858	28736	95782	18
43	458848	70.1	981247	6.3	477601	76.4	522399	28764	95774	17
44	459268	70.0	981209	6.3	478059	76.3	521941	28792	95766	16
45	459688	69.9	981171	6.3	478517	76.3	521483	28820	95757	15
46	460108	69.8	981133	6.4	478975	76.2	521025	28847	95749	14
47	460527	69.8	981095	6.4	479432	76.1	520568	28875	95740	13
48	460946	69.7	981057	6.4	479889	76.1	520111	28903	95732	12
49	461364	69.6	981019	6.4	480345	76.0	519655	28931	95724	11
50	461782	69.5	980981	6.4	480801	75.9	519199	28959	95715	10
51	9.462199	69.5	9.980942	6.4	9.481257	75.9	10.518743	28987	95707	9
52	462616	69.4	980904	6.4	481712	75.8	518288	29015	95698	8
53	463032	69.3	980866	6.4	482167	75.7	517833	29042	95690	7
54	463448	69.3	980827	6.4	482621	75.7	517379	29070	95681	6
55	463864	69.2	980789	6.4	483075	75.6	516925	29098	95673	5
56	464279	69.1	980750	6.4	483529	75.5	516471	29126	95664	4
57	464694	69.0	980712	6.4	483982	75.5	516018	29154	95656	3
58	465108	69.0	980673	6.4	484435	75.4	515565	29182	95647	2
59	465522	68.9	980635	6.4	484887	75.3	515113	29209	95639	1
60	465935		980596		485339		514661	29247	95630	0
	Cosine.		Sine.		Cotang.		Tang.	N. cos.	N. sine.	′

73 Degrees.

′	Sine.	D. 10″	Cosine.	D. 10″	Tang.	D. 10″	Cotang.	N. sine.	N. cos.	
0	9.465935	68.8	9.980596	6.4	9.485339	75.3	10.514661	29237	95630	60
1	466348	68.8	980558	6.4	485791	75.2	514209	29265	95622	59
2	466761	68.7	980519	6.5	486242	75.1	513758	29293	95613	58
3	467173	68.6	980480	6.5	486693	75.1	513307	29321	95605	57
4	467585	68.5	980442	6.5	487143	75.0	512857	29348	95596	56
5	467996	68.5	980403	6.5	487593	74.9	512407	29376	95588	55
6	468407	68.4	980364	6.5	488043	74.9	511957	29404	95579	54
7	468817	68.3	980325	6.5	488492	74.8	511508	29432	95571	53
8	469227	68.3	980286	6.5	488941	74.7	511059	29460	95562	52
9	469637	68.2	980247	6.5	489390	74.7	510610	29487	95554	51
10	470046	68.1	980208	6.5	489838	74.6	510162	29515	95545	50
11	9.470455	68.0	9.980169	6.5	9.490286	74.6	10.509714	29543	95536	49
12	470863	68.0	980130	6.5	490733	74.5	509267	29571	95528	48
13	471271	67.9	980091	6.5	491180	74.4	508820	29599	95519	47
14	471679	67.8	980052	6.5	491627	74.4	508373	29626	95511	46
15	472086	67.8	980012	6.5	492073	74.3	507927	29654	95502	45
16	472492	67.7	979973	6.5	492519	74.3	507481	29682	95493	44
17	472898	67.6	979934	6.5	492965	74.2	507035	29710	95485	43
18	473304	67.6	979895	6.6	493410	74.1	506590	29737	95476	42
19	473710	67.5	979855	6.6	493854	74.0	506146	29765	95467	41
20	474115	67.4	979816	6.6	494299	74.0	505701	29793	95459	40
21	9.474519	67.4	9.979776	6.6	9.494743	74.0	10.505257	29821	95450	39
22	474923	67.3	979737	6.6	495186	73.9	504814	29849	95441	38
23	475327	67.2	979697	6.6	495630	73.8	504370	29876	95433	37
24	475730	67.2	979658	6.6	496073	73.7	503927	29904	95424	36
25	476133	67.1	979618	6.6	496515	73.7	503485	29932	95415	35
26	476536	67.0	979579	6.6	496957	73.6	503043	29960	95407	34
27	476938	66.9	979539	6.6	497399	73.6	502601	29987	95398	33
28	477340	66.9	979499	6.6	497841	73.5	502159	30015	95389	32
29	477741	66.8	979459	6.6	468282	73.4	501718	30043	95380	31
30	478142	66.7	979420	6.6	498722	73.4	501278	30071	95372	30
31	9.478542	66.7	9.979380	6.6	9.499163	73.3	10.500837	30098	95363	29
32	478942	66.6	979340	6.6	499603	73.3	500397	30126	95354	28
33	479342	66.5	979300	6.7	500042	73.2	499958	30154	95345	27
34	479741	66.5	979260	6.7	500481	73.1	499519	30182	95337	26
35	480140	66.4	979220	6.7	500920	73.1	499080	30209	95328	25
36	480539	66.3	979180	6.7	501359	73.0	498641	30237	95319	24
37	480937	66.3	979140	6.7	501797	73.0	498203	30265	95310	23
38	481334	66.2	979100	6.7	502235	72.9	497765	30292	95301	22
39	481731	66.1	979059	6.7	502672	72.8	497328	30320	95293	21
40	482128	66.1	979019	6.7	503109	72.8	496891	30348	95284	20
41	9.482525	66.0	9.978979	6.7	9.503546	72.7	10.496454	30376	95275	19
42	482921	65.9	978939	6.7	503982	72.7	496018	30403	95266	18
43	483316	65.9	978898	6.7	504418	72.6	495582	30431	95257	17
44	483712	65.8	978858	6.7	504854	72.5	495146	30459	95248	16
45	484107	65.7	978817	6.7	505289	72.5	494711	30486	95240	15
46	484501	65.7	978777	6.7	505724	72.4	494276	30514	95231	14
47	484895	65.6	978736	6.7	506159	72.4	493841	30542	95222	13
48	485289	65.5	978696	6.8	506593	72.3	493407	30570	95213	12
49	485682	65.5	978655	6.8	507027	72.2	492973	30597	95204	11
50	486075	65.4	978615	6.8	507460	72.2	492540	30625	95195	10
51	9.486467	66.3	9.978574	6.8	9.507893	72.1	10.492107	30653	95186	9
52	486860	65.3	978533	6.8	508326	72.1	491674	30680	95177	8
53	487251	65.3	978493	6.8	508759	72.0	491241	30708	95168	7
54	487643	65.2	978452	6.8	509191	71.9	490809	30736	95159	6
55	488034	65.1	978411	6.8	509622	71.9	490378	30763	95150	5
56	488424	65.1	978370	6.8	510054	71.8	489946	30791	95142	4
57	488814	65.0	978329	6.8	510485	71.8	489515	30819	95133	3
58	489204	65.0	978288	6.8	510916	71.8	489084	30846	95124	2
59	489593	64.9	978247	6.8	511346	71.7	488654	30874	95115	1
60	489982	64.8	978206		511776	71.6	488224	30902	95106	0
	Cosine.		Sine.		Cotang.		Tang.	N. cos.	N. sine.	′

'	Sine.	D. 10"	Cosine.	D. 10"	Tang.	D. 10"	Cotang.	N. sine.	N. cos.	'
0	9.489982	64.8	9.978206	6.8	9.511776	71.6	10.488224	30902	95106	60
1	490371	64.8	978165	6.8	512206	71.6	487794	30929	95097	59
2	490759	64.7	978124	6.8	512635	71.5	487365	30957	95088	58
3	491147	64.6	978083	6.9	513064	71.4	486936	30985	95079	57
4	491535	64.6	978042	6.9	513493	71.4	486507	31012	95070	56
5	491922	64.5	978001	6.9	513921	71.3	486079	31040	95061	55
6	492308	64.4	977959	6.9	514349	71.3	485651	31068	95052	54
7	492695	64.4	977918	6.9	514777	71.3	485223	31095	95043	53
8	493081	64.3	977877	6.9	515204	71.2	484796	31123	95033	52
9	493466	64.2	977835	6.9	515631	71.2	484369	31151	95024	51
10	493851	64.2	977794	6.9	516057	71.1	483943	31178	95015	50
11	9.494236	64.1	9.977752	6.9	9.516484	71.0	10.483516	31206	95006	49
12	494621	64.1	977711	6.9	516910	71.0	483090	31233	94997	48
13	495005	64.0	977669	6.9	517335	70.9	482665	31261	94988	47
14	495388	63.9	977628	6.9	517761	70.9	482239	31289	94979	46
15	495772	63.9	977586	6.9	518185	70.8	481815	31316	94970	45
16	496154	63.8	977544	7.0	518610	70.8	481390	31344	94961	44
17	496537	63.7	977503	7.0	519034	70.7	480966	31372	94952	43
18	496919	63.7	977461	7.0	519458	70.6	480542	31399	94943	42
19	497301	63.6	977419	7.0	519882	70.6	480118	31427	94933	41
20	497682	63.6	977377	7.0	520305	70.5	479695	31454	94924	40
21	9.498064	63.5	9.977335	7.0	9.520728	70.5	10.479272	31482	94915	39
22	498444	63.5	977293	7.0	521151	70.4	478849	31510	94906	38
23	498825	63.4	977251	7.0	521573	70.3	478427	31537	94897	37
24	499204	63.4	977209	7.0	521995	70.3	478005	31565	94888	36
25	499584	63.3	977167	7.0	522417	70.3	477583	31593	94878	35
26	499963	63.2	977125	7.0	522838	70.2	477162	31620	94869	34
27	500342	63.2	977083	7.0	523259	70.2	476741	31648	94860	33
28	500721	63.1	977041	7.0	523680	70.1	476320	31675	94851	32
29	501099	63.1	976999	7.0	524100	70.1	475900	31703	94842	31
30	501476	63.0	976957	7.0	524520	70.0	475480	31730	94832	30
31	9.501854	62.9	9.976914	7.0	9.524939	69.9	10.475061	31758	94823	29
32	502231	62.9	976872	7.1	525359	69.9	474641	31786	94814	28
33	502607	62.8	976830	7.1	525778	69.8	474222	31813	94805	27
34	502984	62.8	976787	7.1	526197	69.8	473803	31841	94795	26
35	503360	62.7	976745	7.1	526615	69.7	473385	31868	94786	25
36	503735	62.6	976702	7.1	527033	69.7	472967	31896	94777	24
37	504110	62.6	976660	7.1	527451	69.6	472549	31923	94768	23
38	504485	62.5	976617	7.1	527868	69.6	472132	31951	94758	22
39	504860	62.5	976574	7.1	528285	69.5	471715	31979	94749	21
40	505234	62.4	976532	7.1	528702	69.5	471298	32006	94740	20
41	9.505608	62.3	9.976489	7.1	9.529119	69.4	10.470881	32034	94730	19
42	505981	62.3	976446	7.1	529535	69.3	470465	32061	94721	18
43	506354	62.2	976404	7.1	529950	69.3	470050	32089	94712	17
44	506727	62.2	976361	7.1	530366	69.3	469634	32116	94702	16
45	507099	62.1	976318	7.1	530781	69.2	469219	32144	94693	15
46	507471	62.0	976275	7.1	531196	69.1	468804	32171	94684	14
47	507843	62.0	976232	7.1	531611	69.1	468389	32199	94674	13
48	508214	61.9	976189	7.2	532025	69.0	467975	32227	94665	12
49	508585	61.9	976146	7.2	532439	69.0	467561	32250	94656	11
50	508956	61.8	976103	7.2	532853	68.9	467147	32282	94646	10
51	9.509326	61.8	9.976060	7.2	9.533266	68.9	10.466734	32309	94637	9
52	509696	61.7	976017	7.2	533679	68.8	466321	32337	94627	8
53	510065	61.6	975974	7.2	534092	68.8	465908	32364	94618	7
54	510434	61.6	975930	7.2	534504	68.7	465496	32392	94609	6
55	510803	61.5	975887	7.2	534916	68.7	465084	32419	94599	5
56	511172	61.5	975844	7.2	535328	68.6	464672	32447	94590	4
57	511540	61.4	975800	7.2	535739	68.6	464261	32474	94580	3
58	511907	61.3	975757	7.2	536150	68.5	463850	32502	94571	2
59	512275	61.3	975714	7.2	536561	68.5	463439	32529	94561	1
60	512642	61.2	975670		536972	68.4	463028	3255,	94552	0
	Cosine.		Sine.		Cotang.		Tang.	N. cos.	N. sine.	'

′	Sine.	D. 10′	Cosine.	D. 10′	Tang.	D. 10′	Cotang.	N. sine.	N. cos.	
0	9.512642	61.2	9.975670	7.3	9.536972	68.4	10.463028	32557	94552	60
1	513009	61.1	975627	7.3	537382	68.3	462618	32584	94542	59
2	513375	61.1	975583	7.3	537792	68.3	462208	32612	94533	58
3	513741	61.0	975539	7.3	538202	68.2	461798	32639	94523	57
4	514107	60.9	975496	7.3	538611	68.2	461389	32667	94514	56
5	514472	60.9	975452	7.3	539020	68.1	460980	32694	94504	55
6	514837	60.8	975408	7.3	539429	68.1	460571	32722	94495	54
7	515202	60.8	975365	7.3	539837	68.0	460163	32749	94485	53
8	515566	60.7	975321	7.3	540245	68.0	459755	32777	94476	52
9	515930	60.7	975277	7.3	540653	67.9	459347	32804	94466	51
10	516294	60.6	975233	7.3	541061	67.9	458939	32832	94457	50
11	9.516657	60.5	9.975189	7.3	9.541468	67.8	10.458532	32859	94447	49
12	517020	60.5	975145	7.3	541875	67.8	458125	32887	94438	48
13	517382	60.4	975101	7.3	542281	67.7	457719	32914	94428	47
14	517745	60.4	975057	7.3	542688	67.7	457312	32942	94418	46
15	518107	60.3	975013	7.3	543094	67.6	456906	32969	94409	45
16	518468	60.3	974969	7.3	543499	67.6	456501	32997	94399	44
17	518829	60.2	974925	7.4	543905	67.5	456095	33024	94390	43
18	519190	60.1	974880	7.4	544310	67.5	455690	33051	94380	42
19	519551	60.1	974836	7.4	544715	67.4	455285	33079	94370	41
20	519911	60.0	974792	7.4	545119	67.4	454881	33106	94361	40
21	9.520271	60.0	9.974748	7.4	9.545524	67.3	10.454476	33134	94351	39
22	520631	59.9	974703	7.4	545928	67.3	454072	33161	94342	38
23	520990	59.9	974659	7.4	546331	67.2	453669	33189	94332	37
24	521349	59.8	974614	7.4	546735	67.2	453265	33216	94322	36
25	521707	59.8	974570	7.4	547138	67.1	452862	33244	94313	35
26	522066	59.7	974525	7.4	547540	67.1	452460	33271	94303	34
27	522424	59.6	974481	7.4	547943	67.0	452057	33298	94293	33
28	522781	59.6	974436	7.4	548345	67.0	451655	33326	94284	32
29	523138	59.5	974391	7.4	548747	66.9	451253	33353	94274	31
30	523495	59.5	974347	7.5	549149	66.9	450851	33381	94264	30
31	9.523852	59.4	9.974302	7.5	9.549550	66.8	10.450450	33408	94254	29
32	524208	59.4	974257	7.5	549951	66.8	450049	33436	94245	28
33	524564	59.3	974212	7.5	550352	66.7	449648	33463	94235	27
34	524920	59.3	974167	7.5	550752	66.7	449248	33490	94225	26
35	525275	59.2	974122	7.5	551152	66.6	448848	33518	94215	25
36	525630	59.1	974077	7.5	551552	66.6	448448	33545	94206	24
37	525984	59.1	974032	7.5	551952	66.5	448048	33573	94196	23
38	526339	59.0	973987	7.5	552351	66.5	447649	33600	94186	22
39	526693	59.0	973942	7.5	552750	66.5	447250	33627	94176	21
40	527046	58.9	973897	7.5	553149	66.4	446851	33655	94167	20
41	9.527400	58.9	9.973852	7.5	9.553548	66.4	10.446452	33682	94157	19
42	527753	58.8	973807	7.5	553946	66.3	446054	33710	94147	18
43	528105	58.8	973761	7.5	554344	66.3	445656	33737	94137	17
44	528458	58.7	973716	7.6	554741	66.2	445259	33764	94127	16
45	528810	58.7	973671	7.6	555139	66.2	444861	33792	94118	15
46	529161	58.6	973625	7.6	555536	66.1	444464	33819	94108	14
47	529513	58.6	973580	7.6	555933	66.1	444067	33846	94098	13
48	529864	58.5	973535	7.6	556329	66.0	443671	33874	94083	12
49	530215	58.5	973489	7.6	556725	66.0	443275	33901	94078	11
50	530565	58.4	973444	7.6	557121	66.0	442879	33929	94068	10
51	9.530915	58.4	9.973398	7.6	9.557517	65.9	10.442483	33956	94058	9
52	531265	58.3	973352	7.6	557913	65.9	442087	33983	94049	8
53	531614	58.2	973307	7.6	558308	65.9	441692	34011	94039	7
54	531963	58.2	973261	7.6	558702	65.8	441298	34038	94029	6
55	532312	58.1	973215	7.6	559097	65.8	440903	34065	94019	5
56	532661	58.1	973169	7.6	559491	65.7	440509	34093	94009	4
57	533009	58.0	973124	7.6	559885	65.7	440115	34120	93999	3
58	533357	58.0	973078	7.6	560279	65.6	439721	34147	93989	2
59	533704	58.0	973032	7.6	560673	65.6	439327	34175	93979	1
60	534052	57.9	972986	7.7	561066	65.5	438934	34202	93969	0
′	Cosine.		Sine.		Cotang.		Tang.	N. cos.	N. sine.	′

'	Sine.	D. 10"	Cosine.	D. 10"	Tang.	D. 10"	Cotang.	N. sine.	N. cos.	
0	9.534052	57.8	9.972986	7.7	9.561066	65.5	10.438934	34202	93909	60
1	534399	57.7	972940	7.7	561459	65.4	438541	34229	93959	59
2	534745	57.7	972894	7.7	561851	65.4	438149	34257	93949	58
3	535092	57.7	972848	7.7	562244	65.3	437756	34284	93939	57
4	535438	57.6	972802	7.7	562636	65.3	437364	34311	93929	56
5	535783	57.6	972755	7.7	563028	65.3	436972	34339	93919	55
6	536129	57.5	972709	7.7	563419	65.2	436581	34366	93909	54
7	536474	57.4	972663	7.7	563811	65.2	436189	34393	93899	53
8	536818	57.4	972617	7.7	564202	65.1	435798	34421	93889	52
9	537163	57.3	972570	7.7	564592	65.1	435408	34448	93879	51
10	537507	57.3	972524	7.7	564983	65.0	435017	34475	93869	60
11	9.537851	57.2	9.972478	7.7	9.565373	65.0	10.434627	34503	93859	49
12	538194	57.2	972431	7.8	565763	64.9	434237	34530	93849	48
13	538538	57.1	972385	7.8	566153	64.9	433847	34557	93839	47
14	538880	57.1	972338	7.8	566542	64.9	433458	34584	93829	46
15	539223	57.0	972291	7.8	566932	64.9	433068	34612	93819	45
16	539565	57.0	972245	7.8	567320	64.8	432680	34639	93809	44
17	539907	56.9	972198	7.8	567709	64.8	432291	34666	93799	43
18	540249	56.9	972151	7.8	568098	64.7	431902	34694	93789	42
19	540590	56.8	972105	7.8	568486	64.7	431514	34721	93779	41
20	540931	56.8	972058	7.8	568873	64.6	431127	34748	93769	40
21	9.541272	56.7	9.972011	7.8	9.569261	64.6	10.430739	34775	93759	39
22	541613	56.7	971964	7.8	569648	64.5	430352	34803	93748	38
23	541953	56.6	971917	7.8	570035	64.5	429965	34830	93738	37
24	542293	56.6	971870	7.8	570422	64.5	429578	34857	93728	36
25	542632	56.5	971823	7.8	570809	64.4	429191	34884	93718	35
26	542971	56.5	971776	7.8	571195	64.4	428805	34912	93708	34
27	543310	56.5	971729	7.8	571581	64.3	428419	34939	93698	33
28	543649	56.4	971682	7.9	571967	64.3	428033	34966	93688	32
29	543987	56.4	971635	7.9	572352	64.2	427648	34993	93677	31
30	544325	56.3	971588	7.9	572738	64.2	427262	35021	93667	30
31	9.544663	56.3	9.971540	7.9	9.573123	64.2	10.426877	35048	93657	29
32	545000	56.2	971493	7.9	573507	64.1	426493	35075	93647	28
33	545338	56.2	971446	7.9	573892	64.1	426108	35102	93637	27
34	545674	56.1	971398	7.9	574276	64.0	425724	35130	93626	26
35	546011	56.1	971351	7.9	574660	64.0	425340	35157	93616	25
36	546347	56.0	971303	7.9	575044	63.9	424956	35184	93606	24
37	546683	56.0	971256	7.9	575427	63.9	424573	35211	93596	23
38	547019	55.9	971208	7.9	575810	63.9	424190	35239	93585	22
39	547354	55.9	971161	7.9	576193	63.8	423807	35266	93575	21
40	547689	55.8	971113	7.9	576576	63.8	423424	35293	93565	20
41	9.548024	55.8	9.971066	8.0	9.576958	63.7	10.423041	35320	93555	19
42	548359	55.7	971018	8.0	577341	63.7	422659	35347	93544	18
43	548693	55.7	970970	8.0	577723	63.6	422277	35375	93534	17
44	549027	55.6	970922	8.0	578104	63.6	421896	35402	93524	16
45	549360	55.6	970874	8.0	578486	63.6	421514	35429	93514	15
46	549693	55.5	970827	8.0	578867	63.5	421133	35456	93503	14
47	550026	55.5	970779	8.0	579248	63.5	420752	35484	93493	13
48	550359	55.4	970731	8.0	579629	63.4	420371	35511	93483	12
49	550692	55.4	970683	8.0	580009	63.4	419991	35538	93472	11
50	551024	55.3	970635	8.0	580389	63.4	419611	35565	93462	10
51	9.551356	55.3	9.970586	8.0	9.580769	63.3	10.419231	35592	93452	9
52	551687	55.2	970538	8.0	581149	63.3	418851	35619	93441	8
53	552018	55.2	970490	8.0	581528	63.2	418472	35647	93431	7
54	552349	55.2	970442	8.0	581907	63.2	418093	35674	93420	6
55	552680	55.1	970394	8.0	582286	63.2	417714	35701	93410	5
56	553010	55.1	970345	8.1	582665	63.1	417335	35728	93400	4
57	553341	55.0	970297	8.1	583043	63.1	416957	35755	93389	3
58	553670	55.0	970249	8.1	583422	63.0	416578	35782	93379	2
59	554000	54.9	970200	8.1	583800	63.0	416200	35810	93368	1
60	554329	54.9	970152	8.1	584177	62.9	415823	35837	93358	0
	Cosine.		Sine.		Cotang.		Tang.	N. cos.	N.sine.	'

′	Sine.	D. 10″	Cosine.	D. 10″	Tang.	D. 10″	Cotang.	N.sine	N. cos.	′
0	9.554329	54.8	9.970152	8.1	9.584177	62.9	10.415823	35837	93358	60
1	554658	54.8	970103	8.1	584555	62.9	415445	35864	93348	59
2	554987	54.7	970055	8.1	584932	62.8	415068	35891	93337	58
3	555315	54.7	970006	8.1	585309	62.8	414691	35918	93327	57
4	555643	54.6	969957	8.1	585686	62.7	414314	35945	93316	56
5	555971	54.6	969909	8.1	586062	62.7	413938	35973	93306	55
6	556299	54.5	969860	8.1	586439	62.7	413561	36000	93295	54
7	556626	54.5	969811	8.1	586815	62.6	413185	36027	93285	53
8	556953	54.4	969762	8.1	587190	62.6	412810	36054	93274	52
9	557280	54.4	969714	8.1	587566	62.5	412434	36081	93264	51
10	557606	54.3	969665	8.1	587941	62.5	412059	36108	93253	50
11	9.557932	54.3	9.969616	8.2	9.588316	62.5	10.411684	36135	93243	49
12	558258	54.3	969567	8.2	588691	62.4	411309	36162	93232	48
13	558583	54.2	969518	8.2	589066	62.4	410934	36190	93222	47
14	558909	54.2	969469	8.2	589440	62.3	410560	36217	93211	46
15	559234	54.1	969420	8.2	589814	62.3	410186	36244	93201	45
16	559558	54.1	969370	8.2	590188	62.3	409812	36271	93190	44
17	559883	54.0	969321	8.2	590562	62.2	409438	36298	93180	43
18	560207	54.0	969272	8.2	590935	62.2	409065	36325	93169	42
19	560531	53.9	969223	8.2	591308	62.2	408692	36352	93159	41
20	560855	53.9	969173	8.2	591681	62.2	408319	36379	93148	40
21	9.561178	53.8	9.969124	8.2	9.592054	62.1	10.407946	36406	93137	39
22	561501	53.8	969075	8.2	592426	62.1	407574	36434	93127	38
23	561824	53.7	969025	8.2	592798	62.0	407202	36461	93116	37
24	562146	53.7	968976	8.2	593170	62.0	406829	36488	93106	36
25	562468	53.6	968926	8.3	593542	61.9	406458	36515	93095	35
26	562790	53.6	968877	8.3	593914	61.9	406086	36542	93084	34
27	563112	53.6	968827	8.3	594285	61.8	405715	36569	93074	33
28	563433	53.5	968777	8.3	594656	61.8	405344	36596	93063	32
29	563755	53.5	968728	8.3	595027	61.8	404973	36623	93052	31
30	564075	53.4	968678	8.3	595398	61.7	404602	36650	93042	30
31	9.564396	53.4	9.968628	8.3	9.595768	61.7	10.404232	36677	93031	29
32	564716	53.3	968578	8.3	596138	61.7	403862	36704	93020	28
33	565036	53.3	968528	8.3	596508	61.6	403492	36731	93010	27
34	565356	53.2	968479	8.3	596878	61.6	403122	36758	92999	26
35	565676	53.2	968429	8.3	597247	61.6	402753	36785	92988	25
36	565995	53.1	968379	8.3	597616	61.5	402384	36812	92978	24
37	566314	53.1	968329	8.3	597985	61.5	402015	36839	92967	23
38	566632	53.1	968278	8.3	598354	61.5	401646	36867	92956	22
39	566951	53.0	968228	8.3	598722	61.4	401278	36894	92945	21
40	567269	53.0	968178	8.4	599091	61.4	400909	36921	92935	20
41	9.567587	52.9	9.968128	8.4	9.599459	61.3	10.400541	36948	92926	19
42	567904	52.9	968078	8.4	599827	61.3	400173	36975	92913	18
43	568222	52.8	968027	8.4	600194	61.3	399806	37002	92902	17
44	568539	52.8	967977	8.4	600562	61.2	399438	37029	92892	16
45	568856	52.8	967927	8.4	600929	61.2	399071	37056	92881	15
46	569172	52.7	967876	8.4	601296	61.1	398704	37083	92870	14
47	569488	52.7	967826	8.4	601662	61.1	398338	37110	92859	13
48	569804	52.6	967775	8.4	602029	61.1	397971	37137	92849	12
49	570120	52.6	967725	8.4	602395	61.0	397605	37164	92838	11
50	570435	52.5	967674	8.4	602761	61.0	397239	37191	92827	10
51	9.570751	52.5	9.967624	8.4	9.603127	61.0	10.396873	37218	92816	9
52	571066	52.4	967573	8.4	603493	60.9	396507	37245	92805	8
53	571380	52.4	967522	8.4	603858	60.9	396142	37272	92794	7
54	571695	52.3	967471	8.5	604223	60.9	395777	37299	92784	6
55	572009	52.3	967421	8.5	604588	60.8	395412	37326	92773	5
56	572323	52.3	967370	8.5	604953	60.8	395047	37353	92762	4
57	572636	52.2	967319	8.5	605317	60.7	394683	37380	92751	3
58	572950	52.2	967268	8.5	605682	60.7	394318	37407	92740	2
59	573263	52.1	967217	8.5	606046	60.7	393954	37434	92729	1
60	573575		967166		606410	60.6	393590	37461	92718	0
	Cosine.		Sine.		Cotang.		Tang.	N. cos.	N.sine.	′

′	Sine.	D. 10″	Cosine.	D. 10″	Tang.	D. 10″	Cotang.	N. sine.	N. cos.	′
0	9.573575	52.1	9.967166	8.5	9.606410	60.6	10.393590	37461	92718	60
1	573888	52.0	967115	8.5	606773	60.6	393227	37488	92707	59
2	574200	52.0	967064	8.5	607137	60.5	392863	37515	92697	58
3	574512	51.9	967013	8.5	607500	60.5	392500	37542	92686	57
4	574824	51.9	966961	8.5	607863	60.5	392137	37569	92675	56
5	575136	51.9	966910	8.5	608225	60.4	391775	37595	92664	55
6	575447	51.8	966859	8.5	608588	60.4	391412	37622	92653	54
7	575758	51.8	966808	8.5	608950	60.4	391050	37649	92642	53
8	576069	51.7	966756	8.6	609312	60.3	390688	37676	92631	52
9	576379	51.7	966705	8.6	609674	60.3	390326	37703	92620	51
10	576689	51.6	966653	8.6	610036	60.3	389964	37730	92609	50
11	9.576999	51.6	9.966602	8.6	9.610397	60.2	10.389603	37757	92598	49
12	577309	51.6	966550	8.6	610759	60.2	389241	37784	92587	48
13	577618	51.5	966499	8.6	611120	60.2	388880	37811	92576	47
14	577927	51.5	966447	8.6	611480	60.1	388520	37838	92565	46
15	578236	51.4	966395	8.6	611841	60.1	388159	37865	92554	45
16	578545	51.4	966344	8.6	612201	60.1	387799	37892	92543	44
17	578853	51.3	966292	8.6	612561	60.0	387439	37919	92532	43
18	579162	51.3	966240	8.6	612921	60.0	387079	37946	92521	42
19	579470	51.3	966188	8.6	613281	60.0	386719	37973	92510	41
20	579777	51.2	966136	8.6	613641	59.9	386359	37999	92499	40
21	9.580085	51.2	9.966085	8.7	9.614000	59.9	10.386000	38026	92488	39
22	580392	51.1	966033	8.7	614359	59.8	385641	38053	92477	38
23	580699	51.1	965981	8.7	614718	59.8	385282	38080	92466	37
24	581005	51.1	965928	8.7	615077	59.8	384923	38107	92455	36
25	581312	51.0	965876	8.7	615435	59.7	384565	38134	92444	35
26	581618	51.0	965824	8.7	615793	59.7	384207	38161	92432	34
27	581924	50.9	965772	8.7	616151	59.7	383849	38188	92421	33
28	582229	50.9	965720	8.7	616509	59.6	383491	38215	92410	32
29	582535	50.9	965668	8.7	616867	59.6	383133	38241	92399	31
30	582840	50.8	965615	8.7	617224	59.6	382776	38268	92388	30
31	9.583145	50.8	9.965563	8.7	9.617582	59.5	10.382418	38295	92377	29
32	583449	50.7	965511	8.7	617939	59.5	382061	38322	92366	28
33	583754	50.7	965458	8.7	618295	59.5	381705	38349	92355	27
34	584058	50.6	965406	8.7	618652	59.4	381348	38376	92343	26
35	584361	50.6	965353	8.7	619008	59.4	380992	38403	92332	25
36	584665	50.6	965301	8.8	619364	59.4	380636	38430	92321	24
37	584968	50.5	965248	8.8	619721	59.3	380279	38456	92310	23
38	585272	50.5	965195	8.8	620076	59.3	379924	38483	92299	22
39	585574	50.4	965143	8.8	620432	59.3	379568	38510	92287	21
40	585877	50.4	965090	8.8	620787	59.2	379213	38537	92276	20
41	9.586179	50.3	9.965037	8.8	9.621142	59.2	10.378858	38564	92265	19
42	586482	50.3	964984	8.8	621497	59.2	378503	38591	92254	18
43	586783	50.3	964931	8.8	621852	59.1	378148	38617	92243	17
44	587085	50.2	964879	8.8	622207	59.1	377793	38644	92231	16
45	587386	50.2	964826	8.8	622561	59.0	377439	38671	92220	15
46	587688	50.1	964773	8.8	622915	59.0	377085	38698	92209	14
47	587989	50.1	964719	8.8	623269	59.0	376731	38725	92198	13
48	588289	50.1	964666	8.8	623623	58.9	376377	38752	92186	12
49	588590	50.0	964613	8.9	623976	58.9	376024	38778	92175	11
50	588890	50.0	964560	8.9	624330	58.9	375670	38805	92164	10
51	9.589190	49.9	9.964507	8.9	9.624683	58.8	10.375317	38832	92152	9
52	589489	49.9	964454	8.9	625036	58.8	374964	38859	92141	8
53	589789	49.9	964400	8.9	625388	58.8	374612	38886	92130	7
54	590088	49.9	964347	8.9	625741	58.7	374259	38912	92119	6
55	590387	49.8	964294	8.9	626093	58.7	373907	38939	92107	5
56	590686	49.8	964240	8.9	626445	58.7	373555	38966	92096	4
57	590984	49.7	964187	8.9	626797	58.6	373203	38993	92085	3
58	591282	49.7	964133	8.9	627149	58.6	372851	39020	92073	2
59	591580	49.7	964080	8.9	627501	58.6	372499	39046	92062	1
60	591878	49.6	964026		627852	58.5	372148	39073	92050	0
	Cosine.		Sine.		Cotang.		Tang.	N. cos.	N. sine.	′

'	Sine.	D. 10"	Cosine.	D. 10"	Tang.	D. 10"	Cotang.	N. sine.	N. cos.	'
0	9.591878	49.6	9.964026	8.9	9.627852	58.5	10.372148	39073	92050	60
1	592176	49.5	963972	8.9	628203	58.5	371797	39100	92039	59
2	592473	49.5	963919	8.9	628554	58.5	371446	39127	92028	58
3	592770	49.5	963865	9.0	628905	58.4	371095	39153	92016	57
4	593067	49.4	963811	9.0	629255	58.4	370745	39180	92005	56
5	593363	49.4	963757	9.0	629606	58.3	370394	39207	91994	55
6	593659	49.3	963704	9.0	629956	58.3	370044	39234	91982	54
7	593955	49.3	963650	9.0	630306	58.3	369694	39260	91971	53
8	594251	49.3	963596	9.0	630656	58.3	369344	39287	91959	52
9	594547	49.2	963542	9.0	631005	58.2	368995	39314	91948	51
10	594842	49.2	963488	9.0	631355	58.2	368645	39341	91936	50
11	9.595137	49.1	9.963434	9.0	9.631704	58.2	10.368296	39367	91925	49
12	595432	49.1	963379	9.0	632053	58.1	367947	39394	91914	48
13	595727	49.1	963325	9.0	632401	58.1	367599	39421	91902	47
14	596021	49.1	963271	9.0	632750	58.1	367250	39448	91891	46
15	596315	49.0	963217	9.0	633098	58.1	366902	39474	91879	45
16	596609	49.0	963163	9.0	633447	58.0	366553	39501	91868	44
17	596903	48.9	963108	9.0	633795	58.0	366205	39528	91856	43
18	597196	48.9	963054	9.1	634143	58.0	365857	39555	91845	42
19	597490	48.9	962999	9.1	634490	57.9	365510	39581	91833	41
20	597783	48.8	962945	9.1	634838	57.9	365162	39608	91822	40
21	9.598075	48.8	9.962890	9.1	9.635185	57.9	10.364815	39635	91810	39
22	598368	48.7	962836	9.1	635532	57.8	364468	39661	91799	38
23	598660	48.7	962781	9.1	635879	57.8	364121	39688	91787	37
24	598952	48.7	962727	9.1	636226	57.8	363774	39715	91775	36
25	599244	48.6	962672	9.1	636572	57.7	363428	39741	91764	35
26	599536	48.6	962617	9.1	636919	57.7	363081	39768	91752	34
27	599827	48.5	962562	9.1	637265	57.7	362735	39795	91741	33
28	600118	48.5	962508	9.1	637611	57.7	362389	39822	91729	32
29	600409	48.5	962453	9.1	637956	57.6	362044	39848	91718	31
30	600700	48.4	962398	9.2	638302	57.6	361698	39875	91706	30
31	9.600990	48.4	9.962343	9.2	9.638647	57.6	10.361353	39902	91694	29
32	601280	48.4	962288	9.2	638992	57.5	361008	39928	91683	28
33	601570	48.3	962233	9.2	639337	57.5	360663	39955	91671	27
34	601860	48.3	962178	9.2	639682	57.5	360318	39982	91660	26
35	602150	48.2	962123	9.2	640027	57.4	359973	40008	91648	25
36	602439	48.2	962067	9.2	640371	57.4	359629	40035	91636	24
37	602728	48.2	962012	9.2	640716	57.4	359284	40062	91625	23
38	603017	48.1	961957	9.2	641060	57.3	358940	40088	91613	22
39	603305	48.1	961902	9.2	641404	57.3	358596	40115	91601	21
40	603594	48.1	961846	9.2	641747	57.3	358253	40141	91590	20
41	9.603882	48.0	9.961791	9.2	9.642091	57.2	10.357909	40168	91578	19
42	604170	48.0	961735	9.2	642434	57.2	357566	40195	91566	18
43	604457	47.9	961680	9.2	642777	57.2	357223	40221	91555	17
44	604745	47.9	961624	9.3	643120	57.2	356880	40248	91543	16
45	605032	47.9	961569	9.3	643463	57.1	356537	40275	91531	15
46	605319	47.8	961513	9.3	643806	57.1	356194	40301	91519	14
47	605606	47.8	961458	9.3	644148	57.1	355852	40328	91508	13
48	605892	47.8	961402	9.3	644490	57.0	355510	40355	91496	12
49	606179	47.7	961346	9.3	644832	57.0	355168	40381	91484	11
50	606465	47.7	961290	9.3	645174	57.0	354826	40408	91472	10
51	9.606751	47.6	9.961235	9.3	9.645516	56.9	10.354484	40434	91461	9
52	607036	47.6	961179	9.3	645857	56.9	354143	40461	91449	8
53	607322	47.6	961123	9.3	646199	56.9	353801	40488	91437	7
54	607607	47.5	961067	9.3	646540	56.9	353460	40514	91425	6
55	607892	47.5	961011	9.3	646881	56.8	353119	40541	91414	5
56	608177	47.4	960955	9.3	647222	56.8	352778	40567	91402	4
57	608461	47.4	960899	9.3	647562	56.8	352438	40594	91390	3
58	608745	47.4	960843	9.4	647903	56.7	352097	40621	91378	2
59	609029	47.3	960786	9.4	648243	56.7	351757	40647	91366	1
60	609313	47.3	960730	9.4	648583	56.7	351417	40674	91355	0
	Cosine.		Sine.		Cotang.		Tang.	N. cos.	N. sine.	'

′	Sine.	D. 10″	Cosine.	D. 10″	Tang.	D. 10″	Cotang.	N. sine.	N. cos.	
0	9.609313	47.3	9.960730	9.4	9.648583	56.6	10.351417	40674	91355	60
1	609597	47.2	960674	9.4	648923	56.6	351077	40700	91343	59
2	609830	47.2	960618	9.4	649263	56.6	350737	40727	91331	58
3	610164	47.2	960561	9.4	649602	56 6	350398	40753	91319	57
4	610447	47.1	960505	9.4	649942	56.5	350058	40780	91307	56
5	610729	47.1	960448	9.4	650281	56.5	349719	40806	91295	55
6	611012	47.0	960392	9.4	650620	59.5	349380	40833	91283	54
7	611294	47.0	960335	9.4	650959	56.4	349041	40860	91272	53
8	611576	47.0	960279	9.4	651297	56.4	348703	40886	91260	52
9	611858	46.9	960222	9.4	651636	56.4	348364	40913	91248	51
10	612140	46.9	960165	9.4	651974	56.3	348026	40939	91236	50
11	9.612421	46.9	9.960109	9.5	9.652312	56.3	10.347688	40966	91224	49
12	612702	46.8	960052	9.5	652650	56.3	347350	40992	91212	48
13	612983	46.8	959995	9.5	652988	56.3	347012	41019	91200	47
14	613264	46.7	959938	9.5	653326	56.2	346674	41045	91188	46
15	613545	46.7	959882	9.5	653663	56.2	346337	41072	91176	45
16	613825	46.7	959825	9.5	654000	56.2	346000	41098	91164	44
17	614105	46.6	959768	9.5	654337	56.1	345663	41125	91152	43
18	614385	46.6	959711	9.5	654174	56.1	345326	41151	91140	42
19	614665	46.6	959654	9.5	655011	56.1	344989	41178	91128	41
20	614944	46.5	959596	9.5	655348	56.1	344652	41204	91116	40
21	9.615223	46.5	9.959539	9.5	9.655684	56.0	10.344316	41231	91104	39
22	615502	46.5	959482	9.5	656020	56.0	343980	41257	91092	38
23	615781	46.4	959425	9.5	656356	56.0	343644	41284	91080	37
24	616030	46.4	959368	9.5	656692	55.9	343308	41310	91068	36
25	616338	46.4	959310	9.6	657028	55.9	342972	41337	91056	35
26	616616	46.3	959253	9.6	657364	55.9	342636	41363	91044	34
27	616894	46.3	959195	9.6	657699	55.9	342301	41390	91032	33
28	617172	46.2	959138	9.6	658034	55.8	341966	41416	91020	32
29	617450	46.2	959081	9.6	658369	55.8	341631	41443	91008	31
30	617727	46.2	959023	9.6	658704	55.8	341296	41469	90996	30
31	9.618004	46.1	9.958965	9.6	9.659039	55.8	10.340961	41496	90984	29
32	618281	46.1	958908	9.6	659373	55.7	340627	41522	90972	28
33	618558	46.1	958850	9.6	659708	55.7	340292	41549	90960	27
34	618834	46.0	958792	9.6	660042	55.7	339958	41575	90948	26
35	619110	46.0	958734	9.6	660376	55.7	339624	41602	90936	25
36	619386	46.0	958677	9.6	660710	55.6	339290	41628	90924	24
37	619662	45.9	958619	9.6	661043	55.6	338957	41655	90911	23
38	619938	45.9	958561	9.6	661377	55.6	338623	41681	90899	22
39	620213	45.9	958503	9.7	661710	55.5	338290	41707	90887	21
40	620488	45.8	958445	9.7	662043	55.5	337957	41734	90875	20
41	9.620763	45.8	9.958387	9.7	9.662376	55.5	10.337624	41760	90863	19
42	621038	45.7	958329	9.7	662709	55.4	337291	41787	90851	18
43	621313	45.7	958271	9.7	663042	55.4	336958	41813	90839	17
44	621587	45.7	958213	9.7	663375	55.4	336625	41840	90826	16
45	621861	45.6	958154	9.7	663707	55.4	336293	41866	90814	15
46	622135	45.6	958096	9.7	664039	55.3	335961	41892	90802	14
47	622409	45.6	958038	9.7	664371	55.3	335629	41919	90790	13
48	622682	45.5	957979	9.7	664703	55.3	335297	41945	90778	12
49	622956	45.5	957921	9.7	665035	55.3	334965	41972	90766	11
50	623229	45.5	957863	9.7	665366	55.2	334634	41998	90753	10
51	9.623512	45.4	9.957804	9.7	9.665697	55.2	10.334303	42024	90741	9
52	623774	45.4	957746	9.8	666029	55.2	333971	42051	90729	8
53	624047	45.4	957687	9.8	666360	55.1	333620	42077	90717	7
54	624319	45.3	957628	9.8	666691	55.1	333309	42104	90704	6
55	624591	45.3	957570	9.8	667021	55.1	332979	42130	90692	5
56	624863	46.3	957511	9.8	667352	55.1	332648	42156	90680	4
57	625135	45.2	957452	9.8	667682	55.0	332318	42183	90668	3
58	625406	45.2	957393	9.8	668013	55.0	331987	42209	90655	2
59	625677	45.2	957335	9.8	668343	55.0	331657	42235	90643	1
60	625948		957276		668672		331328	42262	90631	0
	Cosine.		Sine.		Cotang.		Tang.	N. cos.	N. sine.	′

65 Degrees.

'	Sine.	D. 10″	Cosine.	D. 10″	Tang.	D. 10″	Cotang.	N. sine.	N. cos.	'
0	9.625948	45.1	9.957276	9.8	9.668673	55.0	10.331327	42262	90631	60
1	626219	45.1	957217	9.8	669002	54.9	330998	42288	90613	59
2	626490	45.1	957158	9.8	669332	54.9	330668	42315	90606	58
3	626760	45.0	957099	9.8	669661	54.9	330339	42341	90594	57
4	627030	45.0	957040	9.8	669991	54.8	330009	42367	90582	56
5	627300	45.0	956981	9.8	670320	54.8	329680	42394	90569	55
6	627570	44.9	956921	9.9	670649	54.8	329351	42420	90557	54
7	627840	44.9	956862	9.9	670977	54.8	329023	42446	90545	53
8	628109	44.9	956803	9.9	671306	54.7	328694	42473	90532	52
9	628378	44.8	956744	9.9	671634	54.7	328366	42499	90520	51
10	628647	44.8	956684	9.9	671963	54.7	328037	42525	90507	50
11	9.628916	44.7	9.956625	9.9	9.672291	54.7	10.327709	42552	90495	49
12	629185	44.7	956566	9.9	672619	54.6	327381	42578	90483	48
13	629453	44.7	956506	9.9	672947	54.6	327053	42604	90470	47
14	629721	44.6	956447	9.9	673274	54.6	326726	42631	90458	46
15	629989	44.6	956387	9.9	673602	54.6	326398	42657	90446	45
16	630257	44.6	956327	9.9	673929	54.5	326071	42683	90433	44
17	630524	44.6	956268	9.9	674257	54.5	325743	42709	90421	43
18	630792	44.5	956208	10.0	674584	54.5	325416	42736	90408	42
19	631059	44.5	956148	10.0	674910	54.4	325090	42762	90396	41
20	631326	44.5	956089	10.0	675237	54.4	324763	42788	90383	40
21	9.631593	44.4	9.956029	10.0	9.675564	54.4	10.324436	42815	90371	39
22	631859	44.4	955969	10.0	675890	54.4	324110	42841	90358	38
23	632125	44.4	955909	10.0	676216	54.3	323784	42867	90346	37
24	632392	44.3	955849	10.0	676543	54.3	323457	42894	90334	36
25	632658	44.3	955789	10.0	676869	54.3	323131	42920	90321	35
26	632923	44.3	955729	10.0	677194	54.3	322806	42946	90309	34
27	633189	44.2	955669	10.0	677520	54.2	322480	42972	90296	33
28	633454	44.2	955609	10.0	677846	54.2	322154	42999	90284	32
29	633719	44.2	955548	10.0	678171	54.2	321829	43025	90271	31
30	633984	44.1	955488	10.0	678496	54.2	321504	43051	90259	30
31	9.634249	44.1	9.955428	10.1	9.678821	54.2	10.321179	43077	90246	29
32	634514	44.0	955368	10.1	679146	54.1	320854	43104	90233	28
33	634778	44.0	955307	10.1	679471	54.1	320529	43130	90221	27
34	635042	44.0	955247	10.1	679795	54.1	320205	43156	90208	26
35	635303	43.9	955186	10.1	680120	54.1	319880	43182	90196	25
36	635570	43.9	955126	10.1	680444	54.0	319556	43209	90183	24
37	635834	43.9	955065	10.1	680768	54.0	319232	43235	90171	23
38	636097	43.8	955005	10.1	681092	54.0	318908	43261	90158	22
39	636360	43.8	954944	10.1	681416	54.0	318584	43287	90146	21
40	636623	43.8	954883	10.1	681740	53.9	318260	43313	90133	20
41	9.636886	43.7	9.954823	10.1	9.682063	53.9	10.317937	43340	90120	19
42	637148	43.7	954762	10.1	682387	53.9	317613	43366	90108	18
43	637411	43.7	954701	10.1	682710	53.9	317290	43392	90095	17
44	637673	43.7	954640	10.1	683033	53.8	316967	43418	90082	16
45	637935	43.6	954579	10.1	683356	53.8	316644	43445	90070	15
46	638197	43.6	954518	10.2	683679	53.8	316321	43471	90057	14
47	638458	43.6	954457	10.2	684001	53.8	315999	43497	90045	13
48	638720	43.5	954396	10.2	684324	53.7	315676	43523	90032	12
49	638981	43.5	954335	10.2	684646	53.7	315354	43549	90019	11
50	639242	43.5	954274	10.2	684968	53.7	315032	43575	90007	10
51	9.639503	43.4	9.954213	10.2	9.685290	53.7	10.314710	43602	89994	9
52	639764	43.4	954152	10.2	685612	53.6	314388	43628	89981	8
53	640024	43.4	954090	10.2	685934	53.6	314066	43654	89968	7
54	640284	43.3	954029	10.2	686255	53.6	313745	43680	89956	6
55	640544	43.3	953968	10.2	686577	53.6	313423	43706	89943	5
56	640804	43.3	953906	10.2	686898	53.5	313102	43733	89930	4
57	641064	43.2	953845	10.2	687219	53.5	312781	43759	89918	3
58	641324	43.2	953783	10.2	687540	53.5	312460	43785	89905	2
59	641584	43.2	953722	10.3	687861	53.4	312139	43811	89892	1
60	641842		953660		688182		311818	43837	89879	0
	Cosine.		Sine.		Cotang.		Tang.	N. cos.	N. sine.	'

64 Degrees.

'	Sine.	D. 10"	Cosine.	D. 10"	Tang.	D. 10"	Cotang.	N. sine	N. cos.	'
0	9.641842	43.1	9.953660	10.3	9.688182	53.4	10.311818	43837	89879	60
1	642101	43.1	953599	10.3	688502	53.4	311498	43863	89867	59
2	642360	43.1	953537	10.3	688823	53.4	311177	43889	89854	58
3	642618	43.1	953475	10.3	689143	53.4	310857	43916	89841	57
4	642877	43.0	953413	10.3	689463	53.3	310537	43942	89828	56
5	643135	43.0	953352	10.3	689783	53.3	310217	43968	89816	55
6	643393	43.0	953290	10.3	690103	53.3	309897	43994	89803	54
7	643650	43.0	953228	10.3	690423	53.3	309577	44020	89790	53
8	643908	42.9	953166	10.3	690742	53.3	309258	44046	89777	52
9	644165	42.9	953104	10.3	691062	53.2	308938	44072	89764	51
10	644423	42.9	953042	10.3	691381	53.2	308619	44098	89752	50
11	9.644680	42.8	9.952980	10.3	9.691700	53.2	10.308300	44124	89739	49
12	644936	42.8	952918	10.4	692019	53.1	307981	44151	89726	48
13	645193	42.8	952855	10.4	692338	53.1	307662	44177	89713	47
14	645450	42.7	952793	10.4	692656	53.1	307344	44203	89700	46
15	645706	42.7	952731	10.4	692975	53.1	307025	44229	89687	45
16	645962	42.7	952669	10.4	693293	53.1	306707	44255	89674	44
17	646218	42.6	952606	10.4	693612	53.0	306388	44281	89662	43
18	646474	42.6	952544	10.4	693930	53.0	306070	44307	89649	42
19	646729	42.6	952481	10.4	694248	53.0	305752	44333	89636	41
20	646984	42.5	952419	10.4	694566	53.0	305434	44359	89623	40
21	9.647240	42.5	9.952356	10.4	9.694883	52.9	10.305117	44385	89610	39
22	647494	42.5	952294	10.4	695201	52.9	304799	44411	89597	38
23	647749	42.4	952231	10.4	695518	52.9	304482	44437	89584	37
24	648004	42.4	952168	10.4	695836	52.9	304164	44464	89571	36
25	648258	42.4	952106	10.5	696153	52.9	303847	44490	89558	35
26	648512	42.4	952043	10.5	696470	52.8	303550	44516	89545	34
27	648766	42.3	951980	10.5	696787	52.8	303213	44542	89532	33
28	649020	42.3	951917	10.5	697103	52.8	302897	44568	89519	32
29	649274	42.3	951854	10.5	697420	52.8	302580	44594	89506	31
30	649527	42.2	951791	10.5	697736	52.7	302264	44620	89493	30
31	9.649781	42.2	9.951728	10.5	9.698053	52.7	10.301947	44646	89480	29
32	650034	42.2	951665	10.5	698369	52.7	301631	44672	89467	28
33	650287	42.2	951602	10.6	698685	52.7	301315	44698	89454	27
34	650539	42.1	951539	10.5	699001	52.6	300999	44724	89441	26
35	650792	42.1	951476	10.5	699316	52.6	300684	44750	89428	25
36	651044	42.1	951412	10.5	699632	52.6	300368	44776	89415	24
37	651297	42.0	951349	10.5	699947	52.6	300053	44802	89402	23
38	651549	42.0	951286	10.6	700263	52.6	299737	44828	89389	22
39	651800	42.0	951222	10.6	700578	52.5	299422	44854	89376	21
40	652052	41.9	951159	10.6	700893	52.5	299107	44880	89363	20
41	9.652304	41.9	9.951096	10.6	9.701208	52.5	10.298792	44906	89350	19
42	652555	41.9	951032	10.6	701523	52.4	298477	44932	89337	18
43	652806	41.8	950968	10.6	701837	52.4	298163	44958	89324	17
44	653057	41.8	950905	10.6	702152	52.4	297848	44984	89311	16
45	653308	41.8	950841	10.6	702466	52.4	297534	45010	89298	15
46	653558	41.8	950778	10.6	702780	52.4	297220	45036	89285	14
47	653808	41.7	950714	10.6	703095	52.3	296905	45062	89272	13
48	654059	41.7	950650	10.6	703409	52.3	296591	45088	89259	12
49	654309	41.7	950586	10.6	703723	52.3	296277	45114	89245	11
50	654558	41.6	950522	10.6	704036	52.3	295964	45140	89232	10
51	9.654808	41.6	9.950458	10.7	9.704350	52.2	10.295650	45166	89219	9
52	655058	41.6	950394	10.7	704663	52.2	295337	45192	89206	8
53	655307	41.6	950330	10.7	704977	52.2	295023	45218	89193	7
54	655556	41.5	950366	10.7	705290	52.2	294710	45243	89180	6
55	655805	41.5	950202	10.7	705603	52.2	294397	45269	89167	5
56	656054	41.5	950138	10.7	705916	52.1	294084	45295	89153	4
57	656302	41.4	950074	10.7	706228	52.1	293772	45321	89140	3
58	656551	41.4	950010	10.7	706541	52.1	293459	45347	89127	2
59	656799	41.4	949945	10.7	706854	52.1	293146	45373	89114	1
60	657047	41.3	949881	10.7	707166	52.1	292834	45399	89101	0
	Cosine.		Sine.		Cotang.		Tang.	N. cos.	N. sine.	'

′	Sine.	D. 10″	Cosine.	D. 10″	Tang.	D. 10″	Cotang.	N. sine.	N. cos.	
0	9.657047	41.3	9.949881	10.7	9.707166	52.0	10.292834	45399	89101	60
1	657295	41.3	949816	10.7	707478	52.0	292522	45425	89087	59
2	657542	41.2	949752	10.7	707790	52.0	292210	45451	89074	58
3	657790	41.2	949688	10.7	708102	52.0	291898	45477	89061	57
4	658037	41.2	949623	10.8	708414	52.0	291586	45503	89048	56
5	658284	41.2	949558	10.8	708726	51.9	291274	45529	89035	55
6	658531	41.2	949494	10.8	709037	51.9	290963	45554	89021	54
7	658778	41.1	949429	10.8	709349	51.9	290651	45580	89008	53
8	659025	41.1	949364	10.8	709660	51.9	290340	45606	88995	52
9	659271	41.1	949300	10.8	709971	51.9	290029	45632	88981	51
10	659517	41.0	949235	10.8	710282	51.8	289718	45658	88968	50
11	9.659763	41.0	9.949170	10.8	9.710593	51.8	10.289407	45684	88955	49
12	660009	41.0	949105	10.8	710904	51.8	289096	45710	88942	48
13	660255	40.9	949040	10.8	711215	51.8	288785	45736	88928	47
14	660501	40.9	948975	10.8	711525	51.8	288475	45762	88915	46
15	660746	40.9	948910	10.8	711836	51.7	288164	45787	88902	45
16	660991	40.9	948845	10.8	712146	51.7	287854	45813	88888	44
17	661236	40.8	948780	10.8	712456	51.7	287544	45839	88875	43
18	661481	40.8	948715	10.9	712766	51.7	287234	45865	88862	42
19	661726	40.8	948650	10.9	713076	51.6	286924	45891	88848	41
20	661970	40.7	948584	10.9	713386	51.6	286614	45917	88835	40
21	9.662214	40.7	9.948519	10.9	9.713696	51.6	10.286304	45942	88822	39
22	662459	40.7	948454	10.9	714005	51.6	285995	45968	88808	38
23	662703	40.7	948388	10.9	714314	51.6	285686	45994	88795	37
24	662946	40.6	948323	10.9	714624	51.5	285376	46020	88782	36
25	663190	40.6	948257	10.9	714933	51.5	285067	46046	88768	35
26	663433	40.6	948192	10.9	715242	51.5	284758	46072	88755	34
27	663677	40.5	948126	10.9	715551	51.5	284449	46097	88741	33
28	663920	40.5	948060	10.9	715860	51.4	284140	46123	88728	32
29	664163	40.5	947995	11.0	716168	51.4	283832	46149	88715	31
30	664406	40.5	947929	11.0	716477	51.4	283523	46175	88701	30
31	9.664648	40.4	9.947863	11.0	9.716785	51.4	10.283215	46201	88688	29
32	664891	40.4	947797	11.0	717093	51.4	282907	46226	88674	28
33	665133	40.4	947731	11.0	717401	51.3	282599	46252	88661	27
34	665375	40.3	947665	11.0	717709	51.3	282291	46278	88647	26
35	665617	40.3	947600	11.0	718017	51.3	281983	46304	88634	25
36	665859	40.3	947533	11.0	718325	51.3	281675	46330	88620	24
37	666100	40.2	947467	11.0	718633	51.3	281367	46355	88607	23
38	666342	40.2	947401	11.0	718940	51.2	281060	46381	88593	22
39	666583	40.2	947335	11.0	719248	51.2	280752	46407	88580	21
40	666824	40.2	947269	11.0	719555	51.2	280445	46433	88566	20
41	9.667065	40.1	9.947203	11.0	9.719862	51.2	10.280138	46458	88553	19
42	667305	40.1	947136	11.1	720169	51.2	279831	46484	88539	18
43	667546	40.1	947070	11.1	720476	51.1	279524	46510	88526	17
44	667786	40.1	947004	11.1	720783	51.1	279217	46536	88512	16
45	668027	40.0	946937	11.1	721089	51.1	278911	46561	88499	15
46	668267	40.0	946871	11.1	721396	51.1	278604	46587	88485	14
47	668506	40.0	946804	11.1	721702	51.1	278298	46613	88472	13
48	668746	39.9	946738	11.1	722009	51.0	277991	46639	88458	12
49	668986	39.9	946671	11.1	722315	51.0	277685	46664	88445	11
50	669225	39.9	946604	11.1	722621	51.0	277379	46690	88431	10
51	9.669464	39.9	9.946538	11.1	9.722927	51.0	10.277073	46716	88417	9
52	669703	39.8	946471	11.1	723232	51.0	276768	46742	88404	8
53	669942	39.8	946404	11.1	723538	50.9	276462	46767	88390	7
54	670181	39.8	946337	11.1	723844	50.9	276156	46793	88377	6
55	670419	39.7	946270	11.1	724149	50.9	275851	46819	88363	5
56	670658	39.7	946203	11.2	724454	50.9	275546	46844	88349	4
57	670896	39.7	946136	11.2	724759	50.9	275241	46870	88336	3
58	671134	39.7	946069	11.2	725065	50.8	274935	46896	88322	2
59	671372	39.6	946002	11.2	725369	50.8	274631	46921	88308	1
60	671609	39.6	945935	11.2	725674	50.8	274326	46947	88295	0
	Cosine.		Sine.		Cotang.		Tang.	N. cos.	N. sine.	

′	Sine.	D. 10″	Cosine.	D. 10″	Tang.	D. 10″	Cotang.	N. sine	N. cos.	′
0	9.671609	39.6	9.945935	11.2	9.725674	50.8	10.274326	46947	88295	60
1	671847	39.5	945868	11.2	725979	50.8	274021	46973	88281	59
2	672034	39.5	945800	11.2	726284	50.7	273716	46999	88267	58
3	672321	39.5	945733	11.2	726588	50.7	273412	47024	88254	57
4	672558	39.5	945666	11.2	726892	50.7	273108	47050	88240	56
5	672795	39.5	945598	11.2	727197	50.7	272803	47076	88226	55
6	673032	39.4	945531	11.2	727501	50.7	272499	47101	88213	54
7	673268	39.4	945464	11.2	727805	50.7	272195	47127	88199	53
8	673505	39.4	945396	11.3	728109	50.6	271891	47153	88185	52
9	673741	39.4	945328	11.3	728412	50.6	271588	47178	88172	51
10	673977	39.3	945261	11.3	728716	50.6	271284	47204	88158	50
11	9.674213	39.3	9.945193	11.3	9.729020	50.6	10.270980	47229	88144	49
12	674448	39.3	945125	11.3	729323	50.6	270677	47255	88130	48
13	674684	39.2	945058	11.3	729626	50.5	270374	47281	88117	47
14	674919	39.2	944990	11.3	729929	50.5	270071	47306	88103	46
15	675155	39.2	944922	11.3	730233	50.5	269767	47332	88089	45
16	675390	39.2	944854	11.3	730535	50.5	269465	47358	88075	44
17	675624	39.1	944786	11.3	730838	50.5	269162	47383	88062	43
18	675859	39.1	944718	11.3	731141	50.4	268859	47409	88048	42
19	676094	39.1	944650	11.3	731444	50.4	268556	47434	88034	41
20	676328	39.1	944582	11.3	731746	50.4	268254	47460	88020	40
21	9.676562	39.0	9.944514	11.4	9.732048	50.4	10.267952	47486	88006	39
22	676796	39.0	944446	11.4	732351	50.4	267649	47511	87993	38
23	677030	39.0	944377	11.4	732653	50.3	267347	47537	87979	37
24	677264	39.0	944309	11.4	732955	50.3	267045	47562	87965	36
25	677498	38.9	944241	11.4	733257	50.3	266743	47588	87951	35
26	677731	38.9	944172	11.4	733558	50.3	266442	47614	87937	34
27	677964	38.9	944104	11.4	733860	50.3	266140	47639	87923	33
28	678197	38.8	944036	11.4	734162	50.2	265838	47665	87909	32
29	678430	38.8	943967	11.4	734463	50.2	265537	47690	87896	31
30	678663	38.8	943899	11.4	734764	50.2	265236	47716	87882	30
31	9.678895	38.8	9.943830	11.4	9.735066	50.2	10.264934	47741	87868	29
32	679128	38.7	943761	11.4	735367	50.2	264633	47767	87854	28
33	679360	38.7	943693	11.4	735668	50.2	264332	47793	87840	27
34	679592	38.7	943624	11.5	735969	50.1	264031	47818	87826	26
35	679824	38.7	943555	11.5	736269	50.1	263731	47844	87812	25
36	680056	38.6	943486	11.5	736570	50.1	263430	47869	87798	24
37	680288	38.6	943417	11.5	736871	50.1	263129	47895	87784	23
38	680519	38.6	943348	11.5	737171	50.1	262829	47920	87770	22
39	680750	38.5	943279	11.5	737471	50.0	262529	47946	87756	21
40	680982	38.5	943210	11.5	737771	50.0	262229	47971	87743	20
41	9.681213	38.5	9.943141	11.5	9.738071	50.0	10.261929	47997	87729	19
42	681443	38.5	943072	11.5	738371	50.0	261629	48022	87715	18
43	681674	38.4	943003	11.5	738671	50.0	261329	48048	87701	17
44	681905	38.4	942934	11.5	738971	49.9	261029	48073	87687	16
45	682135	38.4	942864	11.5	739271	49.9	260729	48099	87673	15
46	682365	38.4	942795	11.6	739570	49.9	260430	48124	87659	14
47	682595	38.3	942726	11.6	739870	49.9	260130	48150	87645	13
48	682825	38.3	942656	11.6	740169	49.9	259831	48175	87631	12
49	683035	38.3	942587	11.6	740468	49.9	259532	48201	87617	11
50	683284	38.3	942517	11.6	740767	49.8	259233	48226	87603	10
51	9.683514	38.2	9.942448	11.6	9.741066	49.8	10.258934	48252	87589	9
52	683743	38.2	942378	11.6	741365	49.8	258635	48277	87575	8
53	683972	38.2	942308	11.6	741664	49.8	258336	48303	87561	7
54	684201	38.2	942239	11.6	741962	49.8	258038	48328	87546	6
55	684430	38.1	942169	11.6	742261	49.7	257739	48354	87532	5
56	684658	38.1	942099	11.6	742559	49.7	257441	48379	87518	4
57	684887	38.1	942029	11.6	742858	49.7	257142	48405	87504	3
58	685115	38.0	941959	11.6	743156	49.7	256844	48430	87490	2
59	685343	38.0	941889	11.6	743454	49.7	256546	48456	87476	1
60	685571	38.0	941819	11.7	743752		256248	48481	87462	0
	Cosine.		Sine.		Cotang.		Tang.	N. cos.	N. sine.	′

61 Degrees.

′	Sine.	D. 10″	Cosine.	D. 10″	Tang.	D. 10″	Cotang.	N. sine.	N. cos.	
0	9.685571	38.0	9.941819	11.7	9.743752	49.6	10.256248	48481	87462	60
1	685799	37.9	941749	11.7	744050	49.6	255950	48506	87448	59
2	686027	37.9	941679	11.7	744348	49.6	255652	48532	87434	58
3	686254	37.9	941609	11.7	744645	49.6	255355	48557	87420	57
4	686482	37.9	941539	11.7	744943	49.6	255057	48583	87406	56
5	686709	37.9	941469	11.7	745240	49.6	254760	48608	87391	55
6	686936	37.8	941398	11.7	745538	49.6	254462	48634	87377	54
7	687163	37.8	941328	11.7	745835	49.5	254165	48659	87363	53
8	687389	37.8	941258	11.7	746132	49.5	253868	48684	87349	52
9	687616	37.8	941187	11.7	746429	49.5	253571	48710	87335	51
10	687843	37.7	941117	11.7	746726	49.5	253274	48735	87321	50
11	9.688069	37.7	9.941046	11.7	9.747023	49.5	10.252977	48761	87306	49
12	688295	37.7	940975	11.8	747319	49.4	252681	48786	87292	48
13	688521	37.7	940905	11.8	747616	49.4	252384	48811	87278	47
14	688747	37.6	940834	11.8	747913	49.4	252087	48837	87264	46
15	688972	37.6	940763	11.8	748209	49.4	251791	48862	87250	45
16	689198	37.6	940693	11.8	748505	49.4	251495	48888	87235	44
17	689423	37.6	940622	11.8	748801	49.3	251199	48913	87221	43
18	689648	37.5	940551	11.8	749097	49.3	250903	48938	87207	42
19	689873	37.5	940480	11.8	749393	49.3	250607	48964	87193	41
20	690098	37.5	940409	11.8	749689	49.3	250311	48989	87178	40
21	9.690323	37.5	9.940338	11.8	9.749985	49.3	10.250015	49014	87164	39
22	690548	37.4	940267	11.8	750281	49.3	249719	49040	87150	38
23	690772	37.4	940196	11.8	750576	49.2	249424	49065	87136	37
24	690996	37.4	940125	11.8	750872	49.2	249128	49090	87121	36
25	691220	37.4	940054	11.9	751167	49.2	248833	49116	87107	35
26	691444	37.3	939982	11.9	751462	49.2	248538	49141	87093	34
27	691668	37.3	939911	11.9	751757	49.2	248243	49166	87079	33
28	691892	37.3	939840	11.9	752052	49.1	247948	49192	87064	32
29	692115	37.3	939768	11.9	752347	49.1	247653	49217	87050	31
30	692339	37.2	939697	11.9	752642	49.1	247358	49242	87036	30
31	9.692562	37.2	9.939625	11.9	9.752937	49.1	10.247063	49268	87021	29
32	692785	37.2	939554	11.9	753231	49.1	246769	49293	87007	28
33	693008	37.1	939482	11.9	753526	49.1	246474	49318	86993	27
34	693231	37.1	939410	11.9	753820	49.1	246180	49344	86978	26
35	693453	37.1	939339	11.9	754115	49.0	245885	49369	86964	25
36	693676	37.1	939267	12.0	754409	49.0	245591	49394	86949	24
37	693898	37.0	939195	12.0	754703	49.0	245297	49419	86935	23
38	694120	37.0	939123	12.0	754997	49.0	245003	49445	86921	22
39	694342	37.0	939052	12.0	755291	49.0	244709	49470	86906	21
40	694564	37.0	938980	12.0	755585	48.9	244415	49495	86892	20
41	9.694786	36.9	9.938908	12.0	9.755878	48.9	10.244122	49521	86878	19
42	695007	36.9	938836	12.0	756172	48.9	243828	49546	86863	18
43	695229	36.9	938763	12.0	756465	48.9	243535	49571	86849	17
44	695450	36.9	938691	12.0	756759	48.9	243241	49596	86834	16
45	695671	36.8	938619	12.0	757052	48.9	242948	49622	86820	15
46	695892	36.8	938547	12.0	757345	48.8	242655	49647	86805	14
47	696113	36.8	938475	12.0	757638	48.8	242362	49672	86791	13
48	696334	36.8	938402	12.1	757931	48.8	242069	49697	86777	12
49	696554	36.7	938330	12.1	758224	48.8	241776	49723	86762	11
50	696775	36.7	938258	12.1	758517	48.8	241483	49748	86748	10
51	9.696995	36.7	9.938185	12.1	9.758810	48.8	10.241190	49773	86733	9
52	697215	36.7	938113	12.1	759102	48.7	240898	49798	86719	8
53	697435	36.6	938040	12.1	759395	48.7	240605	49824	86704	7
54	697654	36.6	937967	12.1	759687	48.7	240313	49849	86690	6
55	697874	36.6	937895	12.1	759979	48.7	240021	49874	86675	5
56	698094	36.6	937822	12.1	760272	48.7	239728	49899	86661	4
57	698313	36.5	937749	12.1	760564	48.7	239436	49924	86646	3
58	698532	36.5	937676	12.1	760856	48.6	239144	49950	86632	2
59	698751	36.5	937604	12.1	761148	48.6	238852	49975	86617	1
60	698970	36.5	937531		761439		238561	50000	86603	0
	Cosine.		Sine.		Cotang.		Tang.	N. cos.	N. sine.	′

′	Sine.	D. 10″	Cosine.	D. 10″	Tang.	D. 10″	Cotang.	N. sine	N. cos.	′
0	9.698970	36.4	9.937531	12.1	9.761439	48.6	10.238561	50000	86603	60
1	699189	36.4	937458	12.2	761731	48.6	238269	50025	86588	59
2	699407	36.4	937385	12.2	762023	48.6	237977	50050	86573	58
3	699626	36.4	937312	12.2	762314	48.6	237686	50076	86559	57
4	699844	36.4	937238	12.2	762606	48.6	237394	50101	86544	56
5	700062	36.3	937165	12.2	762897	48.5	237103	50126	86530	55
6	700280	36.3	937092	12.2	763188	48.5	236812	50151	86515	54
7	700498	36.3	937019	12.2	763479	48.5	236521	50176	86501	53
8	700716	36.3	936946	12.2	763770	48.5	236230	50201	86486	52
9	700933	36.3	936872	12.2	764061	48.5	235939	50227	86471	51
10	701151	36.2	936799	12.2	764352	48.5	235648	50252	86457	50
11	9.701368	36.2	9.936725	12.2	9.764643	48.4	10.235357	50277	86442	49
12	701585	36.2	936652	12.3	764933	48.4	235067	50302	86427	48
13	701802	36.2	936578	12.3	765224	48.4	234776	50327	86413	47
14	702019	36.1	936505	12.3	765514	48.4	234486	50352	86398	46
15	702236	36.1	936431	12.3	765805	48.4	234195	50377	86384	45
16	702452	36.1	936357	12.3	766095	48.4	233905	50403	86369	44
17	702669	36.1	936284	12.3	766385	48.4	233615	50428	86354	43
18	702885	36.0	936210	12.3	766675	48.3	233325	50453	86340	42
19	703101	36.0	936136	12.3	766965	48.3	233035	50478	86325	41
20	703317	36.0	936062	12.3	767255	48.3	232745	50503	86310	40
21	9.703533	36.0	9.935988	12.3	9.767545	48.3	10.232455	50528	86295	39
22	703749	35.9	935914	12.3	767834	48.3	232166	50553	86281	38
23	703964	35.9	935840	12.3	768124	48.3	231876	50578	86266	37
24	704179	35.9	935766	12.4	768413	48.2	231587	50603	86251	36
25	704395	35.9	935692	12.4	768703	48.2	231297	50628	86237	35
26	704610	35.9	935618	12.4	768992	48.2	231008	50654	86222	34
27	704825	35.8	935543	12.4	769281	48.2	230719	50679	86207	33
28	705010	35.8	935469	12.4	769570	48.2	230430	50704	86192	32
29	705254	35.8	935395	12.4	769860	48.2	230140	50729	86178	31
30	705469	35.8	935320	12.4	770148	48.1	229852	50754	86163	30
31	9.705683	35.7	9.935246	12.4	9.770437	48.1	10.229563	50779	86148	29
32	705898	35.7	935171	12.4	770726	48.1	229274	50804	86133	28
33	706112	35.7	935097	12.4	771015	48.1	228985	50829	86119	27
34	706326	35.7	935022	12.4	771303	48.1	228697	50854	86104	26
35	706539	35.6	934948	12.4	771592	48.1	228408	50879	86089	25
36	706753	35.6	934873	12.4	771880	48.1	228120	50904	86074	24
37	706967	35.6	934798	12.4	772168	48.0	227832	50929	86059	23
38	707180	35.6	934723	12.5	772457	48.0	227543	50954	86045	22
39	707393	35.5	934649	12.5	772745	48.0	227255	50979	86030	21
40	707606	35.5	934574	12.5	773033	48.0	226967	51004	86015	20
41	9.707819	35.5	9.934499	12.5	9.773321	48.0	10.226679	51029	86000	19
42	708032	35.5	934424	12.5	773608	48.0	226392	51054	85985	18
43	708245	35.4	934349	12.5	773896	47.9	226104	51079	85970	17
44	708458	35.4	934274	12.5	774184	47.9	225816	51104	85956	16
45	708670	35.4	934199	12.5	774471	47.9	225529	51129	85941	15
46	708882	35.4	934123	12.5	774759	47.9	225241	51154	85926	14
47	709094	35.3	934048	12.5	775046	47.9	224954	51179	85911	13
48	709305	35.3	933973	12.5	775333	47.9	224667	51204	85896	12
49	709518	35.3	933898	12.5	775621	47.9	224379	51229	85881	11
50	709730	35.3	933822	12.6	775908	47.8	224092	51254	85866	10
51	9.709941	35.3	9.933747	12.6	9.776195	47.8	10.223805	51279	85851	9
52	710153	35.2	933671	12.6	776482	47.8	223518	51304	85836	8
53	710364	35.2	933596	12.6	776769	47.8	223231	51329	85821	7
54	710575	35.2	933520	12.6	777055	47.8	222945	51354	85806	6
55	710786	35.2	933445	12.6	777342	47.8	222658	51379	85792	5
56	710967	35.1	933369	12.6	777628	47.8	222372	51404	85777	4
57	711208	35.1	933293	12.6	777915	47.7	222085	51429	85762	3
58	711419	35.1	933217	12.6	778201	47.7	221799	51454	85747	2
59	711629	35.1	933141	12.6	778487	47.7	221512	51479	85732	1
60	711839	35.0	933066		778774	47.7	221226	51504	85717	0
	Cosine.		Sine.		Cotang.		Tang.	N. cos.	N. sine.	′

59 Degrees.

'	Sine.	D. 10"	Cosine.	D. 10"	Tang.	D. 10"	Cotang.	N.sine.	N. cos.	'
0	9.711839	35.0	9.933036	12.6	9.778774	47.7	10.221226	51504	85717	60
1	712050	35.0	932990	12.7	779060	47.7	220940	51529	85702	59
2	712260	35.0	932914	12.7	779346	47.6	220654	51554	85687	58
3	712469	34.9	932838	12.7	779632	47.6	220368	51579	85672	57
4	712679	34.9	932762	12.7	779918	47.6	220082	51604	85657	56
5	712889	34.9	932685	12.7	780203	47.6	219797	51628	85642	55
6	713098	34.9	932609	12.7	780489	47.6	219511	51653	85627	54
7	713308	34.9	932533	12.7	780775	47.6	219225	51678	85612	53
8	713517	34.9	932457	12.7	781060	47.6	218940	51703	85597	52
9	713726	34.8	932380	12.7	781346	47.6	218654	51728	85582	51
10	713935	34.8	932304	12.7	781631	47.5	218369	51753	85567	50
11	9.714144	34.8	9.932228	12.7	9.781916	47.5	10.218084	51778	85551	49
12	714352	34.8	932151	12.7	782201	47.5	217799	51803	85536	48
13	714561	34.7	932075	12.7	782486	47.5	217514	51828	85521	47
14	714769	34.7	931998	12.8	782771	47.5	217229	51852	85506	46
15	714978	34.7	931921	12.8	783056	47.5	216944	51877	85491	45
16	715186	34.7	931845	12.8	783341	47.5	216659	51902	85476	44
17	715394	34.7	931768	12.8	783626	47.5	216374	51927	85461	43
18	715602	34.6	931691	12.8	783910	47.4	216090	51952	85446	42
19	715809	34.6	931614	12.8	784195	47.4	215805	51977	85431	41
20	716017	34.6	931537	12.8	784479	47.4	215521	52002	85416	40
21	9.716224	34.6	9.931460	12.8	9.784764	47.4	10.215236	52026	85401	39
22	716432	34.5	931383	12.8	785048	47.4	214952	52051	85385	38
23	716639	34.5	931306	12.8	785332	47.3	214668	52076	85370	37
24	716846	34.5	931229	12.8	785616	47.3	214384	52101	85355	36
25	717053	34.5	931152	12.9	785900	47.3	214100	52126	85340	35
26	717259	34.5	931075	12.9	786184	47.3	213816	52151	85325	34
27	717466	34.4	930998	12.9	786468	47.3	213532	52175	85310	33
28	717673	34.4	930921	12.9	786752	47.3	213248	52200	85294	32
29	717879	34.4	930843	12.9	787036	47.3	212964	52225	85279	31
30	718085	34.4	930766	12.9	787319	47.3	212681	52250	85264	30
31	9.718291	34.3	9.930688	12.9	9.787603	47.2	10.212397	52275	85249	29
32	718497	34.3	930611	12.9	787886	47.2	212114	52299	85234	28
33	718703	34.3	930533	12.9	788170	47.2	211830	52324	85218	27
34	718909	34.3	930456	12.9	788453	47.2	211547	52349	85203	26
35	719114	34.3	930378	12.9	788736	47.2	211264	52374	85188	25
36	719320	34.2	930300	12.9	789019	47.2	210981	52399	85173	24
37	719525	34.2	930223	13.0	789302	47.2	210698	52423	85157	23
38	719730	34.2	930145	13.0	789585	47.1	210415	52448	85142	22
39	719935	34.2	930067	13.0	789868	47.1	210132	52473	85127	21
40	720140	34.1	929989	13.0	790151	47.1	209849	52498	85112	20
41	9.720345	34.1	9.929911	13.0	9.790433	47.1	10.209567	52522	85096	19
42	720549	34.1	929833	13.0	790716	47.1	209284	52547	85081	18
43	720754	34.1	929755	13.0	790999	47.1	209001	52572	85066	17
44	720958	34.0	929677	13.0	791281	47.1	208719	52597	85051	16
45	721162	34.0	929599	13.0	791563	47.1	208437	52621	85035	15
46	721366	34.0	929521	13.0	791846	47.0	208154	52646	85020	14
47	721570	34.0	929442	13.0	792128	47.0	207872	52671	85005	13
48	721774	34.0	929364	13.0	792410	47.0	207590	52696	84989	12
49	721978	33.9	929286	13.1	792692	47.0	207308	52720	84974	11
50	722181	33.9	929207	13.1	792974	47.0	207026	52745	84959	10
51	9.722385	33.9	9.929129	13.1	9.793256	47.0	10.206744	52770	84943	9
52	722588	33.9	929050	13.1	793538	47.0	206462	52794	84928	8
53	722791	33.9	928972	13.1	793819	46.9	206181	52819	84913	7
54	722994	33.8	928893	13.1	794101	46.9	205899	52844	84897	6
55	723197	33.8	928815	13.1	794383	46.9	205617	52869	84882	5
56	723400	33.8	928736	13.1	794664	46.9	205336	52893	84866	4
57	723603	33.8	928657	13.1	794945	46.9	205055	52918	84851	3
58	723805	33.7	928578	13.1	795227	46.9	204773	52943	84836	2
59	724007	33.7	928499	13.1	795508	46.9	204492	52967	84820	1
60	724210	33.7	928420	13.1	795789	46.8	204211	52992	84805	0
	Cosine.		Sine.		Cotang.		Tang.	N. cos.	N.sine.	'

′	Sine.	D. 10″	Cosine.	D. 10″	Tang.	D. 10″	Cotang.	N. sine.	N. cos.	′
0	9.724210	33.7	9.928420	13.2	9.795789	46.8	10.204211	52992	84805	60
1	724412	33.7	928342	13.2	796070	46.8	203930	53017	84789	59
2	724614	33.6	928263	13.2	796351	46.8	203649	53041	84774	58
3	724816	33.6	928183	13.2	796632	46.8	203368	53066	84759	57
4	725017	33.6	928104	13.2	796913	46.8	203087	53091	84743	56
5	725219	33.6	928025	13.2	797194	46.8	202806	53115	84728	55
6	725420	33.6	927946	13.2	797475	46.8	202525	53140	84712	54
7	725622	33.5	927867	13.2	797755	46.8	202245	53164	84697	53
8	725823	33.5	927787	13.2	798036	46.8	201964	53189	84681	52
9	726024	33.5	927708	13.2	798316	46.7	201684	53214	84666	51
10	726225	33.5	927629	13.2	798596	46.7	201404	53238	84650	50
11	9.726426	33.5	9.927549	13.2	9.798877	46.7	10.201123	53263	84635	49
12	726626	33.4	927470	13.3	799157	46.7	200843	53288	84619	48
13	726827	33.4	927390	13.3	799437	46.7	200563	53312	84604	47
14	727027	33.4	927310	13.3	799717	46.7	200283	53337	84588	46
15	727228	33.4	927231	13.3	799997	46.7	200003	53361	84573	45
16	727428	33.4	927151	13.3	800277	46.6	199723	53386	84557	44
17	727628	33.3	927071	13.3	800557	46.6	199443	53411	84542	43
18	727828	33.3	926991	13.3	800836	46.6	199164	53435	84526	42
19	728027	33.3	926911	13.3	801116	46.6	198884	53460	84511	41
20	728227	33.3	926831	13.3	801396	46.6	198604	53484	84495	40
21	9.728427	33.3	9.926751	13.3	9.801675	46.6	10.198325	53509	84480	39
22	728626	33.2	926671	13.3	801955	46.6	198045	53534	84464	38
23	728825	33.2	926591	13.3	802234	46.5	197766	53558	84448	37
24	729024	33.2	926511	13.4	802513	46.5	197487	53583	84433	36
25	729223	33.2	926431	13.4	802792	46.5	197208	53607	84417	35
26	729422	33.1	926351	13.4	803072	46.5	196928	53632	84402	34
27	729621	33.1	926270	13.4	803351	46.5	196649	53656	84386	33
28	729820	33.1	926190	13.4	803630	46.5	196370	53681	84370	32
29	730018	33.1	926110	13.4	803908	46.5	196092	53705	84355	31
30	730216	33.0	926029	13.4	804187	46.5	195813	53730	84339	30
31	9.730415	33.0	9.925949	13.4	9.804466	46.5	10.195534	53754	84324	29
32	730613	33.0	925868	13.4	804745	46.4	195255	53779	84308	28
33	730811	33.0	925788	13.4	805023	46.4	194977	53804	84292	27
34	731009	33.0	925707	13.4	805302	46.4	194698	53828	84277	26
35	731206	32.9	925626	13.4	805580	46.4	194420	53853	84261	25
36	731404	32.9	925545	13.5	805859	46.4	194141	53877	84245	24
37	731602	32.9	925465	13.5	806137	46.4	193863	53902	84230	23
38	731799	32.9	925384	13.5	806415	46.4	193585	53926	84214	22
39	731996	32.9	925303	13.5	806693	46.3	193307	53951	84198	21
40	732193	32.8	925222	13.5	806971	46.3	193029	53975	84182	20
41	9.732390	32.8	9.925141	13.5	9.807249	46.3	10.192751	54000	84167	19
42	732587	32.8	925060	13.5	807527	46.3	192473	54024	84151	18
43	732784	32.8	924979	13.5	807805	46.3	192195	54049	84135	17
44	732980	32.8	924897	13.5	808083	46.3	191917	54073	84120	16
45	733177	32.7	924816	13.5	808361	46.3	191639	54097	84104	15
46	733373	32.7	924735	13.6	808638	46.2	191362	54122	84088	14
47	733569	32.7	924654	13.6	808916	46.2	191084	54146	84072	13
48	733765	32.7	924572	13.6	809193	46.2	190807	54171	84057	12
49	733961	32.7	924491	13.6	809471	46.2	190529	54195	84041	11
50	734157	32.6	924409	13.6	809748	46.2	190252	54220	84025	10
51	9.734353	32.6	9.924328	13.6	9.810025	46.2	10.189975	54244	84009	9
52	734549	32.6	924246	13.6	810302	46.2	189698	54269	83994	8
53	734744	32.6	924164	13.6	810580	46.2	189420	54293	83978	7
54	734939	32.5	924083	13.6	810857	46.2	189143	54317	83962	6
55	735135	32.5	924001	13.6	811134	46.2	188866	54342	83946	5
56	735330	32.5	923919	13.6	811410	46.1	188590	54366	83930	4
57	735525	32.5	923837	13.6	811687	46.1	188313	54391	83915	3
58	735719	32.5	923755	13.7	811964	46.1	188036	54415	83899	2
59	735914	32.4	923673	13.7	812241	46.1	187759	54440	83883	1
60	736109	32.4	923591	13.7	812517	46.1	187483	54464	83867	0
	Cosine.		Sine.		Cotang.		Tang.	N. cos.	N. sine.	′

′	Sine.	D. 10″	Cosine.	D. 10″	Tang.	D. 10″	Cotang.	N. sine.	N. cos.	
0	9.736109	32.4	9.923591	13.7	9.812517	46.1	10.187482	54464	83867	60
1	736303	32.4	923509	13.7	812794	46.1	187206	54488	83851	59
2	736498	32.4	923427	13.7	813070	46.1	186930	54513	83835	58
3	736692	32.3	923345	13.7	813347	46.0	186653	54537	83819	57
4	736886	32.3	923263	13.7	813623	46.0	186377	54561	83804	56
5	737080	32.3	923181	13.7	813899	46.0	186101	54586	83788	55
6	737274	32.3	923098	13.7	814175	46.0	185825	54610	83772	54
7	737467	32.3	923016	13.7	814452	46.0	185548	54635	83756	53
8	737661	32.2	922933	13.7	814728	46.0	185272	54659	83740	52
9	737855	32.2	922851	13.7	815004	46.0	184996	54683	83724	51
10	738048	32.2	922768	13.7	815279	46.0	184721	54708	83708	50
11	9.738241	32.2	9.922686	13.8	9.815555	46.0	10.184445	54732	83692	49
12	738434	32.2	922603	13.8	815831	45.9	184169	54756	83676	48
13	738627	32.1	922520	13.8	816107	45.9	183893	54781	83660	47
14	738820	32.1	922438	13.8	816382	45.9	183618	54805	83645	46
15	739013	32.1	922355	13.8	816658	45.9	183342	54829	83629	45
16	739206	32.1	922272	13.8	816933	45.9	183067	54854	83613	44
17	739398	32.1	922189	13.8	817209	45.9	182791	54878	83597	43
18	739590	32.1	922106	13.8	817484	45.9	182516	54902	83581	42
19	739783	32.0	922023	13.8	817759	45.9	182241	54927	83565	41
20	739975	32.0	921940	13.8	818035	45.9	181965	54951	83549	40
21	9.740167	32.0	9.921857	13.8	9.818310	45.8	10.181690	54975	83533	39
22	740359	32.0	921774	13.9	818585	45.8	181415	54999	83517	38
23	740550	32.0	921691	13.9	818860	45.8	181140	55024	83501	37
24	740742	31.9	921607	13.9	819135	45.8	180865	55048	83485	36
25	740934	31.9	921524	13.9	819410	45.8	180590	55072	83469	35
26	741125	31.9	921441	13.9	819684	45.8	180316	55097	83453	34
27	741316	31.9	921357	13.9	819959	45.8	180041	55121	83437	33
28	741508	31.9	921274	13.9	820234	45.8	179766	55145	83421	32
29	741699	31.8	921190	13.9	820508	45.8	179492	55169	83405	31
30	741889	31.8	921107	13.9	820783	45.7	179217	55194	83389	30
31	9.742080	31.8	9.921023	13.9	9.821057	45.7	10.178943	55218	83373	29
32	742271	31.8	920939	13.9	821332	45.7	178668	55242	83356	28
33	742462	31.8	920856	14.0	821606	45.7	178394	55266	83340	27
34	742652	31.7	920772	14.0	821880	45.7	178120	55291	83324	26
35	742842	31.7	920688	14.0	822154	45.7	177846	55315	83308	25
36	743033	31.7	920604	14.0	822429	45.7	177571	55339	83292	24
37	743223	31.7	920520	14.0	822703	45.7	177297	55363	83276	23
38	743413	31.7	920436	14.0	822977	45.7	177023	55388	83260	22
39	743602	31.6	920352	14.0	823250	45.6	176750	55412	83244	21
40	743792	31.6	920268	14.0	823524	45.6	176476	55436	83228	20
41	9.743982	31.6	9.920184	14.0	9.823798	45.6	10.176202	55460	83212	19
42	744171	31.6	920099	14.0	824072	45.6	175928	55484	83195	18
43	744361	31.6	920015	14.0	824345	45.6	175655	55509	83179	17
44	744550	31.5	919931	14.0	824619	45.6	175381	55533	83163	16
45	744739	31.5	919846	14.1	824893	45.6	175107	55557	83147	15
46	744928	31.5	919762	14.1	825166	45.6	174834	55581	83131	14
47	745117	31.5	919677	14.1	825439	45.6	174561	55605	83115	13
48	745306	31.5	919593	14.1	825713	45.5	174287	55630	83098	12
49	745494	31.4	919508	14.1	825986	45.5	174014	55654	83082	11
50	745683	31.4	919424	14.1	826259	45.5	173741	55678	83066	10
51	9.745871	31.4	9.919339	14.1	9.826532	45.5	10.173468	55702	83050	9
52	746059	31.4	919254	14.1	826805	45.5	173195	55726	83034	8
53	746248	31.4	919169	14.1	827078	45.5	172922	55750	83017	7
54	746436	31.3	919085	14.1	827351	45.5	172649	55775	83001	6
55	746624	31.3	919000	14.1	827624	45.5	172376	55799	82985	5
56	746812	31.3	918915	14.1	827897	45.5	172103	55823	82969	4
57	746999	31.3	918830	14.2	828170	45.4	171830	55847	82953	3
58	747187	31.3	918745	14.2	828442	45.4	171558	55871	82936	2
59	747374	31.2	918659	14.2	828715	45.4	171285	55895	82920	1
60	747562	31.2	918574	14.2	828987	45.4	171013	55919	82904	0
	Cosine.		Sine.		Cotang.		Tang.	N. cos.	N. sine.	′

56 Degrees.

′	Sine.	D. 10″	Cosine.	D. 10″	Tang.	D. 10″	Cotang.	N.sine	N. cos.	
0	9.747562	31.2	9.918574	14.2	9.828987	45.4	10.171013	55919	82904	60
1	747749	31.2	918489	14.2	829260	45.4	170740	55943	82887	59
2	747936	31.2	918404	14.2	829532	45.4	170468	55968	82871	58
3	748123	31.2	918318	14.2	829805	45.4	170195	55992	82855	57
4	748310	31.1	918233	14.2	830077	45.4	169923	56016	82839	56
5	748497	31.1	918147	14.2	830349	45.4	169651	56040	82822	55
6	748683	31.1	918062	14.2	830621	45.3	169379	56064	82806	54
7	748870	31.1	917976	14.2	830893	45.3	169107	56088	82790	53
8	749056	31.1	917891	14.3	831165	45.3	168835	56112	82773	52
9	749243	31.1	917805	14.3	831437	45.3	168563	56136	82757	51
10	749426	31.0	917719	14.3	831709	45.3	168291	56160	82741	50
11	9.749615	31.0	9.917634	14.3	9.831981	45.3	10.168019	56184	82724	49
12	749801	31.0	917548	14.3	832253	45.3	167747	56208	82708	48
13	749987	31.0	917462	14.3	832525	45.3	167475	56232	82692	47
14	750172	31.0	917376	14.3	832796	45.3	167204	56256	82675	46
15	750358	30.9	917290	14.3	833068	45.3	166932	56280	82659	45
16	750543	30.9	917204	14.3	833339	45.2	166661	56305	82643	44
17	750729	30.9	917118	14.3	833611	45.2	166389	56329	82626	43
18	750914	30.9	917032	14.4	833882	45.2	166118	56353	82610	42
19	751099	30.9	916946	14.4	834154	45.2	165846	56377	82593	41
20	751284	30.9	916859	14.4	834425	45.2	165575	56401	82577	40
21	9.751469	30.8	9.916773	14.4	9.834696	45.2	10.165304	56425	82561	39
22	751654	30.8	916687	14.4	834967	45.2	165033	56449	82544	38
23	751839	30.8	916600	14.4	835238	45.2	164762	56473	82528	37
24	752023	30.8	916514	14.4	835509	45.2	164491	56497	82511	36
25	752208	30.8	916427	14.4	835780	45.2	164220	56521	82495	35
26	752392	30.7	916341	14.4	836051	45.1	163949	56545	82478	34
27	752576	30.7	916254	14.4	836322	45.1	163678	56569	82462	33
28	752760	30.7	916167	14.4	836593	45.1	163407	56593	82446	32
29	752944	30.7	916081	14.5	836864	45.1	163136	56617	82429	31
30	753128	30.7	915994	14.5	837134	45.1	162866	56641	82413	30
31	9.753312	30.6	9.915907	14.5	9.837405	45.1	10.162595	56665	82396	29
32	753495	30.6	915820	14.5	837675	45.1	162325	56689	82380	28
33	753679	30.6	915733	14.5	837946	45.1	162054	56713	82363	27
34	753862	30.6	915646	14.5	838216	45.1	161784	56736	82347	26
35	754046	30.6	915559	14.5	838487	45.1	161513	56760	82330	25
36	754229	30.5	915472	14.5	838757	45.0	161243	56784	82314	24
37	754412	30.5	915385	14.5	839027	45.0	160973	56808	82297	23
38	754595	30.5	915297	14.5	839297	45.0	160703	56852	82281	22
39	754778	30.5	915210	14.5	839568	45.0	160432	56856	82264	21
40	754960	30.5	915123	14.6	839838	45.0	160162	56880	82248	20
41	9.755143	30.4	9.915035	14.6	9.840108	45.0	10.159892	56904	82231	19
42	755326	30.4	914948	14.6	840378	45.0	159622	56928	82214	18
43	755508	30.4	914860	14.6	840647	45.0	159353	56952	82198	17
44	755690	30.4	914773	14.6	840917	45.0	159083	56976	82181	16
45	755872	30.4	914685	14.6	841187	44.9	158813	57000	82165	15
46	756054	30.3	914598	14.6	841457	44.9	158543	57024	82148	14
47	756236	30.3	914510	14.6	841726	44.9	158274	57047	82132	13
48	756418	30.3	914422	14.6	841996	44.9	158004	57071	82115	12
49	756600	30.3	914334	14.6	842266	44.9	157734	57095	82098	11
50	756782	30.3	914246	14.7	842535	44.9	157465	57119	82082	10
51	9.756963	30.2	9.914158	14.7	9.842805	44.9	10.157195	57143	82065	9
52	757144	30.2	914070	14.7	843074	44.9	156926	57167	82048	8
53	757326	30.2	913982	14.7	843343	44.9	156657	57191	82032	7
54	757507	30.2	913894	14.7	843612	44.9	156388	57215	82015	6
55	757688	30.2	913806	14.7	843882	44.9	156118	57238	81999	5
56	757869	30.1	913718	14.7	844151	44.8	155849	57262	81982	4
57	758050	30.1	913630	14.7	844420	44.8	155580	57286	81965	3
58	758230	30.1	913541	14.7	844689	44.8	155311	57310	81949	2
59	758411	30.1	913453	14.7	844958	44.8	155042	57334	81932	1
60	758591	30.1	913365		845227	44.8	154773	57358	81915	0
	Cosine.		Sine.		Cotang.		Tang.	N. cos.	N sine.	′

'	Sine.	D. 10'	Cosine.	D. 10"	Tang.	D. 10"	Cotang.	N. sine.	N. cos.	
0	9.758591	30.1	9.913365	14.7	9.845227	44.8	10.154773	57358	81915	60
1	758772	30.0	913276	14.7	845496	44.8	154504	57381	81899	59
2	758952	30.0	913187	14.8	845764	44.8	154236	57405	81882	58
3	759132	30.0	913099	14.8	846033	44.8	153967	57429	81865	57
4	759312	30.0	913010	14.8	846302	44.8	153698	57453	81848	56
5	759492	30.0	912922	14.8	846570	44.8	153430	57477	81832	55
6	759672	30.0	912833	14.8	846839	44.7	153161	57501	81815	54
7	759852	29.9	912744	14.8	847107	44.7	152893	57524	81798	53
8	760031	29.9	912655	14.8	847376	44.7	152624	57548	81782	52
9	760211	29.9	912566	14.8	847644	44.7	152356	57572	81765	51
10	760390	29.9	912477	14.8	847913	44.7	152087	57596	81748	50
11	9.760569	29.9	9.912388	14.8	9.848181	44.7	10.151819	57619	81731	49
12	760748	29.8	912299	14.8	848449	44.7	151551	57643	81714	48
13	760927	29.8	912210	14.9	848717	44.7	151283	57667	81698	47
14	761106	29.8	912121	14.9	848986	44.7	151014	57691	81681	46
15	761285	29.8	912031	14.9	849254	44.7	150746	57715	81664	45
16	761464	29.8	911942	14.9	849522	44.7	150478	57738	81647	44
17	761642	29.8	911853	14.9	849790	44.7	150210	57762	81631	43
18	761821	29.7	911763	14.9	850058	44.6	149942	57786	81614	42
19	761999	29.7	911674	14.9	850325	44.6	149675	57810	81597	41
20	762177	29.7	911584	14.9	850593	44.6	149407	57833	81580	40
21	9.762356	29.7	9.911495	14.9	9.850861	44.6	10.149139	57857	81563	39
22	762534	29.7	911405	14.9	851129	44.6	148871	57881	81546	38
23	762712	29.6	911315	14.9	851396	44.6	148604	57904	81530	37
24	762889	29.6	911226	15.0	851664	44.6	148336	57928	81513	36
25	763067	29.6	911136	15.0	851931	44.6	148069	57952	81496	35
26	763245	29.6	911046	15.0	852199	44.6	147801	57976	81479	34
27	763422	29.6	910956	15.0	852466	44.6	147534	57999	81462	33
28	763600	29.6	910866	15.0	852733	44.6	147267	58023	81445	32
29	763777	29.5	910776	15.0	853001	44.5	146999	58047	81428	31
30	763954	29.5	910686	15.0	853268	44.5	146732	58070	81412	30
31	9.764131	29.5	9.910596	15.0	9.853535	44.5	10.146465	58094	81395	29
32	764308	29.5	910506	15.0	853802	44.5	146198	58118	81378	28
33	764485	29.5	910415	15.0	854069	44.5	145931	58141	81361	27
34	764662	29.4	910325	15.0	854336	44.5	145664	58165	81344	26
35	764838	29.4	910235	15.1	854603	44.5	145397	58189	81327	25
36	765015	29.4	910144	15.1	854870	44.5	145130	58212	81310	24
37	765191	29.4	910054	15.1	855137	44.5	144863	58236	81293	23
38	765367	29.4	909963	15.1	855404	44.5	144596	58260	81276	22
39	765544	29.4	909873	15.1	855671	44.5	144329	58283	81259	21
40	765720	29.3	909782	15.1	855938	44.4	144062	58307	81242	20
41	9.765896	29.3	9.909691	15.1	9.856204	44.4	10.143796	58330	81225	19
42	766072	29.3	909601	15.1	856471	44.4	143529	58354	81208	18
43	766247	29.3	909510	15.1	856737	44.4	143263	58378	81191	17
44	766423	29.3	909419	15.1	857004	44.4	142996	58401	81174	16
45	766598	29.3	909328	15.1	857270	44.4	142730	58425	81157	15
46	766774	29.2	909237	15.2	857537	44.4	142463	58449	81140	14
47	766949	29.2	909146	15.2	857803	44.4	142197	58472	81123	13
48	767124	29.2	909055	15.2	858069	44.4	141931	58496	81106	12
49	767300	29.2	908964	15.2	858336	44.4	141664	58519	81089	11
50	767475	29.2	908873	15.2	858602	44.4	141398	58543	81072	10
51	9.767649	29.1	9.908781	15.2	9.858868	44.3	10.141132	58567	81055	9
52	767824	29.1	908690	15.2	859134	44.3	140866	58590	81038	8
53	767999	29.1	908599	15.2	859400	44.3	140600	58614	81021	7
54	768173	29.1	908507	15.2	859666	44.3	140334	58637	81004	6
55	768348	29.1	908416	15.2	859932	44.3	140068	58661	80987	5
56	768522	29.0	908324	15.3	860198	44.3	139802	58684	80970	4
57	768697	29.0	908233	15.3	860464	44.3	139536	58708	80953	3
58	768871	29.0	908141	15.3	860730	44.3	139270	58731	80936	2
59	769045	29.0	908049	15.3	860995	44.3	139005	58755	80919	1
60	769219	29.0	907958	15.3	861261	44.3	138739	58779	80902	0
	Cosine.		Sine.		Cotang.		Tang.	N. cos.	N. sine.	'

′	Sine.	D. 10″	Cosine.	D. 10″	Tang.	D. 10″	Cotang.	N. sine.	N. cos.	
0	9.769219	29.0	9.907958	15.3	9.861261	44.3	10.138739	58779	80902	60
1	769393	28.9	907866	15.3	861527	44.3	138473	58802	80885	59
2	769566	28.9	907774	15.3	861792	44.2	138208	58826	80867	58
3	769740	28.9	907682	15.3	862038	44.2	137942	58849	80850	57
4	769913	28.9	907590	15.3	862323	44.2	137677	58873	80833	56
5	770087	28.9	907498	15.3	862589	44.2	137411	58896	80816	55
6	770260	28.9	907406	15.3	862854	44.2	137146	58920	80799	54
7	770433	28.8	907314	15.3	863119	44.2	136881	58943	80782	53
8	770606	28.8	907222	15.4	863385	44.2	136615	58967	80765	52
9	770779	28.8	907129	15.4	863650	44.2	136350	58990	80748	51
10	770952	28.8	907037	15.4	863915	44.2	136085	59014	80730	50
11	9.771125	28.8	9.906945	15.4	9.864180	44.2	10.135820	59037	80713	49
12	771298	28.7	906852	15.4	864445	44.2	135555	59061	80696	48
13	771470	28.7	906760	15.4	864710	44.2	135290	59084	80679	47
14	771643	28.7	906667	15.4	864975	44.2	135025	59108	80662	46
15	771815	28.7	906575	15.4	865240	44.1	134760	59131	80644	45
16	771987	28.7	906482	15.4	865505	44.1	134495	59154	80627	44
17	772159	28.7	906389	15.5	865770	44.1	134230	59178	80610	43
18	772331	28.6	906296	15.5	866035	44.1	133965	59201	80593	42
19	772503	28.6	906204	15.5	866300	44.1	133700	59225	80576	41
20	772675	28.6	906111	15.5	866564	44.1	133436	59248	80558	40
21	9.772847	28.6	9.906018	15.5	9.866829	44.1	10.133171	59272	80541	39
22	773018	28.6	905925	15.5	867094	44.1	132906	59295	80524	38
23	773190	28.6	905832	15.5	867358	44.1	132642	59318	80507	37
24	773361	28.5	905739	15.5	867623	44.1	132377	59342	80489	36
25	773533	28.5	905645	15.5	867887	44.1	132113	59365	80472	35
26	773704	28.5	905552	15.5	868152	44.0	131848	59389	80455	34
27	773875	28.5	905459	15.5	868416	44.0	131584	59412	80438	33
28	774046	28.5	905366	15.6	868680	44.0	131320	59436	80422	32
29	774217	28.5	905272	15.6	868945	44.0	131055	59459	80403	31
30	774388	28.4	905179	15.6	869209	44.0	130791	59482	80386	30
31	9.774558	28.4	9.905085	15.6	9.869473	44.0	10.130527	59506	80368	29
32	774729	28.4	904992	15.6	869737	44.0	130263	59529	80351	28
33	774899	28.4	904898	15.6	870001	44.0	129999	59552	80334	27
34	775070	28.4	904804	15.6	870265	44.0	129735	59576	80316	26
35	775240	28.4	904711	15.6	870529	44.0	129471	59599	80299	25
36	775410	28.3	904617	15.6	870793	44.0	129207	59622	80282	24
37	775580	28.3	904523	15.6	871057	44.0	128943	59646	80264	23
38	775750	28.3	904429	15.7	871321	44.0	128679	59669	80247	22
39	775920	28.3	904335	15.7	871585	44.0	128415	59693	80230	21
40	776090	28.3	904241	15.7	871849	43.9	128151	59716	80212	20
41	9.776259	28.3	9.904147	15.7	9.872112	43.9	10.127888	59739	80195	19
42	776429	28.2	904053	15.7	872376	43.9	127624	59763	80178	18
43	776598	28.2	903959	15.7	872640	43.9	127360	59786	80160	17
44	776768	28.2	903864	15.7	872903	43.9	127097	59809	80143	16
45	776937	28.2	903770	15.7	873167	43.9	126833	59832	80125	15
46	777106	28.2	903676	15.7	873430	43.9	126570	59856	80108	14
47	777275	28.1	903581	15.7	873694	43.9	126306	59879	80091	13
48	777444	28.1	903487	15.7	873957	43.9	126043	59902	80073	12
49	777613	28.1	903392	15.8	874220	43.9	125780	59926	80056	11
50	777781	28.1	903298	15.8	874484	43.9	125516	59949	80038	10
51	9.777950	28.1	9.903203	15.8	9.874747	43.9	10.125253	59972	80021	9
52	778119	28.1	903108	15.8	875010	43.9	124990	59995	80003	8
53	778287	28.0	903014	15.8	875273	43.9	124727	60019	79986	7
54	778455	28.0	902919	15.8	875536	43.8	124464	60042	79968	6
55	778624	28.0	902824	15.8	875800	43.8	124200	60065	79951	5
56	778792	28.0	902729	15.8	876063	43.8	123937	60089	79934	4
57	778960	28.0	902634	15.8	876326	43.8	123674	60112	79916	3
58	779128	28.0	902539	15.9	876589	43.8	123411	60135	79899	2
59	779295	27.9	902444	15.9	876851	43.8	123149	60158	79881	1
60	779463		902349		877114		122886	60182	79864	0
	Cosine.		Sine.		Cotang.		Tang.	N. cos.	N. sine.	′

53 Degrees.

′	Sine.	D. 10″	Cosine.	D. 10″	Tang.	D. 10″	Cotang.	N.sine.	N. cos.	
0	9.779463		9.902349		9.877114		10.122886	60182	79864	60
1	779631	27.9	902253	15.9	877377	43.8	122623	60205	79846	59
2	779798	27.9	902158	15.9	877640	43.8	122360	60228	79829	58
3	779966	27.9	902063	15.9	877903	43.8	122097	60251	79811	57
4	780133	27.9	901967	15.9	878165	43.8	121835	60274	79793	56
5	780300	27.9	901872	15.9	878428	43.8	121572	60298	79776	55
6	780467	27.8	901776	15.9	878691	43.8	121309	60321	79758	54
7	780634	27.8	901681	15.9	878953	43.8	121047	60344	79741	53
8	780801	27.8	901585	15.9	879216	43.7	120784	60367	79723	52
9	780968	27.8	901490	15.9	879478	43.7	120522	60390	79706	51
10	781134	27.8	901394	15.9	879741	43.7	120259	60414	79688	50
11	9.781301	27.8	9.901298	16.0	9.880003	43.7	10.119997	60437	79671	49
12	781468	27.7	901202	16.0	880265	43.7	119735	60460	79658	48
13	781634	27.7	901106	16.0	880528	43.7	119472	60483	79635	47
14	781800	27.7	901010	16.0	880790	43.7	119210	60506	79618	46
15	781966	27.7	900914	16.0	881052	43.7	118948	60529	79600	45
16	782132	27.7	900818	16.0	881314	43.7	118686	60553	79583	44
17	782298	27.7	900722	16.0	881576	43.7	118424	60576	79565	43
18	782464	27.6	900626	16.0	881839	43.7	118161	60599	79547	42
19	782630	27.6	900529	16.0	882101	43.7	117899	60622	79530	41
20	782796	27.6	900433	16.0	882363	43.7	117637	60645	79512	40
21	9.782961	27.6	9.900337	16.1	9.882625	43.6	10.117375	60668	79494	39
22	783127	27.6	900242	16.1	882887	43.6	117113	60691	79477	38
23	783292	27.6	900144	16.1	883148	43.6	116852	60714	79459	37
24	783458	27.5	900047	16.1	883410	43.6	116590	60738	79441	36
25	783623	27.5	899951	16.1	883672	43.6	116328	60761	79424	35
26	783788	27.5	899854	16.1	883934	43.6	116066	60784	79406	34
27	783953	27.5	899757	16.1	884196	43.6	115804	60807	79388	33
28	784118	27.5	899660	16.1	884457	43.6	115543	60830	79371	32
29	784282	27.5	899564	16.1	884719	43.6	115281	60853	79353	31
30	784447	27.4	899467	16.1	884980	43.6	115020	60876	79335	30
31	9.784612	27.4	9.899370	16.2	9.885242	43.6	10.114758	60899	79318	29
32	784776	27.4	899273	16.2	885503	43.6	114497	60922	79300	28
33	784941	27.4	899176	16.2	885765	43.6	114235	60945	79282	27
34	785105	27.4	899078	16.2	886026	43.6	113974	60968	79264	26
35	785269	27.4	898981	16.2	886288	43.6	113712	60991	79247	25
36	785433	27.3	898884	16.2	886549	43.6	113451	61015	79229	24
37	785597	27.3	898787	16.2	886810	43.5	113190	61038	79211	23
38	785761	27.3	898689	16.2	887072	43.5	112928	61061	79193	22
39	785925	27.3	898592	16.2	887333	43.5	112667	61084	79176	21
40	786089	27.3	898494	16.2	887594	43.5	112406	61107	79158	20
41	9.786252	27.3	9.898397	16.3	9.887855	43.5	10.112145	61130	79140	19
42	786416	27.2	898299	16.3	888116	43.5	111884	61153	79122	18
43	786579	27.2	898202	16.3	888377	43.5	111623	61176	79105	17
44	786742	27.2	898104	16.3	888639	43.5	111361	61199	79087	16
45	786906	27.2	898006	16.3	888900	43.5	111100	61222	79069	15
46	787069	27.2	897908	16.3	889160	43.5	110840	61245	79051	14
47	787232	27.2	897810	16.3	889421	43.5	110579	61268	79033	13
48	787395	27.1	897712	16.3	889682	43.5	110318	61291	79016	12
49	787557	27.1	897614	16.3	889943	43.5	110057	61314	78998	11
50	787720	27.1	897516	16.3	890204	43.5	109796	61337	78980	10
51	9.787883	27.1	9.897418	16.3	9.890465	43.4	10.109535	61360	78962	9
52	788045	27.1	897320	16.4	890725	43.4	109275	61383	78944	8
53	788208	27.1	897222	16.4	890986	43.4	109014	61406	78926	7
54	788370	27.1	897123	16.4	891247	43.4	108753	61429	78908	6
55	788532	27.0	897025	16.4	891507	43.4	108493	61451	78891	5
56	788694	27.0	896926	16.4	891768	43.4	108232	61474	78873	4
57	788856	27.0	896828	16.4	892028	43.4	107972	61497	78855	3
58	789018	27.0	896729	16.4	892289	43.4	107711	61520	78837	2
59	789180	27.0	896631	16.4	892549	43.4	107451	61543	78819	1
60	789342	27.0	896532	16.4	892810	43.4	107190	61566	78801	0
	Cosine.		Sine.		Cotang.		Tang.	N. cos.	N.sine.	′

52 Degrees.

′	Sine.	D. 10″	Cosine.	D. 10″	Tang.	D. 10″	Cotang.	N. sine.	N. cos.	
0	9.789342	26.9	9.896532	16.4	9.892810	43.4	10.107190	61566	78801	60
1	789504	26.9	896433	16.5	893070	43.4	106930	61589	78783	59
2	789665	26.9	896335	16.5	893331	43.4	106669	61612	78765	58
3	789827	26.9	896236	16.5	893591	43.4	106409	61635	78747	57
4	789988	26.9	896137	16.5	893851	43.4	106149	61658	78729	56
5	790149	26.9	896038	16.5	894111	43.4	105889	61681	78711	55
6	790310	26.9	895939	16.5	894371	43.4	105629	61704	78694	54
7	790471	26.8	895840	16.5	894632	43.4	105368	61726	78676	53
8	790632	26.8	895741	16.5	894892	43.3	105108	61749	78658	52
9	790793	26.8	895641	16.5	895152	43.3	104848	61772	78640	51
10	790954	26.8	895542	16.5	895412	43.3	104588	61795	78622	50
11	9.791115	26.8	9.895443	16.6	9.895672	43.3	10.104328	61818	78604	49
12	791275	26.7	895343	16.6	895932	43.3	104068	61841	78586	48
13	791436	26.7	895244	16.6	896192	43.3	103808	61864	78568	47
14	791596	26.7	895145	16.6	896452	43.3	103548	61887	78550	46
15	791757	26.7	895045	16.6	896712	43.3	103288	61909	78532	45
16	791917	26.7	894945	16.6	896971	43.3	103029	61932	78514	44
17	792077	26.7	894846	16.6	897231	43.3	102769	61955	78496	43
18	792237	26.7	894746	16.6	897491	43.3	102509	61978	78478	42
19	792397	26.6	894646	16.6	897751	43.3	102249	62001	78460	41
20	792557	26.6	894546	16.6	898010	43.3	101990	62024	78442	40
21	9.792716	26.6	9.894446	16.6	9.898270	43.3	10.101730	62046	78424	39
22	792876	26.6	894346	16.7	898530	43.3	101470	62069	78405	38
23	793035	26.6	894246	16.7	898789	43.3	101211	62092	78387	37
24	793195	26.5	894146	16.7	899049	43.3	100951	62115	78369	36
25	793354	26.5	894046	16.7	899308	43.2	100692	62138	78351	35
26	793514	26.5	893946	16.7	899568	43.2	100432	62160	78333	34
27	793673	26.5	893846	16.7	899827	43.2	100173	62183	78315	33
28	793832	26.5	893745	16.7	900086	43.2	099914	62206	78297	32
29	793991	26.5	893645	16.7	900346	43.2	099654	62229	78279	31
30	794150	26.5	893544	16.7	900605	43.2	099395	62251	78261	30
31	9.794308	26.4	9.893444	16.7	9.900864	43.2	10.099136	62274	78243	29
32	794467	26.4	893343	16.8	901124	43.2	098876	62297	78225	28
33	794626	26.4	893243	16.8	901383	43.2	098617	62320	78206	27
34	794784	26.4	893142	16.8	901642	43.2	098358	62342	78188	26
35	794942	26.4	893041	16.8	901901	43.2	098099	62365	78170	25
36	795101	26.4	892940	16.8	902160	43.2	097840	62388	78152	24
37	795259	26.3	892839	16.8	902419	43.2	097581	62411	78134	23
38	795417	26.3	892739	16.8	902679	43.2	097321	62433	78116	22
39	795575	26.3	892638	16.8	902938	43.2	097062	62456	78098	21
40	795733	26.3	892536	16.8	903197	43.1	096803	62479	78079	20
41	9.795891	26.3	9.892435	16.9	9.903455	43.1	10.096545	62502	78061	19
42	796049	26.3	892334	16.9	903714	43.1	096286	62524	78043	18
43	796206	26.3	892233	16.9	903973	43.1	096027	62547	78025	17
44	796364	26.3	892132	16.9	904232	43.1	095768	62570	78007	16
45	796521	26.2	892030	16.9	904491	43.1	095509	62592	77988	15
46	796679	26.2	891929	16.9	904750	43.1	095250	62615	77970	14
47	796836	26.2	891827	16.9	905008	43.1	094992	62638	77952	13
48	796993	26.2	891726	16.9	905267	43.1	094733	62660	77934	12
49	797150	26.2	891624	16.9	905526	43.1	094474	62683	77916	11
50	797307	26.1	891523	16.9	905784	43.1	094216	62706	77897	10
51	9.797464	26.1	9.891421	17.0	9.906043	43.1	10.093957	62728	77879	9
52	797621	26.1	891319	17.0	906302	43.1	093698	62751	77861	8
53	797777	26.1	891217	17.0	906560	43.1	093440	62774	77843	7
54	797934	26.1	891115	17.0	906819	43.1	093181	62796	77824	6
55	798091	26.1	891013	17.0	907077	43.1	092923	62819	77806	5
56	798247	26.1	890911	17.0	907336	43.1	092664	62842	77788	4
57	798403	26.1	890809	17.0	907594	43.1	092406	62864	77769	3
58	798560	26.0	890707	17.0	907852	43.1	092148	62887	77751	2
59	798716	26.0	890605	17.0	908111	43.1	091889	62909	77733	1
60	798872	26.0	890503	17.0	908369	43.0	091631	62932	77715	0
	Cosine.		Sine.		Cotang.		Tang.	N. cos.	N. sine.	′

'	Sine.	D. 10"	Cosine.	D. 10"	Tang.	D. 10"	Cotang.	N. sine.	N. cos.	'
0	9.798772	26.0	9.890503	17.0	9.908369	43.0	10.091631	62932	77715	60
1	799028	26.0	890400	17.1	908628	43.0	091372	62955	77696	59
2	799184	26.0	890298	17.1	908886	43.0	091114	62977	77678	58
3	799339	25.9	890195	17.1	909144	43.0	090856	63000	77660	57
4	799495	25.9	890093	17.1	909402	43.0	090598	63022	77641	56
5	799651	25.9	889990	17.1	909660	43.0	090340	63045	77623	55
6	799806	25.9	889888	17.1	909918	43.0	090082	63058	77605	54
7	799962	25.9	889785	17.1	910177	43.0	089823	63090	77586	53
8	800117	25.9	889682	17.1	910435	43.0	089565	63113	77568	52
9	800272	25.8	889579	17.1	910693	43.0	089307	63135	77550	51
10	800427	25.8	889477	17.1	910951	43.0	089049	63158	77531	50
11	9.800582	25.8	9.889374	17.1	9.911209	43.0	10.088791	93180	77513	49
12	800737	25.8	889271	17.2	911467	43.0	088533	63203	77494	48
13	800892	25.8	889168	17.2	911724	43.0	088276	63225	77476	47
14	801047	25.8	889064	17.2	911982	43.0	088018	63248	77458	46
15	801201	25.8	888961	17.2	912240	43.0	087760	63271	77439	45
16	801356	25.7	888858	17.2	912498	43.0	087502	63293	77421	44
17	801511	25.7	888755	17.2	912756	43.0	087244	63316	77402	43
18	801665	25.7	888651	17.2	913014	43.0	086986	63338	77384	42
19	801819	25.7	888548	17.2	913271	42.9	086729	63361	77366	41
20	801973	25.7	888444	17.2	913529	42.9	086471	63383	77347	40
21	9.802128	25.7	9.888341	17.3	9.913787	42.9	10.086213	63406	77329	39
22	802282	25.6	888237	17.3	914044	42.9	085956	63428	77310	38
23	802436	25.6	888134	17.3	914302	42.9	085698	63451	77292	37
24	802589	25.6	888030	17.3	914560	42.9	085440	63473	77273	36
25	802743	25.6	887926	17.3	914817	42.9	085183	63496	77255	35
26	802897	25.6	887822	17.3	915075	42.9	084925	63518	77236	34
27	803050	25.6	887718	17.3	915332	42.9	084668	63540	77218	33
28	803204	25.5	887614	17.3	915590	42.9	084410	63563	77199	32
29	803357	25.5	887510	17.3	915847	42.9	084153	63585	77181	31
30	803511	25.5	887406	17.3	916104	42.9	083896	63608	77162	30
31	9.803664	25.5	9.887302	17.4	9.916362	42.9	10.083638	63630	77144	29
32	803817	25.5	887198	17.4	916619	42.9	083381	63653	77125	28
33	803970	25.5	887093	17.4	916877	42.9	083123	63675	77107	27
34	804123	25.5	886989	17.4	917134	42.9	082866	63698	77088	26
35	804276	25.5	886885	17.4	917391	42.9	082609	63720	77070	25
36	804428	25.4	886780	17.4	917648	42.9	082352	63742	77051	24
37	804581	25.4	886676	17.4	917905	42.9	082095	63765	77033	23
38	804734	25.4	886571	17.4	918163	42.9	081837	63787	77014	22
39	804886	25.4	886466	17.4	918420	42.8	081580	63810	76996	21
40	805039	25.4	886362	17.4	918677	42.8	081323	63832	76977	20
41	9.805191	25.4	9.886257	17.5	9.918934	42.8	10.081066	63854	76959	19
42	805343	25.4	886152	17.5	919191	42.8	080809	63877	76940	18
43	805495	25.3	886047	17.5	919448	42.8	080552	63899	76921	17
44	805647	25.3	885942	17.5	919705	42.8	080295	63922	76903	16
45	805799	25.3	885837	17.5	919962	42.8	080038	63944	76884	15
46	805951	25.3	885732	17.5	920219	42.8	079781	63966	76866	14
47	806103	25.3	885627	17.5	920476	42.8	079524	63989	76847	13
48	806254	25.3	885522	17.5	920733	42.8	079267	64011	76828	12
49	806406	25.3	885416	17.5	920990	42.8	079010	64033	76810	11
50	806557	25.2	885311	17.5	921247	42.8	078753	64056	76791	10
51	9.806709	25.2	9.885205	17.6	9.921503	42.8	10.078497	64078	76772	9
52	806860	25.2	885100	17.6	921760	42.8	078240	64100	76754	8
53	807011	25.2	884994	17.6	922017	42.8	077983	64123	76735	7
54	807163	25.2	884889	17.6	922274	42.8	077726	64145	76717	6
55	807314	25.2	884783	17.6	922530	42.8	077470	64167	76698	5
56	807465	25.2	884677	17.6	922787	42.8	077213	64190	76679	4
57	807615	25.1	884572	17.6	923044	42.8	076956	64212	76661	3
58	807766	25.1	884466	17.6	923300	42.8	076700	64234	76642	2
59	807917	25.1	884360	17.6	923557	42.7	076443	64256	76623	1
60	808067		884254		923813		076187	64279	76604	0
	Cosine.		Sine.		Cotang.		Tang.	N. cos.	N. sine.	'

′	Sine.	D. 10″	Cosine.	D. 10″	Tang.	D. 10″	Cotang.	N. sine.	N. cos.	′
0	9.808067	25.1	9.884254	17.7	9.923813	42.7	10.076187	64279	76604	60
1	808218	25.1	884148	17.7	924070	42.7	075930	64301	76586	59
2	808368	25.1	884042	17.7	924327	42.7	075673	64323	76567	58
3	808519	25.0	883936	17.7	924583	42.7	075417	64346	76548	57
4	808669	25.0	883829	17.7	924840	42.7	075160	64368	76530	56
5	808819	25.0	883723	17.7	925096	42.7	074904	64390	76511	55
6	808969	25.0	883617	17.7	925352	42.7	074648	64412	76492	54
7	809119	25.0	883510	17.7	925609	42.7	074391	64435	76473	53
8	809269	25.0	883404	17.7	925865	42.7	074135	64457	76455	52
9	809419	25.0	883297	17.7	926122	42.7	073878	64479	76436	51
10	809569	24.9	883191	17.8	926378	42.7	073622	64501	76417	50
11	9.809718	24.9	9.883084	17.8	9.926634	42.7	10.073366	64524	76398	49
12	809868	24.9	882977	17.8	926890	42.7	073110	64546	76380	48
13	810017	24.9	882871	17.8	927147	42.7	072853	64568	76361	47
14	810167	24.9	882764	17.8	927403	42.7	072597	64590	76342	46
15	810316	24.9	882657	17.8	927659	42.7	072341	64612	76323	45
16	810465	24.8	882550	17.8	927915	42.7	072085	64635	76304	44
17	810614	24.8	882443	17.8	928171	42.7	071829	64657	76286	43
18	810763	24.8	882336	17.8	928427	42.7	071573	64679	76267	42
19	810912	24.8	882229	17.9	928683	42.7	071317	64701	76248	41
20	811061	24.8	882121	17.9	928940	42.7	071060	64723	76229	40
21	9.811210	24.8	9.882014	17.9	9.929196	42.7	10.070804	64746	76210	39
22	811358	24.8	881907	17.9	929452	42.7	070548	64768	76192	38
23	811507	24.7	881799	17.9	929708	42.7	070292	64790	76173	37
24	811655	24.7	881692	17.9	929964	42.7	070036	64812	76154	36
25	811804	24.7	881584	17.9	930220	42.6	069780	64834	76135	35
26	811952	24.7	881477	17.9	930475	42.6	069525	64856	76116	34
27	812100	24.7	881369	17.9	930731	42.6	069269	64878	76097	33
28	812248	24.7	881261	18.0	930987	42.6	069013	64901	76078	32
29	812396	24.7	881153	18.0	931243	42.6	068757	64923	76059	31
30	812544	24.6	881046	18.0	931499	42.6	068501	64945	76041	30
31	9.812692	24.6	9.880938	18.0	9.931755	42.6	10.068245	64967	76022	29
32	812840	24.6	880830	18.0	932010	42.6	067990	64989	76003	28
33	812988	24.6	880722	18.0	932266	42.6	067734	65011	75984	27
34	813135	24.6	880613	18.0	932522	42.6	067478	65033	75965	26
35	813283	24.6	880505	18.0	932778	42.6	067222	65055	75946	25
36	813430	24.6	880397	18.0	933033	42.6	066967	65077	75927	24
37	813578	24.5	880289	18.0	933289	42.6	066711	65100	75908	23
38	813725	24.5	880180	18.1	933545	42.6	066455	65122	75889	22
39	813872	24.5	880072	18.1	933800	42.6	066200	65144	75870	21
40	814019	24.5	879963	18.1	934056	42.6	065944	65166	75851	20
41	9.814166	24.5	9.879855	18.1	9.934311	42.6	10.065689	65188	75832	19
42	814313	24.5	879746	18.1	934567	42.6	065433	65210	75813	18
43	814460	24.5	879637	18.1	934823	42.6	065177	65232	75794	17
44	814607	24.4	879529	18.1	935078	42.6	064922	65254	75775	16
45	814753	24.4	879420	18.1	935333	42.6	064667	65276	75756	15
46	814900	24.4	879311	18.1	935589	42.6	064411	65298	75738	14
47	815046	24.4	879202	18.1	935844	42.6	064156	65320	75719	13
48	815193	24.4	879093	18.2	936100	42.6	063900	65342	75700	12
49	815339	24.4	878984	18.2	936355	42.6	063645	65364	75680	11
50	815485	24.4	878875	18.2	936610	42.6	063390	65386	75661	10
51	9.815631	24.3	9.878766	18.2	9.936866	42.6	10.063134	65408	75642	9
52	815778	24.3	878656	18.2	937121	42.5	062879	65430	75623	8
53	815924	24.3	878547	18.2	937376	42.5	062624	65452	75604	7
54	816069	24.3	878438	18.2	937632	42.5	062368	65474	75585	6
55	816215	24.3	878328	18.2	937887	42.5	062113	65496	75566	5
56	816361	24.3	878219	18.2	938142	42.5	061858	65518	75547	4
57	816507	24.3	878109	18.3	938398	42.5	061602	65540	75528	3
58	816652	24.2	877999	18.3	938653	42.5	061347	65562	75509	2
59	816798	24.2	877890	18.3	938908	42.5	061092	65584	75490	1
60	816943	24.2	877780	18.3	939163	42.5	060837	65606	75471	0
	Cosine.		Sine.		Cotang.		Tang.	N. cos.	N. sine.	′

′	Sine.	D. 10″	Cosine.	D. 10″	Tang.	D. 10″	Cotang.	N. sine.	N. cos.	′
0	9.816943	24.2	9.877780	18.3	9.939163	42.5	10.060837	65606	75471	60
1	817088	24.2	877670	18.3	939418	42.5	060582	65628	75452	59
2	817233	24.2	877560	18.3	939673	42.5	060327	65650	75433	58
3	817379	24.2	877450	18.3	939928	42.5	060072	65672	75414	57
4	817524	24.2	877340	18.3	940183	42.5	059817	65694	75395	56
5	817668	24.1	877230	18.3	940438	42.5	059562	65716	75375	55
6	817813	24.1	877120	18.4	940694	42.5	059306	65738	75356	54
7	817958	24.1	877010	18.4	940949	42.5	059051	65759	75337	53
8	818103	24.1	876899	18.4	941204	42.5	058796	65781	75318	52
9	818247	24.1	876789	18.4	941458	42.5	058542	65803	75299	51
10	818392	24.1	876678	18.4	941714	42.5	058286	65825	75280	50
11	9.818536	24.0	9.876568	18.4	9.941968	42.5	10.058032	65847	75261	49
12	818681	24.0	876457	18.4	942223	42.5	057777	65869	75241	48
13	818825	24.0	876347	18.4	942478	42.5	057522	65891	75222	47
14	818969	24.0	876236	18.4	942733	42.5	057267	65913	75203	46
15	819113	24.0	876125	18.5	942988	42.5	057012	65935	75184	45
16	819257	24.0	876014	18.5	943243	42.5	056757	65956	75165	44
17	819401	24.0	875904	18.5	943498	42.5	056502	65978	75146	43
18	819545	24.0	875793	18.5	943752	42.5	056248	66000	75126	42
19	819689	23.9	875682	18.5	944007	42.5	055993	66022	75107	41
20	819832	23.9	875571	18.5	944262	42.5	055738	66044	75088	40
21	9.819976	23.9	9.875459	18.5	9.944517	42.5	10.055483	66066	75069	39
22	820120	23.9	875348	18.5	944771	42.4	055229	66088	75050	38
23	820263	23.9	875237	18.5	945026	42.4	054974	66109	75030	37
24	820406	23.9	875126	18.5	945281	42.4	054719	66131	75011	36
25	820550	23.8	875014	18.6	945535	42.4	054465	66153	74992	35
26	820693	23.8	874903	18.6	945790	42.4	054210	66175	74973	34
27	820836	23.8	874791	18.6	946045	42.4	053955	66197	74953	33
28	820979	23.8	874680	18.6	946299	42.4	053701	66218	74934	32
29	821122	23.8	874568	18.6	946554	42.4	053446	66240	74915	31
30	821265	23.8	874456	18.6	946808	42.4	053192	66262	74896	30
31	9.821407	23.8	9.874344	18.6	9.947063	42.4	10.052937	66284	74876	29
32	821550	23.8	874232	18.6	947318	42.4	052682	66306	74857	28
33	821693	23.8	874121	18.7	947572	42.4	052428	66327	74838	27
34	821835	23.7	874009	18.7	947826	42.4	052174	66349	74818	26
35	821977	23.7	873896	18.7	948081	42.4	051919	66371	74799	25
36	822120	23.7	873784	18.7	948336	42.4	051664	66393	74780	24
37	822262	23.7	873672	18.7	948590	42.4	051410	66414	74760	23
38	822404	23.7	873560	18.7	948844	42.4	051156	66436	74741	22
39	822546	23.7	873448	18.7	949099	42.4	050901	66458	74722	21
40	822688	23.7	873335	18.7	949353	42.4	050647	66480	74703	20
41	9.822830	23.6	9.873223	18.7	9.949607	42.4	10.050393	66501	74683	19
42	822972	23.6	873110	18.7	949862	42.4	050138	66523	74663	18
43	823114	23.6	872998	18.8	950116	42.4	049884	66545	74644	17
44	823255	23.6	872885	18.8	950370	42.4	049630	66566	74625	16
45	823397	23.6	872772	18.8	950625	42.4	049375	66588	74606	15
46	823539	23.6	872659	18.8	950879	42.4	049121	66610	74586	14
47	823680	23.6	872547	18.8	951133	42.4	048867	66632	74567	13
48	823821	23.5	872434	18.8	951388	42.4	048612	66653	74548	12
49	823963	23.5	872321	18.8	951642	42.4	048358	66675	74522	11
50	824104	23.5	872208	18.8	951896	42.4	048104	66697	74509	10
51	9.824245	23.5	9.872095	18.8	9.952150	42.4	10.047850	66718	74489	9
52	824386	23.5	871981	18.9	952405	42.4	047595	66740	74470	8
53	824527	23.5	871868	18.9	952659	42.4	047341	66762	74451	7
54	824668	23.5	871755	18.9	952913	42.4	047087	66783	74431	6
55	824808	23.4	871641	18.9	953167	42.4	046833	66805	74412	5
56	824949	23.4	871528	18.9	953421	42.3	046579	66827	74392	4
57	825090	23.4	871414	18.9	953675	42.3	046325	66848	74373	3
58	825230	23.4	871301	18.9	953929	42.3	046071	66870	74353	2
59	825371	23.4	871187	18.9	954183	42.3	045817	66891	74334	1
60	825511		871073		954437	42.3	045563	66913	74314	0
	Cosine.		Sine.		Cotang.		Tang.	N. cos.	N. sine.	′

′	Sine.	D. 10″	Cosine.	D. 10″	Tang.	D. 10″	Cotang.	N. sine.	N. cos.	
0	9.825511	23.4	9.871073	19.0	9.954437	42.3	10.045563	66913	74314	60
1	825651	23.3	870960	19.0	954691	42.3	045309	66935	74295	59
2	825791	23.3	870846	19.0	954945	42.3	045055	66956	74276	58
3	825931	23.3	870732	19.0	955200	42.3	044800	66978	74256	57
4	826071	23.3	870618	19.0	955454	42.3	044546	66999	74237	56
5	826211	23.3	870504	19.0	955707	42.3	044293	67021	74217	55
6	826351	23.3	870390	19.0	955961	42.3	044039	67043	74198	54
7	826491	23.3	870276	19.0	956215	42.3	043785	67064	74178	53
8	826631	23.3	870161	19.0	956469	42.3	043531	67086	74159	52
9	826770	23.3	870047	19.0	956723	42.3	043277	67107	74139	51
10	826910	23.2	869933	19.1	956977	42.3	043023	67129	74120	50
11	9.827049	23.2	9.869818	19.1	9.957231	42.3	10.042769	67151	74100	49
12	827189	23.2	869704	19.1	957485	42.3	042515	67172	74080	48
13	827328	23.2	869589	19.1	957739	42.3	042261	67194	74061	47
14	827467	23.2	869474	19.1	957993	42.3	042007	67215	74041	46
15	827606	23.2	869360	19.1	958246	42.3	041754	67237	74022	45
16	827745	23.2	869245	19.1	958500	42.3	041500	67258	74002	44
17	827884	23.2	869130	19.1	958754	42.3	041246	67280	73983	43
18	828023	23.1	869015	19.1	959008	42.3	040992	67301	73963	42
19	828162	23.1	868900	19.2	959262	42.3	040738	67323	73944	41
20	828301	23.1	868785	19.2	959516	42.3	040484	67344	73924	40
21	9.828439	23.1	9.868670	19.2	9.959769	42.3	10.040231	67366	73904	39
22	828578	23.1	868555	19.2	960023	42.3	039977	67387	73885	38
23	828716	23.1	868440	19.2	960277	42.3	039723	67409	73865	37
24	828855	23.1	868324	19.2	960531	42.3	039469	67430	73846	36
25	828993	23.0	868209	19.2	960784	42.3	039216	67452	73826	35
26	829131	23.0	868093	19.2	961038	42.3	038962	67473	73806	34
27	829269	23.0	867978	19.2	961291	42.3	038709	67495	73787	33
28	829407	23.0	867862	19.3	961545	42.3	038455	67516	73767	32
29	829545	23.0	867747	19.3	961799	42.3	038201	67538	73747	31
30	829683	23.0	867631	19.3	962052	42.3	037948	67559	73728	30
31	9.829821	23.0	9.867515	19.3	9.962306	42.3	10.037694	67580	73708	29
32	829959	22.9	867399	19.3	962560	42.3	037440	67602	73688	28
33	830097	22.9	867283	19.3	962813	42.3	037187	67623	73669	27
34	830234	22.9	867167	19.3	963067	42.3	036933	67645	73649	26
35	830372	22.9	867051	19.3	963320	42.3	036680	67666	73629	25
36	830509	22.9	866935	19.3	963574	42.3	036426	67688	73610	24
37	830646	22.9	866819	19.4	963827	42.3	036173	67709	73590	23
38	830784	22.9	866703	19.4	964081	42.3	035919	67730	73570	22
39	830921	22.9	866586	19.4	964335	42.3	035665	67752	73551	21
40	831058	22.8	866470	19.4	964588	42.2	035412	67773	73531	20
41	9.831195	22.8	9.866353	19.4	9.964842	42.2	10.035158	67795	73511	19
42	831332	22.8	866237	19.4	965095	42.2	034905	67816	73491	18
43	831469	22.8	866120	19.4	965349	42.2	034651	67837	73472	17
44	831606	22.8	866004	19.4	965602	42.2	034398	67859	73452	16
45	831742	22.8	865887	19.5	965855	42.2	034145	67880	73432	15
46	831879	22.8	865770	19.5	966109	42.2	033891	67901	73413	14
47	832015	22.8	865653	19.5	966362	42.2	033638	67923	73393	13
48	832152	22.7	865536	19.5	966616	42.2	033384	67944	73373	12
49	832288	22.7	865419	19.5	966869	42.2	033131	67965	73353	11
50	832425	22.7	865302	19.5	967123	42.2	032877	67987	73333	10
51	9.832561	22.7	9.865185	19.5	9.967376	42.2	10.032624	68008	73314	9
52	832697	22.7	865068	19.5	967629	42.2	032371	68029	73294	8
53	832833	22.7	864950	19.6	967883	42.2	032117	68051	73274	7
54	832969	22.7	864833	19.6	968136	42.2	031864	68072	73254	6
55	833105	22.6	864716	19.6	968389	42.2	031611	68093	73234	5
56	833241	22.6	864598	19.6	968643	42.2	031357	68115	73215	4
57	833377	22.6	864481	19.6	968896	42.2	031104	68136	73195	3
58	833512	22.6	864363	19.6	969149	42.2	030851	68157	73175	2
59	833648	22.6	864245	19.6	969403	42.2	030597	68179	73155	1
60	833783	22.6	864127		969656	42.2	030344	68200	73135	0
	Cosine.		Sine.		Cotang.		Tang.	N. cos.	N. sine.	′

′	Sine.	D. 10″	Cosine.	D. 10″	Tang.	D. 10″	Cotang.	N.sine.	N. cos.	′
0	9.833783	22.6	9.864127	19.6	9.969656	42.2	10.030344	68200	73135	60
1	833919		864010	19.6	969909	42.2	030091	68221	73116	59
2	834054	22.5	863892		970162	42.2	029838	68242	73096	58
3	834189	22.5	863774	19.7	970416	42.2	029584	68264	73076	57
4	834325	22.5	863656	19.7	970669	42.2	029331	68285	73056	56
5	834460	22.5	863538	19.7	970922	42.2	029078	68306	73036	55
6	834595	22.5	863419	19.7	971175	42.2	028825	68327	73016	54
7	834730	22.5	863301	19.7	971429	42.2	028571	68349	72996	53
8	834865	22.5	863183	19.7	971682	42.2	028318	68370	72976	52
9	834999	22.4	863064	19.7	971935	42.2	028065	68391	72957	51
10	835134	22.4	862946	19.8	972188	42.2	027812	68412	72937	50
11	9.835269	22.4	9.862827	19.8	9.972441	42.2	10.027559	68434	72917	49
12	835403	22.4	862709	19.8	972694	42.2	027306	68455	72897	48
13	835538	22.4	862590	19.8	972948	42.2	027052	68476	72877	47
14	835672	22.4	862471	19.8	973201	42.2	026799	68497	72857	46
15	835807	22.4	862353	19.8	973454	42.2	026546	68518	72837	45
16	835941	22.4	862234	19.8	973707	42.2	026293	68539	72817	44
17	836075	22.4	862115	19.8	973960	42.2	026040	68561	72797	43
18	836209	22.3	861996	19.8	974213	42.2	025787	68582	72777	42
19	836343	22.3	861877	19.8	974466	42.2	025534	68603	72757	41
20	836477	22.3	861758	19.8	974719	42.2	025281	68624	72737	40
21	9.836611	22.3	9.861638	19.9	9.974973	42.2	10.025027	68645	72717	39
22	836745	22.3	861519	19.9	975226	42.2	024774	68666	72697	38
23	836878	22.3	861400	19.9	975479	42.2	024521	68688	72677	37
24	837012	22.3	861280	19.9	975732	42.2	024268	68709	72657	36
25	837146	22.2	861161	19.9	975985	42.2	024015	68730	72637	35
26	837279	22.2	861041	19.9	976238	42.2	023762	68751	72617	34
27	837412	22.2	860922	19.9	976491	42.2	023509	68772	72597	33
28	837546	22.2	860802	19.9	976744	42.2	023256	68793	72577	32
29	837679	22.2	860682	19.9	976997	42.2	023003	68814	72557	31
30	837812	22.2	860562	20.0	977250	42.2	022750	68835	72537	30
31	9.837945	22.2	9.860442	20.0	9.977503	42.2	10.022497	68857	72517	29
32	838078	22.2	860322	20.0	977756	42.2	022244	68878	72497	28
33	838211	22.1	860202	20.0	978000	42.2	021991	68899	72477	27
34	838344	22.1	860082	20.0	978262	42.2	021738	68920	72457	26
35	838477	22.1	859962	20.0	978515	42.2	021485	68941	72437	25
36	838610	22.1	859842	20.0	978768	42.2	021232	68962	72417	24
37	838742	22.1	859721	20.0	979021	42.2	020979	68983	72397	23
38	838875	22.1	859601	20.1	979274	42.2	020726	69004	72377	22
39	839007	22.1	859480	20.1	979527	42.2	020473	69025	72357	21
40	839140	22.1	859360	20.1	979780	42.2	020220	69046	72337	20
41	9.839272	22.0	9.859239	20.1	9.980033	42.2	10.019967	69067	72317	19
42	839404	22.0	859119	20.1	980286	42.2	019714	69088	72297	18
43	839536	22.0	858998	20.1	980538	42.2	019462	69109	72277	17
44	839668	22.0	858877	20.1	980791	42.2	019209	69130	72257	16
45	839800	22.0	858756	20.1	981044	42.1	018956	69151	72236	15
46	839932	22.0	858635	20.2	981297	42.1	018703	69172	72216	14
47	840064	22.0	858514	20.2	981550	42.1	018450	69193	72196	13
48	840196	21.9	858393	20.2	981803	42.1	018197	69214	72176	12
49	840328	21.9	858272	20.2	982056	42.1	017944	69235	72156	11
50	840459	21.9	858151	20.2	982309	42.1	017691	69256	72136	10
51	9.840591	21.9	9.858029	20.2	9.982562	42.1	10.017438	69277	72116	9
52	840722	21.9	857908	20.2	982814	42.1	017186	69298	72095	8
53	840854	21.9	857786	20.2	983067	42.1	016933	69319	72075	7
54	840985	21.9	857665	20.2	983320	42.1	016680	69340	72055	6
55	841116	21.9	857543	20.3	983573	42.1	016427	69361	72035	5
56	841247	21.8	857422	20.3	983826	42.1	016174	69382	72015	4
57	841378	21.8	857300	20.3	984079	42.1	015921	69403	71995	3
58	841509	21.8	857178	20.3	984331	42.1	015669	69424	71974	2
59	841640	21.8	857056	20.3	984581	42.1	015416	69445	71954	1
60	841771	21.8	856934	20.3	984837		015163	69466	71934	0
′	Cosine.		Sine.		Cotang.		Tang.	N. cos.	N. sine.	′

46 Degrees.

′	Sine.	D. 10″	Cosine.	D. 10″	Tang.	D. 10″	Cotang.	N. sine.	N. cos.	
0	9.841771	21.8	9.856934	20.3	9.984837	42.1	10.015163	69466	71934	60
1	841902	21.8	856812	20.3	985090	42.1	014910	69487	71914	59
2	842033	21.8	856690	20.4	985343	42.1	014657	69508	71894	58
3	842163	21.8	856568	20.4	985596	42.1	014404	69529	71873	57
4	842294	21.7	856446	20.4	985848	42.1	014152	69549	71853	56
5	842424	21.7	856323	20.4	986101	42.1	013899	69570	71833	55
6	842555	21.7	856201	20.4	986354	42.1	013646	69591	71813	54
7	842685	21.7	856078	20.4	986607	42.1	013393	69612	71792	53
8	842815	21.7	855956	20.4	986860	42.1	013140	69633	71772	52
9	842946	21.7	855833	20.4	987112	42.1	012888	69654	71752	51
10	843076	21.7	855711	20.4	987365	42.1	012635	69675	71732	50
11	9.843206	21.7	9.855588	20.5	9.987618	42.1	10.012382	69696	71711	49
12	843336	21.6	855465	20.5	987871	42.1	012129	69717	71691	48
13	843466	21.6	855342	20.5	988123	42.1	011877	69737	71671	47
14	843595	21.6	855219	20.5	988376	42.1	011624	69758	71650	46
15	843725	21.6	855096	20.5	988629	42.1	011371	69779	71630	45
16	843855	21.6	854973	20.5	988882	42.1	011118	69800	71610	44
17	843984	21.6	854850	20.5	989134	42.1	010866	69821	71590	43
18	844114	21.6	854727	20.5	989387	42.1	010613	69842	71569	42
19	844243	21.5	854603	20.6	989640	42.1	010360	69862	71549	41
20	844372	21.5	854480	20.6	989893	42.1	010107	69883	71529	40
21	9.844502	21.5	9.854356	20.6	9.990145	42.1	10.009855	69904	71508	39
22	844631	21.5	854233	20.6	990398	42.1	009602	69925	71488	38
23	844760	21.5	854109	20.6	990651	42.1	009349	69946	71468	37
24	844889	21.5	853986	20.6	990903	42.1	009097	69966	71447	36
25	845018	21.5	853862	20.6	991156	42.1	008844	69987	71427	35
26	845147	21.5	853738	20.6	991409	42.1	008591	70008	71407	34
27	845276	21.5	853614	20.7	991662	42.1	008338	70029	71386	33
28	845405	21.4	853490	20.7	991914	42.1	008086	70049	71366	32
29	845533	21.4	853366	20.7	992167	42.1	007833	70070	71345	31
30	845662	21.4	853242	20.7	992420	42.1	007580	70091	71325	30
31	9.845790	21.4	9.853118	20.7	9.992672	42.1	10.007328	70112	71305	29
32	845919	21.4	852994	20.7	992925	42.1	007075	70132	71284	28
33	846047	21.4	852869	20.7	993178	42.1	006822	70153	71264	27
34	846175	21.4	852745	20.7	993430	42.1	006570	70174	71243	26
35	846304	21.4	852620	20.7	993683	42.1	006317	70195	71223	25
36	846432	21.3	852496	20.8	993936	42.1	006064	70215	71203	24
37	846560	21.3	852371	20.8	994189	42.1	005811	70236	71182	23
38	846688	21.3	852247	20.8	994441	42.1	005559	70257	71162	22
39	846816	21.3	852122	20.8	994694	42.1	005306	70277	71141	21
40	846944	21.3	851997	20.8	994947	42.1	005053	70298	71121	20
41	9.847071	21.3	9.851872	20.8	9.995199	42.1	10.004801	70319	71100	19
42	847199	21.3	851747	20.8	995452	42.1	004548	70339	71080	18
43	847327	21.3	851622	20.8	995705	42.1	004295	70360	71059	17
44	847454	21.3	851497	20.8	995957	42.1	004043	70381	71039	16
45	847582	21.2	851372	20.9	996210	42.1	003790	70401	71019	15
46	847709	21.2	851246	20.9	996463	42.1	003537	70422	70998	14
47	847836	21.2	851121	20.9	996715	42.1	003285	70443	70978	13
48	847964	21.2	850996	20.9	996968	42.1	003032	70463	70957	12
49	848091	21.2	850870	20.9	997221	42.1	002779	70484	70937	11
50	848218	21.2	850745	20.9	997473	42.1	002527	70505	70916	10
51	9.848345	21.2	9.850619	20.9	9.997726	42.1	10.002274	70525	70896	9
52	848472	21.2	850493	20.9	997979	42.1	002021	70546	70875	8
53	848599	21.1	850368	21.0	998231	42.1	001769	70567	70855	7
54	848726	21.1	850242	21.0	998484	42.1	001516	70587	70834	6
55	848852	21.1	850116	21.0	998737	42.1	001263	70608	70813	5
56	848979	21.1	849990	21.0	998989	42.1	001011	70628	70793	4
57	849106	21.1	849864	21.0	999242	42.1	000758	70649	70772	3
58	849232	21.1	849738	21.0	999495	42.1	000505	70670	70752	2
59	849359	21.1	849611	21.0	999748	42.1	000253	70690	70731	1
60	849485	21.1	849485	21.0	10.000000	42.1	000000	70711	70711	0
	Cosine.		Sine.		Cotang.		Tang.	N. cos.	N. sine.	′

TABLE III.

LOGARITHMS OF NUMBERS.

From 1 to 200,

INCLUDING TWELVE DECIMAL PLACES.

N.	Log.	N.	Log.	N.	Log.
1	000000 000000	41	612783 856720	81	908485 018879
2	301029 995664	42	623249 290398	82	913813 852384
3	477121 254720	43	633468 455580	83	919078 092376
4	602059 991328	44	643452 676486	84	924279 286062
5	698970 004336	45	653212 513775	85	929418 925714
6	778151 250384	46	662757 831682	86	934498 451244
7	845098 040014	47	672097 857926	87	939519 252619
8	903089 986992	48	681241 237376	88	944482 672150
9	954242 509439	49	690196 080028	89	949390 006645
10	Same as to 1.	50	Same as to 5.	90	Same as to 9.
11	041392 685158	51	707570 176098	91	959041 392321
12	079181 246048	52	716003 343635	92	963787 827346
13	113943 352307	53	724275 869601	93	968482 948554
14	146128 035678	54	732393 759823	94	973127 853600
15	176091 259056	55	740362 689494	95	977723 605889
16	204119 982656	56	748188 027006	96	982271 233040
17	230448 921378	57	755874 855672	97	986771 734266
18	255272 505103	58	763427 993563	98	991226 075692
19	278753 600953	59	770852 011642	99	995635 194598
20	Same as to 2.	60	Same as to 6.	100	Same as to 10,
21	322219 2947	61	785329 835011	101	004321 373783
22	342422 680822	62	792391 699498	102	008600 171762
23	361727 836018	63	799340 549453	103	012837 224705
24	380211 241712	64	806179 973984	104	017033 339299
25	397940 008672	65	812913 356643	105	021189 299070
26	414973 347971	66	819543 935542	106	025305 865265
27	431363 764159	67	826074 802701	107	029383 777685
28	447158 031342	68	832508 912706	108	033423 755487
29	462397 997899	69	838849 090737	109	037426 497941
30	Same as to 3.	70	Same as to 7.	110	Same as to 11.
31	491361 693834	71	851258 348719	111	045322 978787
32	505149 978320	72	857332 496431	112	049218 022670
33	518513 939878	73	863322 860120	113	053078 443483
34	531478 917042	74	869231 719731	114	056904 851336
35	544068 044350	75	875061 263392	115	060397 840354
36	556302 500767	76	880813 592281	116	064457 989227
37	568201 724067	77	886490 725172	117	068185 861746
38	579783 596617	78	892094 602690	118	071882 007306
39	591064 607026	79	897627 091290	119	075546 961393
40	Same as to 4.	80	Same as to 8.	120	Same as to 12.

N.	Log.	N.	Log.	N.	Log
121	082785 370316	148	170261 715395	175	243038 048686
122	086359 830675	149	173186 268412	176	245512 667814
123	089905 111439	150	176091 259056	177	247973 266362
124	093421 685162	151	178976 947293	178	250420 002309
125	096910 013008	152	181843 587945	179	252853 030980
126	100370 545118	153	184691 430818	180	255272 505103
127	103803 720956	154	187520 720836	181	257678 574869
128	107209 969648	155	190331 698170	182	260071 387985
129	110589 710299	156	193124 588354	183	262451 089730
130	· Same as to 13.	157	195899 652409	184	264817 823010
131	117271 295656	158	198657 086954	185	267171 728403
132	120573 931206	159	201397 124320	186	269512 944218
133	123851 640967	160	204119 982656	187	271841 606536
134	127104 798365	161	206825 876032	188	274157 849264
135	130333 768495	162	209515 014543	189	276461 804173
136	133538 908370	163	212187 604404	190	278753 600953
137	136720 567156	164	214843 848048	191	281033 367248
138	139879 086401	165	217483 944214	192	283301 228704
139	143014 800254	166	220108 088040	193	285557 309008
140	146128 035678	167	222716 471148	194	287801 729930
141	149219 112655	168	225309 281726	195	290034 611362
142	152288 344383	169	227886 704614	196	292256 071356
143	155336 037465	170	230448 921378	197	294466 226162
144	158362 492095	171	232996 110392	198	296665 190262
145	161368 002235	172	235528 446908	199	298853 076410
146	164352 855784	173	238046 103129		
147	167317 334748	174	240549 248283		

LOGARITHMS OF THE PRIME NUMBERS

From 200 to 1543,

INCLUDING TWELVE DECIMAL PLACES.

N.	Log.	N.	Log.	N.	Log.
201	303196 057420	277	442479 769064	379	578639 209968
203	307496 037913	281	448706 319905	383	583198 773968
207	315970 345457	283	451786 435524	389	589949 601326
209	320146 286111	293	466867 620354	397	598790 506763
211	324282 455298	307	487138 375477	401	603144 372620
223	348304 863048	311	492760 389027	409	611723 308007
227	356025 857193	313	495544 337546	419	622214 022966
229	359835 482340	317	501059 262218	421	624282 095836
233	367355 921026	331	519827 993776	431	634477 270161
239	378397 900948	337	527629 900871	433	636487 896353
241	3820'7 042575	347	540329 474791	439	642424 520242
251	399673 721481	349	542825 426959	443	646403 726223
257	409933 123331	353	547774 705388	449	652246 341003
263	419955 748490	359	555094 448578	457	659916 200070
269	429752 280002	367	564666 064252	461	663700 925390
271	432969 290874	373	571708 831809	463	665580 991018

N.	Log.	N.	Log.	N.	Log.
467	6.9310 88050.6	821	914343 157119	1171	068556 895072
479	680335 513414	823	915399 835212	1181	072249 807613
487	687528 961215	827	917505 500553	1187	074450 718955
491	691081 492123	829	918554 530550	1193	076640 443670
499	698100 545623	839	923761 960829	1201	079543 007385
503	701567 985056	853	930949 031168	1213	083860 800845
509	706717 782337	857	932980 821923	1217	085290 578210
521	716837 723300	859	933993 163831	1223	087426 458017
523	718501 688867	863	936010 795715	1229	089551 882866
541	733197 265107	877	942099 593356	1231	090258 052912
547	737987 326333	881	944975 908412	1237	092369 699609
557	745855 195174	883	945960 703578	1249	096562 438356
563	750508 394851	887	947923 619832	1259	100025 729204
569	755112 266395	907	957607 287060	1277	106190 896808
571	756636 108246	911	959518 376973	1279	106870 542460
577	761175 813156	919	963315 511386	1283	108226 656362
587	768638 101248	929	968015 713994	1289	110252 917337
593	773054 693364	937	971739 590888	1291	110926 242517
599	777426 822389	941	973589 623427	1297	112939 986066
601	778874 472002	947	976349 979003	1301	114277 296540
607	783138 691075	953	979092 900638	1303	114944 415712
613	787460 474518	967	985426 474083	1307	116275 587564
617	790285 164033	971	987219 229908	1319	120244 795568
619	791690 649020	977	989894 563719	1321	120902 817604
631	800029 359244	983	992553 517832	1327	122870 922849
641	806858 029519	991	996073 654485	1361	133858 125188
643	808210 972924	997	998695 158312	1367	135768 514554
647	810904 280669	1009	003891 166237	1373	137670 537223
653	814913 181275	1013	005609 445360	1381	140193 678544
659	818885 414594	1019	008174 184006	1399	145817 714122
661	810201 459486	1021	009025 742087	1409	148910 994096
673	828015 064224	1031	013258 665284	1423	153204 896557
677	830588 668685	1033	014100 321520	1427	154424 012366
683	834420 703682	1039	016615 547557	1429	155032 228774
691	839478 047374	1049	020775 488194	1433	156246 402184
701	845718 017967	1051	021602 716028	1439	158060 793919
709	850646 235183	1061	025715 383901	1447	160468 531109
719	856728 890383	1063	026533 264523	1451	161667 412427
727	861534 410859	1069	028977 705209	1453	162265 614286
733	865103 974742	1087	036229 544086	1459	164055 291883
739	868644 488395	1091	037824 750588	1471	167612 672629
743	870988 813761	1093	038620 161950	1481	170555 058512
751	855639 937004	1097	040206 627575	1483	171141 151014
757	879005 879500	1103	042595 512440	1487	172310 968489
761	881384 656771	1109	044931 546149	1489	172894 731332
769	885926 339801	1117	048053 173116	1493	174059 807708
773	888179 493918	1123	050379 756261	1499	175801 632866
787	895974 732359	1129	052693 941925	1511	179264 464329
797	901458 321396	1151	061075 323630	1523	182699 903324
809	907948 521612	1153	061829 307295	1531	184975 190807
811	909020 854211	1163	065579 714728	1543	188365 926053

AUXILIARY LOGARITHMS.

N.	Log.		N.	Log.	
1.009	003891166237		1.0009	000390689248	
1.008	003460532110		1.0008	000347296684	
1.007	003029470554		1.0007	000303899784	
1.006	002598080685	A	1.0006	000260498547	B
1.005	002166061756		1.0005	000217092970	
1.004	001733712775		1.0004	000173683057	
1.003	001300933020		1.0003	000130268801	
1.002	000867721529		1.0002	000086850211	
1.001	000434077479		1.0001	000043427277	

C

N.	Log.	N.	Log.
1.00009	000039083266	1.000009	000003908628
1.00008	000034740691	1.000008	000003474338
1.00007	000030398072	1.000007	000003040047
1.03006	000026055410	1.000006	000002605756
1.00005	000021712704	1.000005	000002171464
1.00004	000017371430	1.000004	000001737173
1.00003	000013028638	1.000003	000001302880
1.00002	000008685802	1.000002	000000868587
1.00001	000004342923	1.000001	000000434294

N.	Log.	
1.0000001	000000043429	(n)
1.00000001	000000004343	(o)
1.000000001	000000000434	(p)
1.0000000001	000000000043	(q)

$$m = 0.4342944819 \qquad \log. -1.637784298.$$

By the preceding tables—and the auxiliaries A, B, and C, we can find the logarithm of any number, true to at least ten decimal places.

But some may prefer to use the following direct formula, which may be found in any of the standard works on algebra:

$$\text{Log. } (z+1) = \log. z + 0.8685889638\left(\frac{1}{2z+1}\right)$$

The result will be true to twelve decimal places, if z be over 2000.

The log. of composite numbers can be determined by the combination of logarithms, already in the table, and the prime numbers from the formula.

Thus, the number 3083 is a prime number, find its logarithm.

We first find the log. of the number 3082. By factoring, we discover that this is the product of 46 into 67.

Log. 46,	1.6627578316
Log. 67,	1.8260748027
Log. 3082	3.4888326343

$$\text{Log. } 3083 = 3.4888326343 + \frac{0.8685889638}{6165}$$

NUMBERS AND THEIR LOGARITHMS,

OFTEN USED IN COMPUTATIONS.

		Log.
Circumference of a circle to dia. 1 ⎫		
Surface of a sphere to diameter 1 ⎬ =3.14159265		0.4971499
Area of a circle to *radius* 1 ⎭		
Area of a circle to diameter 1	= .7853982	—1.8950899
Capacity of a sphere to diameter 1	= .5235988	—1.7189986
Capacity of a sphere to radius 1	=4.1887902	0.6220886
Arc of any circle equal to the radius	=57°29578	1.7581226
Arc equal to radius expressed in sec.	=206264″8	5.3144251
Length of a degree, (radius unity)	=.01745329	—2.2418773
12 hours expressed in seconds,	= 43200	4.6354837
Complement of the same,	=0.00002315	—5.3645163
360 degrees expressed in seconds,	= 1296000	6.1126050

A gallon of distilled water, when the temperature is 62° Fahrenheit, and Barometer 30 inches, is 277.$\frac{274}{1000}$ cubic inches.

$$\sqrt{277.274} = 16.651542 \text{ nearly.}$$

$$\sqrt{\frac{277.274}{.775398}} = 18.78925284 \qquad \sqrt{231} = 15.198684.$$

$$\sqrt{282} = 16.792855.$$

$$\sqrt{\frac{282.}{.785398}} = 18.948708.$$

The French Metre=3.2808992, English *feet* linear measure, =39.3707904 inches, the length of a pendulum vibrating seconds.